Pr

Career Development in Bic ... *gy*

"The book offers those contemplating careers in ... a *vade mecum* that should be of great help—a virtual career mentor outlining authoritatively the relevant terrain, values, and pathways to entry and accomplishment."

—Jeremiah Barondess, MD, President Emeritus, New York Academy of Medicine, USA

"This excellent book admirably succeeds in its stated purpose of serving as a professional resource for career development, and goes well beyond it. It offers priceless insights into a fast-moving and far-reaching field that plays an important part in the development of economies and societies everywhere, and touches all our lives in many ways."

—Walter Truett Anderson, PhD, President Emeritus, World Academy of Art & Science, USA; Author of *Reality Isn't What It Used To Be: Theatrical Politics, Ready-to-Wear Religion, Global Myths, Primitive Chic, and Other Wonders of the Postmodern World* (HarperCollins) and *All Connected Now: Life in the First Global Civilization* (Westview Press)

". . . an invaluable reference. I have not seen any similar compendium of such important information in any other field. The editors are to be congratulated for making an outstanding contribution."

—Russell Lefevre, PhD, 2008 President and Member of the Board of Directors, IEEE-USA, USA

"An outstanding resource for the community. . . .the book addresses (professional and social) issues important to individuals in a wide-array of scientific fields that we should all address as scientists and citizens of the world."

—Martin Frank, PhD, Executive Director, The American Physiological Society, USA

". . . a good book is supposed to educate as well as entertain. This book certainly manages to educate and to entertain by providing personal experiences shared by stalwarts of bioengineering. A good book is also supposed to evoke debate. By covering a broad international cross-section of opinions – conventional, contrarian, or, sometimes, opposing – this book is guaranteed to incite discussion, challenge *status quo* and help readers in their pursuit of career development in bioengineering and biotechnology."

—Dorin Panescu, PhD, Principal Staff Scientist, St. Jude Medical, USA

"The material covered in *Career Development in Bioengineering and Biotechnology* is unlike any of the standard publications related to these fields of activity. I found the material to be very broad and appropriate for individuals interested in or already participating in these disciplines. It is clearly a good reference book, one that can be read and reread at different stages of one's career."

—Joseph Bronzino, PhD, PE, The Vernon Roosa Professor of Applied Science, Trinity College; President of Biomedical Engineering Alliance and Consortium; Editor-in-Chief of *The Biomedical Engineering Handbook* (CRC Press); Past-President, IEEE Engineering in Medicine and Biology Society, USA

"I am impressed by the powerful group of authors and the very wide and pertinent range of subjects that this book covers."

—Ralph Benjamin, CB, PhD, DSc, DEng, Professor, Imperial College London and University of Bristol; Past Director of Science and Technology, British Government Communications Headquarters, United Kingdom

"An outstanding contribution! This book not only describes the spectrum of topics in bioengineering but also provides important advice for career development. The field is put into perspective by framing it in terms of the reach of bioengineering beyond the profession. It is an impressive work that will be useful for students considering bioengineering as a career, practicing engineers, and the general public."

—Donna Hudson, PhD, President, IEEE Engineering in Medicine and Biology Society; Professor and Director of Academic Research and Technology, University of California, San Francisco; Past-President, International Society for Computers and Their Applications; Co-Author of *Neural Networks and Artificial Intelligence for Biomedical Engineering* (Wiley-IEEE Press), USA

"This book will be essential reading for all those seeking career guidance in bioengineering and biotechnology."

—Tony Bradshaw, PhD, Director bioProcessUK – BioIndustry Association (BIA); Chairman, The Royal Academy of Engineering/BIA Life Scientists' Career Seminars, United Kingdom

159609
FO-C

"This book fills a void in the availability of realistic career advice for students looking at potential careers and older professionals seeking new opportunities in bioengineering and biotechnology. Its comprehensive chapters are filled with practical advice on curricula, case studies and inspiring stories by experienced professionals who share their knowledge and recommendations for creating successful and fulfilling careers in these domains."

—Edward Perkins, 2008 Chair IEEE-USA Career and Workforce Policy Committee, USA

". . . .the general understanding of the nature of bioengineering and biotechnology, and the notions about what the related professionals actually do, as well as ideas on potential careers one could expect to develop, have all been unclear. The wide variety of the terminology used has not helped in presenting a clear picture of the subject area. Questions that I have personally encountered from students and the general public indicate the real need for authentic and authoritative information to clarify these important issues. *Career Development in Bioengineering and Biotechnology* addresses this pressing need. I admire the collective knowledge and experience that has gone into the writing of this book."

—Rangaraj Rangayyan, PhD, PE, University Professor; Professor of Electrical and Computer Engineering, and Adjunct Professor of Radiology and Surgery, University of Calgary, Alberta, Canada; Author of *Biomedical Signal Analysis: A Case-Study Approach* (Wiley-IEEE Press) and *Biomedical Image Analysis* (CRC Press)

". . . once I started reading [the book], I could not put it down. In less than three days, I read it all, absorbing the stories and details as if I was consumed by watching a high action movie. Reading the book felt like sitting with your best friend, or your favorite professor, listening to the voice of wisdom answering all the questions that you may have about entering a professional life and positioning yourself for great success. The breadth and depth of the wisdom is phenomenal, and the stories shared by the writers are moving, inspiring, and shine of intelligence in seizing one's own passion and talents and turning them into stellar professional careers. What I valued best was the candidness about the reality of entering bioengineering and biotechnology as a profession and the demands it places on agility of mind, highest level of ethics, need to research the field thoroughly for sound decision making, the necessity to work well with other experts, and overall, the imperative to reinvent yourself frequently. I wish for all students to discover this book and use it frequently during their journey as students and as professionals."

—Nathalie Gosset, MS, MBA, Head of Marketing and Business Development, Alfred Mann Institute for Biomedical Engineering, University of Southern California, USA

"Bioengineering and biotechnology are emerging as distinct disciplines amid the biological revolution and during a period of rapid globalization. These interesting times offer us unprecedented opportunities for professional and personal growth. . . .The legacy of our times will include how well we used our rapidly advancing technologies to improve the world around us. This book provides a roadmap for the contributions of bioengineering and biotechnology in this quest."

—James Moore, PhD, Professor, Department of Biomedical Engineering, Texas A&M University, USA

"I am very excited about this book. As a bioengineering educator, I am always looking for information that can provide guidance for students as they prepare for their careers. The contributors in this book are so enthusiastic about their careers that many of the chapters made we want to switch careers on the spot! I believe that engineering students do not receive enough guidance about alternative career paths. This book will very much help fill the void."

—Judy Cezeaux, PhD, Professor, Department of Biomedical Engineering, Western New England College, USA

"In my position as the international student representative for the IEEE Engineering in Medicine and Biology Society, I have observed repeatedly that professional development is a primary concern of bioengineering students. This book is not only timely in addressing this concern, but also in providing such a sophisticated and comprehensive view of the role of bioengineering in society. Every student should read this book!"

—Jennifer Flexman, PhD, 2005–2007 Student Representative, IEEE Engineering in Medicine and Biology Society, USA

"This book is a must read—it contains a great deal of essential information for junior as well as senior professionals."

—Paolo Bonato, PhD, Director, Motion Analysis Laboratory, Spaulding Rehabilitation Hospital, and Assistant Professor, Harvard Medical School and Harvard-MIT Health Sciences and Technology, USA

"This is an exciting undertaking! The book is very well thought through and balanced. I enjoyed very much reading the chapters I have reviewed. Congratulations to all contributors and the editors of this book."

—Gudrun Zahlmann, PhD, Director of Business Development, Siemens Medical Systems, Germany

"This is a *functional* book with immediate impact, and is very helpful to those who *need* and desperately *want* help in making a career choice."

—Jonathan Newman, NSF/IGERT Predoctoral Fellow in Biomedical Engineering, Georgia Institute of Technology and Emory University, USA

"This book covers a comprehensive list of career development topics of particular interest to postdoctoral scholars. The authors provide diverse perspectives on a broad range of career issues of value to early-career bioengineers and scientists. The book is a rich new resource for the postdoctoral community."

—Alyson Reed, Executive Director, National Postdoctoral Association, USA

". . . It is the most comprehensive look at career development in bioengineering and biotechnology I have seen. A great resource that I plan to recommend to students and faculty!"

—Charla Triplett, President, Biomedical Engineering Career Alliance, USA

"The book has quite a full-range of contributors. I had no comments on the contents except "wow"!!! While targeted to the bioengineering and biotechnology fields, any technology professional would benefit from the information in this marvelous book on career development."

—John Paserba, Manager, Mitsubishi Electric Power Products, Inc; IEEE Fellow and Member, Board of Governors, IEEE Power Engineering Society, Past-Chair, IEEE-USA Student Professional Awareness Committee, USA

"What's special about bioengineering is that practitioners can make of it what they wish. This book shows how true this is."

—Maurice Klee, PhD, JD, Attorney-at-Law, Fairfield, Connecticut, USA

"A timely and important publication . . . This should be a required reference book for all those working at the interface of engineering and biology for meeting challenges of the 21st century related to health, energy, environment and sustainability."

—Brahm Verma, PhD, Professor and Founder, Faculty of Engineering, University of Georgia; Founding President, The Institute of Biological Engineering, USA

". . . incredibly interesting. It is not easy to find this kind of 'straight talk' and I'm glad to see that finally such a reference is now available."

—Jennifer Jackson, 2008 President-Elect, American College of Clinical Engineering; Clinical Engineer, Brigham and Women's Hospital (Teaching Affiliate of Harvard Medical School), USA

"A career in bioengineering and biotechnology is not about searching for jobs, but *creating* innovative opportunities; more so, pioneering not just in science or engineering, but also extending into politics, management, governance, transparency, law, and social development. Recent years have shown how bioengineering and biotechnology can shield society and human life from untoward situations be it biological weapons, pandemics or even terrorism at large. And with contributions from luminaries, this volume should just serve as an inspiration for us to make the world a better place."

—Basheerhamad Shadrach, PhD, Senior Program Officer–Asia, The International Development Research Centre (Centre de recherches pour le développement international), India

Career Development in Bioengineering and Biotechnology

Series in Biomedical Engineering

Editor-in-Chief

Prof. Dr. Joachim H. Nagel
Institute of Biomedical Engineering
University of Stuttgart
Seidenstrasse 36
70174 Stuttgart
Germany
E-mail: jn@bmt.uni-stuttgart.de

The Series in Biomedical Engineering is an official publication of the International Federation for Medical and Biological Engineering.

The International Federation for Medical and Biological Engineering (IFMBE) is a federation of national and transnational organizations representing internationally the interests of medical and biological engineering and sciences. The IFMBE is a non-profit organization fostering the creation, dissemination and application of medical and biological engineering knowledge and the management of technology for improved health and quality of life. Its activities include participation in the formulation of public policy and the dissemination of information through publications and forums. Within the field of medical, clinical, and biological engineering, IFMBE's aims are to encourage research and the application of knowledge, and to disseminate information and promote collaboration. The objectives of the IFMBE are scientific, technological, literary, and educational.

The IFMBE is a World Health Organization accredited non-governmental organization covering the full range of biomedical and clinical engineering, healthcare, and healthcare technology and management. It is representing through its 58 member societies some 120,000 professionals involved in the various issues of improving health and healthcare delivery.

IFMBE Officers
President: Makoto Kikuchi, Vice-President: Herbert Voigt, Past-President: Joachim H. Nagel, Treasurer: Shankar M. Krishnan, Secretary-General: Ratko Magjarevic, www.ifmbe.org

Career Development in Bioengineering and Biotechnology

Edited by

Guruprasad Madhavan
State University of New York
Binghamton, New York, USA

Barbara Oakley
Oakland University
Rochester, Michigan, USA

Luis Kun
National Defense University and
American Institute of Medical and Biological Engineering
Washington, District of Columbia, USA

Editorial by **Joachim Nagel**
Foreword by **Robert Langer**
Introduction by **Bruce Alberts**
Afterword by **Shu Chien**

Editors
Guruprasad Madhavan
Department of Bioengineering
Thomas J. Watson School of Engineering
and Applied Science
State University of New York
Binghamton New York, USA
www.binghamton.edu

Barbara Oakley
Department of Industrial and Systems
Engineering
School of Engineering and Computer Science
Oakland University
Rochester, Michigan, USA
www.oakland.edu

Luis Kun
Information Resources Management College
National Defense University and
American Institute of Medical and Biological Engineering
Washington, District of Columbia, USA
www.ndu.edu; www.aimbe.org

ISBN: 978-0-387-76494-8 e-ISBN: 978-0-387-76495-5
DOI: 10.1007/978-0-387-76495-5

Library of Congress Control Number: 2008930852

© 2008 Springer Science+Business Media, LLC
All rights reserved. This work may not be translated or copied in whole or in part without the written permission of the publisher (Springer Science+Business Media, LLC, 233 Spring Street, New York, NY 10013, USA), except for brief excerpts in connection with reviews or scholarly analysis. Use in connection with any form of information storage and retrieval, electronic adaptation, computer soft-ware, or by similar or dissimilar methodology now known or hereafter developed is forbidden.
The use in this publication of trade names, trademarks, service marks, and similar terms, even if they are not identified as such, is not to be taken as an expression of opinion as to whether or not they are subject to proprietary rights.

Printed on acid-free paper

9 8 7 6 5 4 3 2 1

springer.com

IFMBE

The International Federation for Medical and Biological Engineering (IFMBE) was established in 1959 to provide medical and biological engineering with a vehicle for international collaboration in research and practice of the profession. The Federation has a long history of encouraging and promoting international cooperation and collaboration in the use of science and engineering for improving health and quality of life.

The IFMBE is an organization with membership of national and transnational societies and an International Academy. At present there are 53 national members and 5 transnational members representing a total membership in excess of 120,000 professionals worldwide. An observer category is provided to groups or organizations considering formal affiliation. Personal membership is possible for individuals living in countries without a member society. The IFMBE International Academy includes individuals who have been recognized for their outstanding contributions to biomedical engineering.

Objectives

The objectives of the International Federation for Medical and Biological Engineering are scientific, technological, literary, and educational. Within the field of medical, clinical and biological engineering its aims are to encourage research and the application of knowledge, to disseminate information and promote collaboration.

In pursuit of these aims the Federation engages in the following activities: sponsorship of national and international meetings, publication of official journals, cooperation with other societies and organizations, appointment of commissions on special problems, awarding of prizes and distinctions, establishment of professional standards and ethics within the field as well as other activities which in the opinion of the General Assembly or the Administrative Council would further the cause of medical, clinical or biological engineering. It promotes the formation of regional, national, international or specialized societies, groups or boards, the coordination of bibliographic or informational services and the improvement of standards in terminology, equipment, methods and safety practices, and the delivery of health care.

The Federation works to promote improved communication and understanding in the world community of engineering, medicine and biology.

Activities

Publications of the IFMBE include: the journal *Medical and Biological Engineering and Computing*, the electronic magazine *IFMBE News*, and the *Book Series in Biomedical Engineering*. In cooperation with its international and regional conferences, IFMBE also publishes the *IFMBE Proceedings Series*. All publications of the IFMBE are published by Springer.

Every three years the IFMBE hosts a *World Congress on Medical Physics and Biomedical Engineering* in cooperation with the IOMP and the IUPESM. In addition, annual, milestone and regional conferences are organized in different regions of the world, such as Asia-Pacific, Europe, the Nordic-Baltic and Mediterranean regions, Africa and Latin America.

The administrative council of the IFMBE meets once a year and is the steering body for the IFMBE. The council is subject to the rulings of the General Assembly, which meets every three years.

Information on the activities of the IFMBE General Assembly can be at found www.ifmbe.org.

Do not go where the path may lead, go instead where there is no path and leave a trail.

Ralph Waldo Emerson
(1803–1882)

Dedicated to the past, present, and future engineers, technologists, and science professionals committed to better health and human welfare.

Biographical Information

Guruprasad Madhavan (Editor) received his BE degree (Honors with Distinction) in instrumentation and control engineering from the University of Madras, Chennai, India (2001), and MS degree in biomedical engineering from the State University of New York at Stony Brook (2002). Following his medical device industry experience as a research scientist at AFx, Inc. and Guidant Corporation in Fremont, California, Madhavan completed his MBA in leadership and healthcare management from the State University of New York at Binghamton (2007) where he is currently a PhD candidate in biomedical engineering. His professional interests broadly range from conceptualization to commercialization of medical devices, healthcare management and leadership, improving healthcare delivery and access in the developing world, and public policy.

Barbara Oakley (Editor) earned a BA in Slavic languages and literature in 1976 from the University of Washington in Seattle, and a BS in Electrical Engineering from the same institution in 1986. She earned an MS in electrical and computer engineering in 1995, and a PhD in systems engineering in 1998, both from Oakland University in Rochester, Michigan. Oakley is currently an associate professor of engineering at Oakland University and recently served as vice president of the Institute of Electrical and Electronics Engineers' (IEEE) Engineering in Medicine and Biology Society—the world's largest bioengineering society. Oakley's research and teaching interests are in the area of bioelectronics, medical sensors and instrumentation, and the effects of electromagnetic fields on biological cells. An award-winning teacher with a witty way in the classroom, she is also the author of *Hair of the Dog: Tales from Aboard a Russian Trawler* (WSU Press, 1996), and her most recent *Evil Genes: Why Rome Fell, Hitler Rose, Enron Failed, and My Sister Stole My Mother's Boyfriend* (Prometheus Books, 2007).

Luis Kun (Editor), an information technology consultant and professor of systems management in the Information Resources Management College at National Defense University in Washington, District of Columbia, is a medical informatics expert and consultant in the area of healthcare, public health, and scientific computing. Kun's teaching and research interests are in the realm of homeland security, infrastructure control systems, supervisory control and data acquisition, which focus on awareness and protection of the government's infrastructure devices and control centers. Kun spent 14 years at IBM, where he developed the first six clinical applications for the IBM personal computer; was one of the pioneers on bedside terminals for intensive care units; and a developer of a semi-expert, real-time, clinical decision support system. He was also the technical manager of the Nursing Point of Care System at IBM. Kun was a biomedical engineer in the team of four that developed the first teleradiology system, and the first picture archival and communications systems to run on an IBM platform. He was also director of medical systems technology and strategic planning at Cedars Sinai Medical Center in Los Angeles, California.

Kun is a fellow of the Institute of Electrical and Electronic Engineers, American Institute of Medical and Biological Engineering (in which he recently served as the Secretary-Treasurer), and serves on the executive committee or the board of directors of the American Medical Informatics Association and the American Association of Engineering Societies. He advises several federal departments, as well as the legislative and the executive branches of the US government on information technology infrastructure related to healthcare, terrorism, e-government and homeland cybersecurity. In the past 25 years, Kun has written a large number of articles, book chapters/sections, and served as the editor-in-chief of the *Handbook of Biomedical Information Technology* (Elsevier Press, 2007). He has lectured on medical and public health informatics, information technology and bioengineering in over 50 countries. Kun received his BS and MS in electrical engineering and a PhD in biomedical engineering—all from the University of California, Los Angeles.

Joachim Nagel (Editorial) has been a professor and chairman of the department of biomedical engineering at the University of Stuttgart since 1996. He received his MS degree in physics and electronics at the University of Saarbruecken in 1973, and his DSc at the University of Erlangen-Nuremberg, Germany in 1979. Following appointments in industry and as a faculty member in the department of biomedical engineering of the University of Erlangen-Nuremberg, Professor Nagel joined the University of Miami in 1986, where he served as a professor of biomedical engineering (1986–96), radiology (1990–96), and clinical psychophysiology (1988–1996). He also served as director of the medical imaging and instrumentation lab (1986–1996), and director of biomedical engineering in the Behavioral Medicine Research Center (1986–1996).

Professor Nagel is the current president of the International Union for Physical and Engineering Sciences in Medicine (IUPESM) and the immediate past-president of the International Federation for Medical and Biological Engineering (IFMBE). He is a co-founder of the European Alliance for Medical and Biological Engineering and Science and served as the founding chairman of its Division of Academic Programs and Research Institutes. Professor Nagel is a member of the Scientific Council of the International Centre for Bio-Cybernetics of the Polish Academy of Sciences, a fellow of the Institute of Physics (IOP), a fellow of AIMBE and BMES, and an academician member of the UNESCO/UATI World Academy of Biomedical Technologies. He is an honorary member of the Romanian and the Czech Societies for Biomedical Engineering. Professor Nagel is the leader of the European project BIOMEDEA–Biomedical Engineering Preparing for the European Higher Education Area, in charge of the IFMBE participation in the WHO World Alliances on Patient Safety, and the Health Workforce and co-chair of the IUPESM Health Technology and Training Task Force, which is supporting developing countries with regard to health technologies. Professor Nagel is the series editor of the IFMBE/Springer book series on biomedical engineering, and the continuation of the IOP book series in medical physics and biomedical engineering for which he previously served as a series editor (2001–05). His main research interests are in the fields of cardiovascular monitoring, instrumentation and physiology, medical image acquisition and image processing, physiological signals, Bio-MEMS, neurosciences, and biological effects, as well as therapeutic applications of ultrasound. Professor Nagel has published more than 250 scientific papers, books, book chapters, patents, and conference papers.

Robert Langer (Foreword) is one of 13 Institute Professors (the highest honor awarded to a faculty member) at the Massachusetts Institute of Technology, in Cambridge. Langer has written nearly 1,000 articles, and also has over 600 issued or pending patents worldwide, one of which was cited as the outstanding patent in Massachusetts in 1988 and one of 20 outstanding patents in the United States. Langer's patents have been licensed or sublicensed to over 200 pharmaceutical, chemical, biotechnology, and medical device companies. A number of these companies were launched on the basis of these patent licenses. He served as a member of the US Food and Drug Administration's Science Board—the agency's highest advisory board—from 1995–2002 and as its Chairman from 1999–2002. Langer's work is at the interface of biotechnology and materials science. His major focus is in the research and development of polymeric drug delivery systems, particularly genetically engineered proteins, DNA, and RNAi, continuously at controlled rates for prolonged periods of time.

Langer has received over 150 major awards. In 2007, he received the 2006 United States National Medal of Science. In 2002, he received the Charles Stark Draper Prize—considered the equivalent of the Nobel Prize for engineers and the world's most prestigious engineering prize—from the National Academy of Engineering. He is also the only engineer to receive the Gairdner Foundation International Award; 70 recipients of this award have subsequently received a Nobel Prize. Among numerous other awards, Langer has received the Dickson Prize for Science (2002), the Heinz Award for Technology, Economy and Employment (2003), the Harvey Prize (2003), the John Fritz Award (2003) (given previously to inventors such as Thomas Edison and Orville Wright), the General Motors Kettering Prize for Cancer Research (2004), the Dan David Prize in Materials Science (2005) and the Albany Medical Center Prize in Medicine and Biomedical Research (2005)—the largest prize in the United States for medical research. In 2006, he was inducted into the National Inventors Hall of Fame. In 1998, he received the Lemelson-MIT prize—the world's largest prize for invention—for being "one of history's most prolific inventors in medicine." In 1989, Dr. Langer was elected to the Institute of Medicine of the National Academy of Sciences, and in 1992 he was elected to both the National Academy of Engineering and the National Academy of Sciences. He is one of very few people ever elected to all three US National Academies and the youngest in history (at age 43) to ever receive this distinction. *Time Magazine* and CNN (2001) named Langer as one of the 100 most important people in America and one of the 18 top people in science or medicine in America. *Discover Magazine* (2002) named him as one of the 20 most important people in biotechnology. *Forbes Magazine* (2002) selected Langer as one of the 15 innovators world wide who will reinvent our future. Langer has served, at various times, on 15 boards of directors, 30 scientific advisory boards, and received 11 honorary doctorates. He received his BS Degree from Cornell University in 1970 and his ScD from the Massachusetts Institute of Technology in 1974—both in Chemical Engineering.

Bruce Alberts (Introduction) is a professor in the department of biochemistry and biophysics at the University of California, San Francisco, and the editor-in-chief of *Science*, the world's preeminent weekly journal—published by the American Association for the Advancement of Science. He completed two six-year terms as the president of the US National Academy of Sciences, in addition to being the chair of the National Research Council, which, with the National Academy of Engineering and the Institute of Medicine, form the National Academies of the United States. He is presently the co-chair of the InterAcademy Council—a new organization governed by the presidents of science academies from more than 15 nations, designed to provide science advice to the United Nations and other international organizations. A respected scientist recognized for his work in biochemistry and molecular biology, Alberts is known for his extensive molecular analyses of the protein complexes that allow chromosomes to be replicated, and is one of the original authors of *The Molecular Biology of the Cell*, considered to be the world's leading textbook in this field. Alberts received his PhD from Harvard University, and has earned over 15 honorary degrees for his prolific contributions and leadership to science, education, and policy. He currently serves on the boards of more than 15 nonprofit institutions, including the Carnegie Corporation of New York, the Gordon and Betty Moore Foundation, and the Lawrence Berkeley National Laboratory, and also as the past-president of the American Society of Cell Biology.

Shu Chien (Afterword) received his MD in 1953 from National Taiwan University and his PhD in physiology from Columbia University in New York in 1957. Subsequently, Chien was employed at Columbia University as an assistant to associate professor of physiology. In 1969 he was promoted to professor of physiology at Columbia University, and in 1974 became the director of the Division of Circulatory Physiology and Biophysics. In 1987, Chien established Taiwan's Institute of Biomedical Sciences in Academia Sinica. He returned to the United States in 1988 and joined the University of California, San Diego (UCSD). While working at UCSD, he founded the Whitaker Institute of Biomedical Engineering in 1991, and the department of bioengineering in 1994. In 2002, Chien was appointed as a university professor, a unique distinction held currently by only 20 faculty members in the entire University of California system, comprising nearly 9000 faculty members.

As the author of nearly 500 peer-reviewed journal articles, and editor of nine books, Chien's research and teaching interests include the effects of mechanical forces on gene expression and signal transduction, molecular, cellular and tissue bioengineering, blood rheology, and microcirculatory dynamics in health and disease. For his research contributions and leadership excellence, Chien has received numerous honors and awards, including the notable Melville Medal (twice), the Fahraeus Medal, the National Institutes of Health Merit Award, the Landis Award, the ALZA Award, the Zweifach Award, the Poiseuille Gold Medal, and the Galletti Award. In 2005, Chien was honored with the Distinguished Lifetime Achievement Award from the Asian American Engineer of the Year Awards Committee.

Chien is a member of all three US National Academies, the Institute of Medicine, the National Academy of Engineering, and the National Academy of Sciences, and is currently one of only nine members who hold membership to all three academies. He is also a member of the Academia Sinica, a founding fellow of the American Institute for Medical and Biological Engineering, an elected member of the American Academy of Arts and Sciences, and a foreign member of the Chinese Academy of Science.

Chien has served as president of the Microcirculatory Society (1980–81), the American Physiological Society (1990–1991), the Federation of American Societies of Experimental Biology (1992–93), the American Institute of Medical and Biological Engineering (2001–2002), and currently serves as the president of the Biomedical Engineering Society and the International Society of Biorheology.

Any opinion, findings, conclusions or recommendations expressed in this book are those of the individual authors, and do not necessarily reflect the views of the institutes, institutions, organizations, corporations, agencies, or governments they represent.

All Internet links presented in the book were last accessed in December, 2007.

Preface

Career Development in Bioengineering and Biotechnology is an ambitious undertaking that blends a "careers" book with something far more reaching—an understanding of the relation and impact of those careers on human society worldwide. Each of the five sections in this book, totaling seventy-one chapters, is designed to provide balanced, practice-oriented viewpoints that acknowledge international differences in approach. We have attempted to keep our focus on time-tested methods and a professional outlook for success. It is our hope that this book conveys the dynamic and influential nature of bioengineers and biotechnologists in society. More importantly, we hope this book serves as a stimulus for further progress.

Finally, if this volume contains any merit, it is due to the efforts and experiences of our devoted contributors and scrupulous reviewers. Any remaining errors are our own.

We wish you all the very best, and a pleasant journey through this book and your career.

Binghamton, New York
Rochester, Michigan
Washington, District of Columbia

Guruprasad Madhavan
Barbara Oakley
Luis Kun

Acknowledgments

Editing *Career Development in Bioengineering and Biotechnology* was like conducting a world class symphony orchestra. We are gratefully indebted to the many contributors from across the globe who took time from their busy lives to help make this groundbreaking volume a reality and success.

We wish to express our deepest gratitude to our families, students, friends, mentors, and institutions for the support, enthusiasm, and inspiration they have provided while this book was in development.

This book would not have been possible without David Packer, PhD and his army of wonderful colleagues and associates at Springer in New York and Germany. We are very thankful to them and to the partnership series of Springer and the pre-eminent International Federation of Medical and Biological Engineering.

"The worth of a book is to be measured by what you can carry away from it."

—*James Bryce*

Binghamton, New York
Rochester, Michigan
Washington, District of Columbia

Guruprasad Madhavan
Barbara Oakley
Luis Kun

Contents

External Reviewers

The editors and contributors wish to greatly appreciate and applaud the review efforts and enthusiasm of the following individuals, who have played a significant role in improving the quality of this book.

Aimee Betker, MSc
Department of Electrical and Computer Engineering, University of Manitoba, Winnipeg, Manitoba, Canada

Susan Blanchard, PhD
Department of Bioengineering, U. A. Whitaker School of Engineering, Florida Gulf Coast University, Fort Myers, Florida

Paolo Bonato, PhD
Department of Physical Medicine and Rehabilitation, Spaulding Rehabilitation Hospital Harvard Medical School, and Harvard-MIT Health Sciences and Technology, Boston, Massachusetts, USA

Judy Cezeaux, PhD
Department of Biomedical Engineering, School of Engineering, Western New England College, Springfield, Massachusetts, USA

Eric Chua, MS
Department of Electronic and Electrical Engineering, University College Dublin, Belfield, Dublin, Ireland

Jennifer Flexman, PhD
Department of Bioengineering, University of Washington, Seattle, Washington, USA

Emmanuel Gonzalez, MS
Department of Electronics and Communications Engineering, De La Salle University, Manila, Philippines

Faustina Hwang, PhD
School of Systems Engineering University of Reading, Whiteknights, Reading, United Kingdom

Vincent Ling, MHSc
Institute of Biomaterials and Biomedical Engineering, University of Toronto, Toronto, Ontario, Canada

Cristian Linte, MESc
Robarts Research Institute, University of Western Ontario, London, Ontario, Canada

Angela Love, BS
School of Medicine, Veterinary Medicine, and Arts and Sciences, St George's University, Grenada, West Indies

Jonathan Newman, BS
The Wallace H. Coulter Department of Biomedical Engineering, Georgia Institute of Technology and Emory University, Atlanta, Georgia, USA

Sheila Saia, BS
Higashi-Funabashi, Funabashi-shi, Chiba-Ken, Japan

David Schaffer, PhD
Philips Research North America, Briarcliff Manor, New York, USA

V.R.Singh, PhD
Instrumentation, Sensors and Biomedical Measurements & Standards, National Physical Laboratory, New Delhi, India

Winston Tran, MS
Department of Biomedical Engineering Viterbi School of Engineering, University of Southern California, Los Angeles, California, USA

Vishwak Vajendar
Department of Electrical and Electronics Engineering, VEC—Anna University, Chennai, Tamil Nadu, India

Wei Yin, PhD
School of Mechanical and Aerospace Engineering, Oklahoma State University, Stillwater, Oklahoma, USA

Yunfeng Wu, BS
School of Information Engineering, Beijing University of Posts and Telecommunications, Beijing, China

Gudrun Zahlmann, PhD
Siemens Medical Systems, Erlangen, Germany

Contributors

Guruprasad Madhavan, MS, MBA
Department of Bioengineering, Thomas J. Watson School of Engineering and Applied Science, State University of New York, Binghamton, New York, USA

Barbara A. Oakley, PhD, PE
Department of Industrial and Systems Engineering, School of Engineering and Computer Science, Oakland University, Rochester, Michigan, USA

Luis G. Kun, PhD
Information Resources Management College, National Defense University; American Institute of Medical and Biological Engineering, Washington, District of Columbia, USA

Joachim H. Nagel, DSc
International Federation of Medical and Biological Engineering; International Union for Physical and Engineering Sciences in Medicine; Department of Biomedical Engineering, University of Stuttgart, Stuttgart, Germany

Robert S. Langer, ScD
Department of Chemical Engineering and Harvard-MIT Division of Health Sciences and Technology, Massachusetts Institute of Technology, Cambridge, Massachusetts, USA

Bruce M. Alberts, PhD
Department of Biochemistry and Biophysics, University of California, San Francisco, California; The American Society of Cell Biology; National Academy of Sciences and National Research Council, Washington, District of Columbia, USA; and the InterAcademy Council, Amsterdam, The Netherlands

Robert A. Linsenmeier, PhD
Biomedical Engineering Department and Department of Neurobiology and Physiology, Northwestern University, Evanston, Illinois, USA

David W. Gatchell, PhD
Pritzker Institute of Biomedical Science and Engineering and Biomedical Engineering Department, Illinois Institute of Technology, Chicago, Illinois, USA

Makoto Kikuchi, PhD
Department of Medical Engineering, National Defense Medical College, Saitama, Japan

James C.H. Goh, PhD
Department of Orthopaedic Surgery and Division of Bioengineering, National University of Singapore, Singapore

John D. Enderle, PhD
Department of Biomedical Engineering, School of Engineering, University of Connecticut, Storrs, Connecticut, USA

Nitish V. Thakor, PhD
Department of Biomedical Engineering, The Whitaker Biomedical Engineering Institute and the Neuroengineering Training Initiative, School of Medicine, The Johns Hopkins University School of Medicine, Baltimore, Maryland, USA

Maria E. Squire, PhD
Department of Biology, The University of Scranton, Scranton, Pennsylvania, USA

Mark W. Kroll, PhD
Taser International, Inc, Scottsdale, Arizona; Mark Kroll & Associates, Crystal Bay, Minnesota; Department of Biomedical Engineering, California Polytechnic State University, San Luis Obispo, California; Department of Biomedical Engineering, University of Minnesota, Minneapolis, Minnesota; and Anderson School of Management, University of California, Los Angeles, California, USA

David J. Schlyer, PhD
Brookhaven National Laboratory, US Department of Energy, Upton, New York, USA

Jove Graham, PhD
Food and Drug Administration, US Department of Health and Human Services, Center for Devices and Radiological Health, Office of Device Evaluation, Division of General, Neurological and Restorative Devices, Orthopedic Spinal Devices Branch, Rockville, Maryland; and Geisinger Center for Health Research, Danville, Pennsylvania, USA

Leann M. Lesperance, MD, PhD
Department of Pediatrics, State University of New York – Upstate Medical University, Syracuse, New York; Department of Bioengineering, Thomas J. Watson School of Engineering and Applied Science, State University of New York, Binghamton, New York; and United Health Services Hospitals, Johnson City, New York, USA

Kenneth H. Sonnenfeld, PhD, JD
King and Spalding, LLP, New York, New York, USA

Jennifer McGill, MEng
ECRI Institute (Health Devices) and American College of Clinical Engineering, Plymouth Meeting, Pennsylvania, USA

Dany Bérubé, PhD
MicroCube, LLC, Fremont, California, USA

Rabbi Robert G.L. Shorr, PhD, DIC
Cornerstone Pharmaceuticals, Cranbury, New Jersey and Stony Brook, New York; Altira Capital and Consulting, LLC., Edison, New Jersey; Center for Biotechnology, Stony Brook, New York; Metabolic Research, Inc., Montgomery, Texas, USA; and BrainStorm Cell Therapeutics, Tel Aviv, Israel

Joseph H. Schulman, PhD
Alfred E. Mann Foundation for Biomedical Engineering, Valencia, California; Department of Biomedical Engineering, Viterbi School of Engineering, University of Southern California, Los Angeles, California, USA

Charles H. Kachmarik, Jr., MS
Salem Solutions Corporation, Salem, New Hampshire; Paradigm Physician Partners, LLC, Fairfield, Connecticut, USA

Kristi A. Tange, MS
Derivatives Operations, Goldman, Sachs & Co., New York, New York, USA

Ronald A. Guido, MS
Global Research and Development, Pfizer, Inc., New York, New York, USA

Alan V. McEmber, MS
Global Research and Development, Pfizer, Inc., New York, New York, USA

Kathi G. Durdon, MA
Welch Allyn, Inc., Skaneateles Falls, New York, and The Society of Clinical Research Associates, Chalfont, Pennsylvania, USA

Diana M. Falkenbach, PhD
Department of Psychology, John Jay College of Criminal Justice, The City University of New York, New York, New York, USA

Mary E. Reidy, EDM, PE
National Grid USA, Syracuse, New York, USA

Eugene B. Krentsel, PhD
Office of Technology Transfer and Innovation Partnerships, State University of New York, Binghamton, New York, USA

Assemblyman David R. Koon
135th Assembly District, Fairport, New York, and Legislative Offices, New York State Capitol and Assembly, Albany, New York, USA

Robert A. Malkin, PhD
Engineering World Health and Department of Biomedical Engineering, Pratt School of Engineering, Duke University, Durham, North Carolina, USA

John G. Webster, PhD
Department of Biomedical Engineering, University of Wisconsin-Madison, Madison, Wisconsin, USA

Cynthia Isaac, PhD
Healthcare Practice, Ogilvy Public Relations Worldwide, New York, New York, USA

Jason M. Alter, PhD
Aureon Laboratories, Yonkers, New York, USA

Celeste Baine, MAEd
Engineering Education Service Center, Springfield, Oregon, USA

Jennifer A. Flexman, PhD
Department of Bioengineering, College of Engineering, University of Washington, Seattle, Washington, USA

Aimee L. Betker, MSc
Department of Electrical and Computer Engineering, University of Manitoba, Winnipeg, Manitoba, Canada

Domenico Grasso, PhD, PE, DEE
College of Engineering and Mathematical Sciences, University of Vermont, Burlington, Vermont, USA

David Martinelli, PhD
Department of Civil and Environmental Engineering, West Virginia University, Morgantown, West Virginia, USA

Rick L. Smyre
Center for Communities of Future, Gastonia, North Carolina, USA

Steven Kerno, Jr.
Deere & Company, Moline, Illinois and St. Ambrose University, Davenport, Iowa, USA

Jean Houston, PhD
Human Capacities Institute and International Institute for Social Artistry, Ashland, Oregon, USA

Bala S. Prasanna, MS
IBM Corporation, Middletown, New Jersey, USA

Jerry C. Collins, PhD
Department of Biomedical Engineering, School of Engineering, Vanderbilt University, Nashville, Tennessee, USA

Joseph O. Malo, PhD
University of Nairobi and the Kenya National Academy of Sciences, Nairobi, Kenya

Elizabeth M. Whelan, ScD, MPH
The American Council on Science and Health, New York, New York, USA

John C. Polanyi, PhD
Nobel Prize Laureate, Department of Chemistry, University of Toronto, Ontario, Canada

Subrata Saha, PhD
Departments of Orthopaedic Surgery & Rehabilitation Medicine; Neurosurgery; Physiology & Pharmacology, State University of New York – Downstate Medical Center, Brooklyn, New York, USA

Pamela Saha, MD
Department of Psychiatry, State University of New York – Downstate Medical Center, Brooklyn, New York, USA

Semahat S. Demir, PhD
National Science Foundation, Arlington, Virginia, USA

Richard A. Baird, PhD
National Institute for Biomedical Imaging and Bioengineering, National Institutes of Health, US Department of Health & Human Services, Bethesda, Maryland, USA

Roderic I. Pettigrew, PhD, MD
National Institute for Biomedical Imaging and Bioengineering, National Institutes of Health, US Department of Health & Human Services, Bethesda, Maryland, USA

T.K. Partha Sarathy, MD
Sri Ramachandra Medical University and Harvard Medical International Alliance, Chennai, India; Division of Academic Affairs, Gulf Medical College, Ajman, United Arab Emirates

Nigel H. Lovell, PhD
Graduate School of Biomedical Engineering, University of New South Wales, Sydney, and National Information and Communications Technology Australia, Eveleigh, New South Wales, Australia

Pradeep Ray, PhD
Asia-Pacific Ubiquitous Healthcare Research Centre, University of New South Wales, Australia

Dhanjoo Ghista, PhD
School of Mechanical and Aerospace Engineering, Nanyang Technological University, Singapore

Xiaofei F. Teng, PhD
The Joint Research Center for Biomedical Engineering, Department of Electronic Engineering, the Chinese University of Hong Kong, Shatin, N. T., Hong Kong SAR

Yuan-ting Zhang, PhD
The Joint Research Center for Biomedical Engineering, Department of Electronic Engineering, The Chinese University of Hong Kong, Shatin, N. T., Hong Kong SAR; Institute of Biomedical and Health Engineering, Shenzhen Institute of Advanced Technology, Chinese Academy of Sciences, China

Rajendra K. Pachauri, PhD
Intergovernmental Panel on Climate Change, Geneva, Switzerland and The Energy and Resources Institute, New Delhi, India

Arun M. Gandhi
M. K. Gandhi Institute for Nonviolence and University of Rochester, Rochester, New York, USA

M.S. Swaminathan, PhD
World Food Prize Laureate and UNESCO-Cousteau Ecotechnie Chair M.S. Swaminathan Research Foundation, Chennai; Rajya Sabha, Parliament House, Government of India and National Academy of Agricultural Sciences, New Delhi, India; United Nations Millennium Development Goals, Project Task Force on Hunger, USA; and Pugwash Conference on Science and World Affairs, Pugwash, Nova Scotia, Canada

Norman E. Borlaug, PhD
Nobel Prize, Presidential Medal of Freedom, and Congressional Gold Medal Laureate, Department of Soil and Crop Sciences, Texas A&M University, College Station, Texas, USA; International Maize and Wheat Improvement Center, Mexico; Sasakawa Africa Association, The Nippon Foundation, Tokyo, Japan

Raghav Narayanan, MS
Department of Civil and Environmental Engineering, Carnegie Mellon University, Pittsburgh, Pennsylvania, USA

Ashbindu Singh, PhD
United Nations Environment Programme Division of Early Warning and Assessment-North America, Washington, District of Columbia, USA

Andrei Issakov, PhD
Health Technology and Facilities Planning, Department for Health System Governance and Service Delivery, World Health Organization, Geneva, Switzerland

S. Yunkap Kwankam, PhD
E-Health, Department of Health Statistics and Informatics, World Health Organization, Geneva, Switzerland

Vinton G. Cerf, PhD
Google, Inc. Herndon, Virginia, USA

John P. Holdren, PhD
American Association for Advancement of Science, Washington, District of Columbia; Woods Hole Research Center, Falmouth, Massachusetts; Program on Science, Technology, and Public Policy, Belfer Center for Science and International Affairs, John F. Kennedy School of Government, Harvard University, Cambridge, Massachusetts; and Lawrence Livermore National Laboratory, Livermore, California, USA

Sarah Hall Gueldner, RN, DSN
Frances Payne Bolton School of Nursing, Case Western Reserve University, Cleveland, Ohio; and School of Nursing, College of Health and Human Development, The Pennsylvania State University, University Park, Pennsylvania, USA

Yunfeng Wu
School of Information Engineering, Beijing University of Posts and Telecommunications, Beijing, P.R. China

Yachao Zhou
Department of Computer Science and Technology, Tsinghua University, Beijing, P.R. China

Metin Akay, PhD
Harrington Department of Bioengineering, Fulton School of Engineering, Arizona State University, Tempe, Arizona, USA

Yaneer Bar-Yam, PhD
New England Complex Systems Institute Cambridge, Massachusetts, USA

Raymond C. Tallis, D. Litt
The University of Manchester, Hope Hospital Clinical Academic Group, Manchester and Salford Royal Hospitals National Health Services Trust, Salford, UK

Gail D. Baura, PhD
Keck Graduate Institute of Applied Life Sciences, Claremont, California, USA

Dieter Falkenhagen, Dr. Med. Habil., Dipl. Phys.
Department for Environmental and Medical Sciences, Center for Biomedical Technology and Christian Doppler Laboratory,Danube University, Krems, Austria; International Faculty for Artificial Organs, University of Bologna, Italy; International Faculty for Artificial Organs, University of Strathclyde, Glasgow, UK

Joaquin Azpiroz Leehan, PhD
Center for Medical Imaging and Instrumentation, Department of Electrical Engineering, Universidad Autonoma Metropolitana-Iztapalapa, Col. Vicentina, Mexico D.F, Mexico

Ralph W. Wyndrum Jr., ScD, MBA
The Institute of Electrical and Electronics Engineers-USA Innovation Institute, Washington, District of Columbia, USA

David Sloan Wilson, PhD
Departments of Biology and Anthropology, State University of New York, Binghamton, New York, USA

Max E. Valentinuzzi, PhD
Universidad Nacional de Tucumán and Instituto Superior de Investigaciones Biológicas, Tucumán, Argentina

Colonel Barry L. Shoop, PhD
Joint Improvised Explosive Device Defeat Organization, Office of the Deputy Secretary of Defense, Army Pentagon, US Department of Defense, Washington, District of Columbia; and Department of Electrical Engineering and Computer Science, United States Military Academy, West Point, New York, USA

Guy Kawasaki, MBA
Garage Technology Ventures and Nononina, Inc., Palo Alto, California, USA

Reverend John C. Maxwell, PhD
The INJOY Group and Maximum Impact, Atlanta, Georgia, USA

Shu Chien, MD, PhD
Biomedical Engineering Society; International Society of Biorheology; Whitaker Institute of Biomedical Engineering, University of California, San Diego, California, USA

Editorial

Over the past few decades, bioengineering and biotechnology have revolutionized medicine and healthcare, providing tremendous possibilities for the prevention, diagnosis, and treatment of disease, thereby accelerating the generation of innovative industries and numerous career opportunities. Despite this amazing success, public awareness is lagging. All of us working in the field are still frequently confronted with questions about what bioengineering and biotechnology actually is, which training pathways exist, and how career development opportunities compare to those in other science and engineering disciplines.

This book provides all the answers and can be highly recommended as the ultimate guide to anyone interested in bioengineering and biotechnology. The book arrives at a crucial time, and catapults bioengineering and biotechnology to the forefront of disciplines and to a rightly held pinnacle of inspiration for engineers, scientists, and technologists. Anyone thinking about, or involved with, bioengineering or biotechnology would benefit from this extraordinary volume, including students, their instructors and advisors, political, educational, and healthcare decision makers, and even professionals in the field.

Beyond explanations of bioengineering and biotechnology and various educational options, the volume presents comprehensive information on career and professional development through wide-ranging material written by a truly extraordinary team of experts, and diligently put together by competent editors who deserve the credit for not only recognizing the need for such a book, but satisfying that need in such an excellent and unprecedented way.

As the editor of the biomedical engineering series, a publication of the International Federation for Medical and Biological Engineering (IFMBE), I am particularly pleased to include *Career Development in Bioengineering and Biotechnology* in this successful series of predominantly scientific texts, which was originally established in 1994 in cooperation with the International Organization for Medical Physics and the Institute of Physics Publishing (IOPP) as the Series in Medical Physics and Biomedical Engineering, and relaunched in 2007 by the IFMBE with Springer as the new publisher. *Career Development in Bioengineering and Biotechnology* is strongly supportive of the renewed IFMBE philosophy to put the highest priority on services to its constituent societies and their 120,000 members, as well as to students and professionals entering the field, thus furthering the broader interests of the bioengineering and biotechnology community—and the health and safety of everyone.

Joachim H. Nagel, DSc
President, International Union for Physical and Engineering Sciences in Medicine; Past President, International Federation for Medical and Biological Engineering; and Professor and Chair, Institute of Biomedical Engineering, University of Stuttgart, Germany

Foreword

It is a pleasure to be able to write a foreword for such a unique book involving career development in bioengineering and biotechnology. The editors have gone to great effort to discuss a variety of critical topics in the burgeoning areas of bioengineering and biotechnology. The book is divided up into five important sections.

In the first section, introductory information is provided about what makes a bioengineer or a biotechnologist, and the employment prospects for these individuals. The second section goes over traditional types of careers in bioengineering and biotechnology. It covers such areas as intellectual property; clinical engineering; entrepreneurship in the medical device, biotechnology and pharmaceutical industries; work in university research, industry research, and independent research laboratories; and public sector research.

The third section describes innovative alternative careers in biotechnology and engineering. A great many exciting areas are discussed, ranging from finance and banking to regulatory affairs, science and technology policy, sales and marketing, sports engineering, technology transfer, forensic psychology, venture capitalism, being an expert witness, and many others.

The fourth section covers key career development and success strategies and examines developing new "genes," leadership skills, and life management skills for career and success, including perspectives on ethical development.

The final section covers growth and humanitarian responsibilities of bioengineers and biotechnologists beyond the profession. This segment features stimulating contributions on issues such as patient safety, ethics, affordable and accessible healthcare systems, humanistic science and technology for environmental conservation, hunger, poverty, and sustainable development in the developing world, and many other areas.

This book contains a wealth of information and should serve as an excellent resource to high school students, undergraduates, graduates, postdoctoral fellows, and various people thinking about career development in bioengineering and biotechnology.

Robert S. Langer, ScD
Institute Professor
Massachusetts Institute of Technology
Cambridge, Massachusetts

Introduction

In the 1950s, Prime Minister Jawaharlal Nehru emphasized the importance of what he called a "scientific temper" for the new India. He clearly saw the rationality, openness, and tolerance inherent to science as a requirement for his highly populated and diverse nation. As the world becomes an ever more crowded place, threatened by the lethal rise of both nuclear proliferation and dogmatism, it is clear to me that we will need a scientific temper for *every* nation if we are to live together peacefully and celebrate—rather than disparage—our diversity.

By exploring the many career possibilities open to scientists, the present volume on *Career Development in Bioengineering and Biotechnology*—edited by Guruprasad Madhavan, Barbara Oakley, and Luis Kun—makes an important contribution to this broad goal. In my mind, it is essential that students of science—people with "science in their soul"—spread far and wide to adopt many different careers in society, thereby promoting both Nehru's "scientific temper" and what the great American educator John Dewey called the "scientific habit of mind."

A wonderful little book was written in 1956, entitled *Science and Human Values*, in which the physicist Jacob Bronowski looks back over the course of human history to examine the effect of science on human societies [1]. His conclusion continues to ring in my ears, and it seems even more critical today than it was 50 years ago:

> "The society of scientists is simple because it has a directing purpose: to explore the truth. Nevertheless, it has to solve the problem of every society, which is to find a compromise between the individual and the group. It must encourage the single scientist to be independent, and the body of scientists to be tolerant. From these basic conditions, which form the prime values, there follows step by step a range of values: dissent, freedom of thought and speech, justice, honor, human dignity and self respect.... Science has humanized our values. Men have asked for freedom, justice and respect precisely as the scientific spirit has spread among them."

This is why scientists all around the world must now band together to help spread the "scientific habits of mind"—not just to make better living standards possible in a world of increasingly strained resources, but also to create more rational, scientifically based societies that reject the dogmatism that threatens the world today with deadly conflicts.

Science Education: Producing Citizens with Scientific Habits of Mind

Universal education in science, starting at a young age, stands out as an obvious and important element of the movement toward more rational societies. The culture of science is based on a strong respect for evidence and logic and an openness to new ideas. These are habits of mind that can be learned when children study science in school. And they must spread everywhere to produce the tolerance and rationality that our world so urgently needs.

Scientists are optimists who see remarkable progress being made in their own field of work—be it biology or astrophysics—and therefore believe in the possibility of progress more broadly. Imparting this optimism to young people is a critical component of the type of science education programs I advocate.

Science education is not memorizing facts about thirty kinds of whales and then taking a multiple-choice exam. Yet most American schools still focus on having students learn what science has already discovered, rather than having them take part in the process of discovery, so that they come to understand science as a special way of knowing about the world.

A brief anecdote may help. A few years ago, my daughter was distressed when her son reached the second grade without any sign that science would ever be part of his curriculum. She therefore volunteered to teach a few hands-on science lessons to the class. On the first day, she gave each child a hand lens and three different types of soil, and she asked them to describe what they observed in each sample. To her dismay, the class soon became paralyzed, with no one willing to write the requested descriptions.

Why? She discovered that, after two and a half years of formal schooling, these 7-year-old students had been trained to memorize facts and regurgitate the "correct" answers. Since the exercise called for thinking outside this pattern, and offered no "correct" answer, a fear of making a mistake prevented them from writing anything.

An education that aims to fill the heads of students with correct answers is a disaster for many reasons. For one, different cultures will have different answers, and our diverse societies will suffer greatly from intolerance. Instead, all students must learn how to learn, so that they can solve new problems and overcome the many challenges that they will encounter in their adult lives.

How do we teach a five-year old scientific ways of thinking? In some kindergarten classes in San Francisco today, the teacher gives her children a clean white sock to put on while walking around the schoolyard. Back in the classroom, the children are told to collect all of the black specks stuck to their socks and sort them into two piles: they are to figure out which are seeds and which are dirt. To help with this task, each child is given a $3.00 plastic "microscope" and asked to make a drawing of each speck.

Some children will eventually notice that some specks have regular shapes. Are these the seeds? After an extensive discussion of this idea, and with coaching from the teacher, they finally decide to plant both those specks believed to be dirt and those believed to be seeds, thereby testing the class's idea that the regularly shaped ones are seeds.

Imagine an education that includes solving hundreds of such challenges over the course of the 13 years of schooling that lead to high school graduation—challenges that elevate in difficulty as the children age. I believe that children prepared for life in this way will be great problem-solvers in the workplace, with the abilities and the can-do attitude that are needed to be competitive in the global economy. Even more importantly, they will be more rational human beings—people who are able to make wise judgments for their families, their communities, and their nations.

Last but by no means least, if we really care about creating a "scientific temper" for the world, we must do more than catalyze new teaching in each of our nations at the pre-college level. All of us here today who are scientists and technologists will also need to completely rethink how we teach the introductory college courses taken by all students. Right now most of us are "teaching a little about a lot" in these courses. For example, in our first course in biology, we try to touch on nearly everything that scientists have learned about biology in a single year. I claim that it is urgent that we instead focus on a different type of goal: that of conveying an understanding and appreciation of science, and its relation to society, in each of these introductory courses—regardless of the scientific field.

International Science Policy: Promoting a Scientific Temper for the World

From 1993 to 2005, as the president of the US National Academy of Sciences, I had the privilege of working closely with the leading scientists from national scientific academies all over the world, and I have seen the advantages of working with scientifically trained individuals on international policy discussions and negotiations. It was remarkable how easily we could not only communicate with each other but also agree, even when our governments were at loggerheads. As scientists and engineers, we all shared the same respect for evidence and logical analysis. And we were all passionate in our belief that science had enabled humanity to understand the natural world so well that we can accurately predict the effect of an accelerating release of greenhouse gases—or a few parts per billion of arsenic in drinking water—on the world 50 or 100 years hence. Therefore, each of us had been repeatedly frustrated when the decisions of our own government failed to give enough weight to such scientific judgments, whether they were related to future environmental, health, or economic impacts.

It was this joint frustration, and the hope that by banding together we each could be more effective in spreading scientific judgments, that caused the Inter Academy Panel on International Issues (IAP) to be founded. The IAP was envisioned at the first-ever meeting of all of the world's science academies in New Delhi, India in 1993. Today, the IAP membership includes the science academies from more than 90 nations [2]. One primary aim is to help each of its member academies develop a larger role in its own nation—including becoming a respected, independent advisor to its own government. The IAP also promotes science as a "Global Public Good," accelerating the free sharing of information and resources that strengthen both world science and world science education. Thus, for example, the US National Academies publish over 200 reports a year—mostly in response to requests to the US government—applying science to everything from education to detailed evaluations of health and environmental risks. Their website now contains more than 3,000 books, all of which can be accessed for free around the world [3].

Because of their prestige, national academies of science are an obvious place to catalyze major changes in science and technology education at all levels. One of the IAP's most vigorous programs focuses on sharing curricula and other resources for teaching science for students of all ages.

In 2000, the IAP oversaw the foundation of the InterAcademy Council (IAC) in Amsterdam. The IAC is governed by a Board composed of the presidents of 18 science and engineering academies. It was formed to take advantage of the general agreement of scientists regardless of their nation of origin, producing consensus reports that provide world science advice on specific issues to the UN, World Bank, and similar regional (or national) organizations. Four such reports have been published to date: on building capacity in every nation in science and engineering, on African agriculture, on women in science, and on the enormous world energy challenge [4].

Spreading Scientifically and Technically Educated Emissaries Throughout Society

My past experiences have convinced me that we must work on moving young people with scientific and technical expertise into a range of careers throughout general society. Scientific thinking provides clear benefits at all levels of society, and it promotes common understanding where other modes of communication might fail.

Whether in Washington or Africa, it is not enough for academies and others to produce timely reports with sound recommendations—it is also crucial that there be people in the government, the media, and other positions in society with the scientific and technological background necessary to interpret and adapt the advice for the nation. The US National Academies are effective because many people who were originally trained as scientists and engineers are positioned throughout congress and the executive branch of our federal government. These individuals provide an invaluable link between their particular part of the government and the scientific community, acting as "translators" between two very different cultures—that of the political and scientific worlds. These staff often serve as the initial audience for our many policy reports, and it is hard to imagine how the US government could function without them. Likewise, when scientists and engineers try to help their local school system by bringing resources from their industry or university, it is critical that someone inside the school system be able to act as the translators needed between those different cultures to make sure that what is offered or used is appropriate and helpful.

My conclusion is that to effectively spread science and technology throughout our societies, we must also spread scientists and technologists. By this I mean that we will need to mount an intensive effort to make it possible for scientifically and technically trained people to move into a broad range of relevant professions—not just those that we normally define as science or engineering.

From my contacts with large numbers of young scientists, I find a surprising percentage who would be pleased (in fact, eager) to use their science training in any one of a variety of ways. But this requires that we both support their career development choices and continue to maintain contacts after they leave. Thus, we need to treat them as an important part of the broader scientific community—inviting them back to talk to our students and faculty as we currently do for those who pursue standard research paths.

About ten years ago, the national academies produced two important booklets to help enlarge the discussion around this issue. The first, entitled *Careers in Science and Engineering: A Student Planning Guide to Grad School and Beyond* [5], was intended to help students broaden the range of careers that they were considering in thinking about their future. The other booklet, entitled *Adviser, Teacher, Role Model, Friend: On Being a Mentor to Students in Science and Engineering* [6], was produced in response to many students' fear of discrediting themselves with faculty if they even hinted that they might not be aiming at a career as a professor.

In my opinion, we need to do much more to provide pathways from our graduate schools into the wide variety of nonresearch professions that are of interest to students. This is why I am so enthusiastic about *Career Development in Bioengineering and Biotechnology*. I am very impressed with the enormous dedication and skill that created this major, highly-original contribution—I know of nothing like it. The many heartfelt contributions have been written in different styles, but underlying them all is a deep concern for young people and for the well-being of humanity. The articles range from practical guides on what to do and what not to do, based on carefully distilled life experiences; to inspirational pieces on human motivation; to inspirational outlines of many of the challenges that face a world so urgently in need of young scientific talent. This book needs to reach the broad audience that it so richly deserves.

Bruce M. Alberts, PhD
Co-Chair, The InterAcademy Council, Amsterdam, The Netherlands
Editor-in-Chief, Science
President-Emeritus, US National Academy of Sciences
Chair-Emeritus, The National Research Council
Past-President, The American Society of Cell Biology
Professor, Department of Biochemistry and Biophysics,
University of California, San Francisco, California, USA

References

1. Bronowski, J. *Science and Human Values*. Harper and Row, 1956.
2. The InterAcademy Panel on International Issues, www.interacademies.net
3. The National Academies, www.nationalacademies.org
4. The InterAcademy Council, www.interacademycouncil.net
5. National Research Council. *Careers in Science and Engineering: A Student Planning Guide to Grad School and Beyond*. National Academies Press, 1996.
6. National Research Council. *Adviser, Teacher, Role Model, Friend: On Being a Mentor to Students in Science and Engineering*. National Academies Press, 1997.

Part I

An Introduction to Bioengineering and Biotechnology

1 What Makes a Bioengineer and a Biotechnologist?

Robert A. Linsenmeier, PhD

Biomedical Engineering Department and Department of Neurobiology and Physiology, Northwestern University, Evanston, Illinois, USA

David W. Gatchell, PhD

Pritzker Institute of Biomedical Science and Engineering and Biomedical Engineering Department, Illinois Institute of Technology, Chicago, Illinois, USA

Bioengineering and biotechnology are broad fields with many career options across industry, academia, government, and alternative sectors. Work in these areas offers outstanding possibilities for contributing directly to human health and social welfare. This chapter examines careers in bioengineering and biotechnology, and the educational preparation necessary for those careers. The "bio"-related terminology is somewhat confusing, and we will first try to provide some clarification. As we will explain, our focus is on bio*medical* engineering and on bio*logical* engineering, two of the engineering specializations that prepare individuals for employment in biotechnology, rather than on training in the sciences of biology and chemistry, which are also required in the biotechnology industry.

Characteristics that lead to career success in bioengineering and biotechnology are in some respects very clear, and in other respects cannot be specified in detail. Finding the best way to prepare students for work in these fields is a topic of ongoing discussion among academics and industry personnel, but there is some consensus. The fuzzier aspects concern the appropriate breadth and depth of *domain knowledge*; that is, the knowledge and skills that engineers need to be successful in these fields. The knowledge base of a bioengineer or biotechnologist cannot be completely specified, as there are many subspecialties that fall under these headings, and also because the required knowledge changes as the understanding of biology and the ability to manipulate living organisms in a predictable way improve. The attributes of career preparation that can be more clearly specified are a range of professional skills, or "core competencies," in areas such as

G. Madhavan et al. (eds.), *Career Development in Bioengineering and Biotechnology*, DOI: 10.1007/978-0-387-76495-5_1, © Springer Science+Business Media, LLC 2008

communication, teamwork, and management. These skills are also important for successful careers in many other fields, and have little to do with the individual's technical knowledge base. Here we will touch on both the domain knowledge aspects of education in bioengineering and biotechnology, and on the professional skills applicable to these fields and beyond.

The information in this chapter is derived from several sources. The primary source is the experience and research of the VaNTH (Vanderbilt-Northwestern-Texas-Harvard/MIT) Engineering Research Center on Bioengineering Educational Technologies,[1] the largest project that the US National Science Foundation (NSF) has ever funded to study bioengineering education. Its purview extends considerably beyond educational technologies to consider pedagogy and bioengineering curriculum. A second source is discussion with industry representatives, as we have attempted at Northwestern and other institutions to build relationships that can benefit our students. A third source is the developing literature on biological engineering and its similarities and differences with biomedical engineering.

As biomedical engineers, we take the view that curriculum in bioengineering and biotechnology should be driven largely by the needs of industry, even though we recognize that many students will go on to graduate school to focus on research, or to professional school to train for careers in healthcare. This topic is explored further in a previous article [1], but, briefly, the rationale for focusing on industry is that graduate and medical schools generally have fewer specific engineering requirements than industry, and educating for industry will also prepare students for medical and graduate school. Post-baccalaureate education requires some undergraduate knowledge base (all of which is science, and not engineering, for medical school), but the most important skills for any graduate education are the abilities to reason and to think critically. For graduate school, this is accompanied by skill in designing and carrying out research. For medical school, it seems another critical skill is being able to digest, remember, and connect a great deal of material about health and disease. Industry values these same skills of research competence and efficiency in drawing from one's knowledge base, but industry often also cares that a student has taken a particular set of courses.

Industry requirements for particular knowledge and course work are somewhat paradoxical, because three fundamental points about careers in industry are widely recognized: First, within a few years, an engineer's responsibilities may change dramatically due to promotions or changes of employer, and require completely different specific skills than the ones learned as an undergraduate student. Second, even if an engineer keeps the same job title, new techniques, new research results, and new regulations ensure that the information from school will need to be updated and revised. Third, much of the specific information (including software packages, computer languages, detailed regulatory issues, etc) needed for a particular job in industry are too specific to teach in university, and will be learned

[1] Website: www.vanth.org

on the job. In the face of all of these limitations, does learning specific material in college have any benefit? The answer seems to be yes, and presumably this is why industry appears to care about specifics of education. First, truly basic engineering knowledge (e.g., circuit analysis, fluid flow properties, and thermodynamics) does not change, but it does differ among engineering fields. If one had to learn all the fundamentals on the job, progress would be very slow. Second, someone who has succeeded as an engineering student has demonstrated an ability to handle the rather difficult conceptual and mathematical material of engineering, as well as a strong motivation to work hard, which are probably good predictors of success on the job. This does not imply that industry values engineers who have mastered particular sets of material to the exclusion of other characteristics. There is also a high value placed on innovation, leadership, adaptability, and strong professional skills, a topic we return to after discussing bioengineering and biotechnology more thoroughly.

What is Bioengineering?

At present, the term *bioengineering* is almost interchangeable with *biomedical engineering*. This is not quite what the word bioengineering would be expected to mean, because "bio" interacts with "engineering" in many ways beyond medical applications. However, the National Institutes of Health (NIH) chose the term bioengineering to mean biomedical engineering, and did not do so accidentally. That bioengineering means biomedical engineering to NIH is clear from the definition of bioengineering adopted in 1997, and from the use of the word bioengineering in the name of the newest NIH Institute – the National Institute for Biomedical Imaging and Bioengineering, initiated at the end of 2002. The NIH definition is as follows:

> Bioengineering integrates physical, chemical, mathematical, and computational sciences and engineering principles to study biology, medicine, behavior, and health. It advances fundamental concepts; creates knowledge from the molecular to the organ systems levels; and develops innovative biologics, materials, processes, implants, devices, and informatics approaches for the prevention, diagnosis, and treatment of disease, for patient rehabilitation, and for improving health. [2]

The Whitaker Foundation, which was the largest private funding agency for biomedical engineering for thirty years [3], defined biomedical engineering, rather than bioengineering, in similar terms. Their definition, like the NIH definition, captured both the medical device orientation of the field and the biological understanding side of the field. In the US, some academic departments are called Bioengineering, while others are called Biomedical Engineering, and they are for the most part indistinguishable. So, at present, bioengineering means biomedical engineering, although this could change. This is distinct from biological engineering, as discussed below.

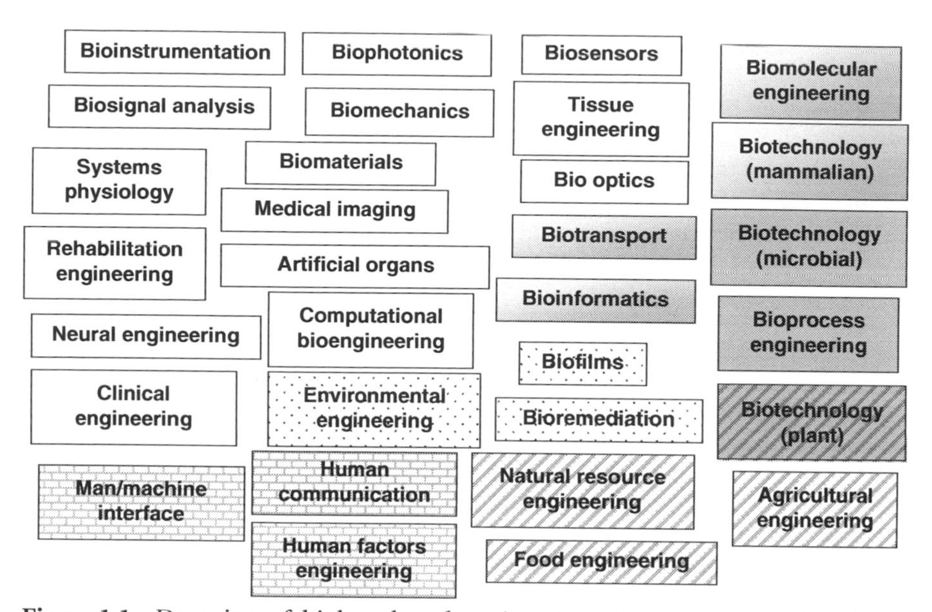

Figure 1.1 Domains of biology-based engineering. Domains with *white background* are generally identified with bioengineering (biomedical engineering), *gray background* with biotechnology and biological engineering, *diagonal background* with agricultural engineering, *stippled background* with environmental engineering, and *brick background* with industrial engineering. Combined backgrounds (*white/gray or diagonal/gray*) are associated with more than one field

Even though the scope of bioengineering is large, it does not nearly exhaust the possible interactions between biology and engineering. Figure 1.1 shows a collection of sub-disciplines that represent many of the areas in which biology and engineering intersect. The ones with white backgrounds would be regarded by most experts as bioengineering. The ones with gray backgrounds are biotechnology, and the ones shading from white to gray are areas important to both bioengineering and biotechnology. In particular, focus areas in bioinformatics and biomolecular engineering are emerging at more biomedical engineering departments. Diagonal hatching represents agricultural engineering, stippling represents environmental engineering, and brick backgrounds are areas frequently associated with industrial or mechanical engineering.

It is difficult to identify a birth date for biomedical engineering. It has existed in a formal way since at least the 1960s, when NIH funded the first Biomedical Engineering Centers, and a few universities began to award graduate degrees in biomedical engineering. Before that, dating back to the 1940s (at least), different types of engineers, physicians, and physiologists collaborated on research projects that we would now call biomedical engineering; however, in the early days, the engineers generally knew little about the biology, and the biologists knew little about the engineering. Medical devices and instruments were invented and manufactured long before that, but the individuals who invented the first systems for X-ray imaging, hemodialysis, and electrophysiological recording, and the first orthopedic and dental prostheses and orthoses, did not consider themselves biomedical engineers, or in many cases engineers at all. What we now call biomedical engineering, however, is a field in which a fully competent engineer is also expected to have a great deal of knowledge of biology, especially at the systems level: the branch called physiology (i.e., the study of how animals or humans work), and a familiarity with clinical problems. Some of the history and important contributions of medical and biological engineering since the 1950s are captured in the "Hall of Fame" of the American Institute of Medical and Biological Engineering [4].

What is Biotechnology?

From its name, biotechnology sounds as though it could almost have the same meaning as biomedical engineering or bioengineering. However, in practice it is rather different. In general, biotechnology is the field in which one manipulates or controls living organisms to generate products, such as drugs. Biotechnologists also devise ways to use (and mimic) strategies from nature to solve, or at least inspire solutions to societal problems that have nothing to do with biology, such as detection of chemical weapons. Many definitions are available on the Internet and elsewhere. Here is one that captures the range of areas within biotechnology:

> Biotechnology is the use of biological processes, organisms, or systems to manufacture products intended to improve the quality of human life. The earliest biotechnologists were farmers who developed improved species of plants and animals by cross pollination or cross breeding.... The science of biotechnology can be broken down into subdisciplines called red, white, green, and blue. Red biotechnology involves medical processes such as getting organisms to produce new drugs, or using stem cells to regenerate damaged human tissues and perhaps re-grow entire organs. White (also called gray) biotechnology involves industrial processes such as the production of new chemicals or the development of new fuels for vehicles. Green biotechnology applies to agriculture and involves such processes as the development of pest-resistant grains or the accelerated evolution of disease-resistant animals. Blue biotechnology, rarely mentioned, encompasses processes in marine and aquatic environments, such as controlling the proliferation of noxious water-borne organisms. [5]

Notice that this definition refers to the "*science* of biotechnology." Biotechnology requires some individuals who are engineers, but the field employs many chemists and biologists as well, and biotechnology *per se* is not an engineering discipline. For the rest of the discussion here, we will confine our attention to engineering careers in the field of biotechnology.

As the above definitions suggest, biotechnology need not be related to human or animal health, but may simply use human, animal, or plant cells, or, in many cases, single celled organisms, such as algae, yeast, or bacteria, to make products. If we refer again to Figure 1.1, we see a few blocks that would be regarded by most experts as biotechnology, but not bioengineering. Much of biotechnology is associated with agricultural engineering and chemical engineering, rather than bioengineering. The interface with agricultural engineering arises in the genetic engineering of plants and animals, production of pesticides and other pest control strategies, and improved processing and safety of food. The interface with chemical engineering occurs because this is the field that deals with scaling up all benchtop chemical reactions so that they work on an industrial scale to produce products, and with computational approaches involving reaction kinetics, heat and mass transfer, and control theory, which come into play in those processes. Just as NIH has a definition for bioengineering, NSF has a statement about Biochemical Engineering and Biotechnology (BEB) that primarily reflects the engineering side of biotechnology. The NSF website states:

> The Biochemical Engineering and Biotechnology programs deal with problems involved in economic processing and manufacturing of products of economic importance by effectively utilizing renewable resources of biological origin and bioinformatics originating from genomic and proteomic information. . . . The BEB programs emphasize basic engineering research that advances the fundamental engineering knowledge base that contributes to better understanding of biomolecular processes (in vivo, in vitro, and/or ex vivo) and eventually to the development of generic enabling technology and practical application. Quantitative assessments of bioprocesses and their rates at the levels of gene regulation and expression, signal transduction pathways, posttranslational protein processing, enzymes in reaction systems, metabolic pathways, cells and tissues in cultivation, biological systems including animal, plant, microbial and insect cells, etc. are considered vital to the successful research projects in the BEB program areas. [6]

Consistent with the above description, the Biotechnology Industry Organization [7] identifies more than 200 new therapies and vaccines currently on the market, 400 more in clinical trials, hundreds of medical diagnostics that keep the blood supply safe and detect changes (such as pregnancy) in the body, several genetically engineered foods, including papaya, soybeans, and corn, microbiological degradation of hazardous waste and microbiological processes in other industries, and biotech production of chemicals, including pesticides and enzymes for detergents.

Agricultural Engineering programs were arguably the first engineering programs that had significant biological content, preceding biomedical engineering programs by decades. However, the movement of agricultural engineering toward a broad based biological foundation occurred largely in the 1980s [8], creating one type of *biological engineering*. While there have long been biochemical engineers, the movement of a large part of the mainstream of chemical engineers into dealing with biology began in the late 1990s, and has created a different type of biological engineering. One now finds a few academic programs called Biological Engineering, and many more called Agricultural *and* Biological (or Biosystems) Engineering, or Chemical *and* Biological (or Biomolecular) Engineering [9]. Unsurprisingly, changes in the curricular content of these programs have not occurred as quickly as the changes in program names [10].

Biological engineering programs arising from agricultural and chemical engineering are distinct from each other at present [8], and from biomedical engineering, as discussed further below. However, a similarity among biological engineering programs, and a key point of difference with biomedical engineering, is the type of biology that biological engineers are expected to know. Consistent with the statements above, an engineer working in biotechnology needs to know cell biology, molecular biology, and microbiology, and generally does not need to know physiology. Biological engineering curricula reflect this [8].

Employment in Bioengineering and Biotechnology

Biomedical Engineering. Biomedical engineers are employed in several industry sectors, all of which are related to health in some way. The largest number are engaged in (1) implantable and extracorporeal (outside the body) medical devices (e.g., pacemakers, dialyzers, intraocular lenses), (2) hospital diagnostic systems and other medical products (e.g. EKG monitors, automated blood analyzers, catheters, surgical tools), (3) medical imaging (e.g. optical, magnetic resonance, computed tomography), and (4) prosthetics, orthotics, and other rehabilitation aids. At this time, the largest segment of the industry is electronic medical devices, although the companies regarded as medical device companies are beginning to diversify in order to deal more directly with cells and drugs. A relatively small number of biomedical engineers, in the authors' experience, are in the large pharmaceutical sector, but the US Bureau of Labor Statistics identified about 15% of biomedical engineers as employed in this sector in 2004 [11].

Data on the placement of biomedical and biological engineers have been collected by the American Institute for Medical and Biological Engineering for several years. As its name suggests, in some respects this organization represents all of the biologically-related engineering fields discussed above, although at present, the members of its Academic Council tend to primarily represent biomedical engineering programs. The 2004–2005 Academic Council survey contains information on more than 1200 baccalaureate graduates during that year from 43 academic programs [12]. Universities tend to collect these data during exit interviews near graduation, when plans may be uncertain, so the data are not complete. However, they show that 56% of bachelor's degree students were planning to pursue a further degree (20% in medical school), 36% of bachelor's level graduates obtained a job, and 8% were seeking employment.

The ability to design systems and products is a key element of any field of engineering, so it is natural to expect that biomedical engineers would be the ones designing medical products and devices. However, biomedical engineering is still a small field relative to electrical, chemical, and mechanical engineering, and relatively few biomedical engineers work on research and development or product design after receiving a bachelor's degree. The number of biomedical engineers engaged in research and development is growing rapidly, however, as industry realizes their unique capabilities. Biomedical engineers will continue to work in teams with mechanical or electrical engineers in much of research and development, because those engineers typically have a somewhat narrower focus and more depth in their areas. In general, a biomedical engineer will not be able to design a circuit as well as an electrical engineer, although he or she may well be better at specifying the requirements of a circuit in a medical device, and understanding the impact of failure.

There are a large number of additional jobs beyond research and development for which biomedical engineers have outstanding qualifications after the bachelor's degree, and for which they are better suited than other types of engineers. Some of these are covered in more detail in other chapters in this book. In general, because of their unique understanding of physiology, biomedical engineers are well qualified to work at the interface between the devices and the delivery of medical care. This means that entry level jobs and advancement in the areas of regulatory affairs, field engineering, quality assurance, and customer training and service are open to them. Consulting is also a frequent career choice for biomedical engineers.

Biotechnology. Because of their focus on cellular processes and microbiology, biological and biochemical engineers may be better suited than biomedical engineers for employment in some areas of biotechnology – for example, the very large pharmaceutical, chemical, and agricultural sectors. Jobs in these sectors include research, design and development of products and processes, management and development of plant processes and control systems, and the wide range of quality assurance, safety, and customer service roles that engineers are called upon to perform in all industries. The Bureau of Labor Statistics does not recognize the category "biological engineer," so statistics on employment are difficult to obtain. However, it is certain that engineering jobs are plentiful. The biotechnology industry is enormous, with both large and small companies participating. At the end of 2005, there were over 1400 biotechnology companies in the US, with a total market capitalization for the publicly traded ones of $410 billion [7]. Revenues in the industry in 2005 were more than $50 billion, more than six times the revenues in 1992 [7]. Overall, in 2004, companies involved in biosciences, a somewhat larger category than biotechnology itself, had a total employment of 1.2 million people [7].

A field that has been attractive to many bioengineering students is tissue engineering, which is in the process of being renamed regenerative engineering. The goal of this field is to build more biologically natural replacements for the function of diseased or damaged organs, by techniques such as gene transfer into damaged cells, or creating materials that are a combination of biocompatible (and usually biodegradable) materials embedded with cells that will proliferate and replace the polymer. This has been an exciting area, but a very difficult one, both scientifically and commercially, for a variety of reasons. At present, tissue engineered materials do not have the mechanical strength of native materials, and regulatory issues concerning safety and reproducibility make production of living products more difficult. At the end of 2002, there had been an investment of approximately $3.5 billion worldwide in tissue engineering, and 89 companies were active, but only four products had been approved and none were commercially viable [13]. Still, these and other types of technology development companies, for instance those devising new drug delivery technologies, will continue to push back the frontiers of what is possible and commercially viable. These companies will remain a good source of biotechnology jobs for biomedical engineers.

New Opportunities. A growing opportunity for both biomedical and biological engineers is to help organize biological facts into more comprehensive systems of knowledge, an area where biology has lagged behind physics and chemistry. This combines computer-based approaches to large datasets (i.e., bioinformatics) and molecular and cellular experiments to create "Systems Biology," an approach that is likely to open up new areas of study and application [14]. Bioinformatics has been embraced as an important subfield in biomedical engineering by only a few universities so far. As the intersection of molecular biology, biotechnology, and computer science, it is often a separate interdisciplinary program.

Another opportunity that is at an early stage is nanotechnology – defined as the investigation and use of materials that are less than one micron in size. Biological materials may interact differently with coatings and agents of this size than with larger materials, and biological materials may also be used to construct nanoassemblies. The range of possibilities is very large [15, 16]. As with systems biology and bioinformatics, the research and applications of bionanotechnology will be partly done by biomedical or biological engineers, but materials scientists, other engineers, chemists, physicists, and structural biologists will be involved.

The Bioengineering Curriculum

Undergraduate Curriculum. The undergraduate curriculum in bioengineering (i.e., biomedical engineering) has developed essentially independently at each of the approximately 80 US universities which offer it. No official body has attempted to prescribe the technical content of an undergraduate bioengineering curriculum. As a result, the curriculum might be expected to vary a great deal from university to university. But to what extent does it actually vary, and what would industry and academia put in an ideal curriculum if they could design one? Is there enough agreement about an ideal curriculum that one could even be defined? In VaNTH, we have taken two approaches to these questions. First, we have characterized the curriculum of the approximately 40 accredited bioengineering and biomedical engineering programs to determine whether there are courses that are taught so frequently in these programs that they constitute a de facto core. We have also characterized the curriculum at non-accredited programs, which have turned out to be similar. The most reliable information about curriculum was obtained from each university's website. Second, we have used online questionnaires (in a format called a Delphi Study) to ask industry and academic representatives to rate approximately 300 concepts for their importance to an undergraduate bioengineering curriculum; that is, what should all undergraduate bioengineers know?

The surveys attempt to define the ideal core at the level of concepts rather than courses, whose names could be misleading. The results have been discussed with members of the bioengineering educational community, and detailed reports on both of these topics are forthcoming. Preliminary results of the work on current curriculum can be found in a proceedings paper [17], and the raw data from the first round of the Delphi Study are all posted on the VaNTH website (www.vanth.org/curriculum). Here we summarize some of the important points.

First, there is good agreement between the actual curriculum and the ideal curriculum. Second, the survey shows remarkably good agreement between industry and academia. Both indicate that the core of bioengineering is, and should be, broad-based. At least 75% of bioengineering programs require courses in physiology, biology other than physiology, mechanics, circuit analysis, computing, materials science, instrumentation, design, and statistics; 71% of the programs require a course in transport phenomena, including some balance among fluid mechanics, heat transfer, and mass transfer. Taking 70% as a cutoff, we can say that the courses mentioned so far, plus the requisite mathematics, physics, and basic chemistry, constitute the functional core at the present time. Thermodynamics and signal analysis are required by just over 50% of programs, and no other course is this common. We found that design (usually at the senior level), mechanics, physiology, and instrumentation are generally taught within bioengineering departments, and will therefore have a strong biology focus, while computing, circuit analysis, and statistics are not generally taught in bioengineering departments, and are more generic. Most of these required subjects are covered in one 3 to 4 credit hour course. The only topics receiving substantially more attention in the required curricula, approximately 6 credit hours on average, or two courses, are mechanics, design, physiology, and biology other than physiology. Our analysis extended only to the core because we recognized that beyond the fundamentals, most undergraduate bioengineers specialize in a subfield (see Figure 1.1). After accounting for all the required courses, the curricular time left for engineering and bioengineering specialization courses was an average of 18 credit hours, or about 6 courses [17]. Beyond this are the courses in humanities and social sciences, which are required at all engineering schools, but which we did not analyze.

The Delphi study of "ideal" content matches the current content remarkably well, with the most highly rated concepts coming from all of the domains in which courses are actually taught most frequently.[2] Perhaps this is not surprising from academic participants in the survey, who are teaching the currently required courses, but industry falls into the same pattern with few deviations. The most highly rated concepts by industry are in the areas of statistical testing and experimental design, followed by concepts from instrumentation, mechanics, materials, and fluids. The top seventeen concepts aggregated across academia and industry are shown in Table 1.1.

[2] See www.vanth.org/curriculum

Table 1.1 Top seventeen concepts from the VaNTH Delphi Study, Round 1. Columns A and I represent the ranking of the importance of each concept by academia and industry, respectively. Concepts are ordered in the table according to the sum of these rankings, from highest to lowest

Concept	Domain	A	I	Sum
Descriptive Statistics (e.g., mean, median, variance, standard deviation)	Statistics	3	1	4
Hypothesis Testing (e.g., paired and un-paired t-tests; chi-squared test)	Statistics	1	3	4
Principles of Statics (e.g., forces; moments; couples; torques; free-body diagrams)	Mechanics	2	8	10
Measurement Concepts (e.g., accuracy, precision, sensitivity; error analysis – sources, propagation of error)	General Engineering	9	2	11
Strength of Materials (e.g., stress, strain; models of material behavior)	Mechanics	8	5	13
Mathematical Descriptions of Physical Systems (e.g., functional relationships, logarithmic, exponential, power-law; ordinary differential equations; partial differential equations)	Modeling	6	10	16
Probability Distributions (e.g., normal, Poisson, binomial)	Statistics	19	4	23
Circuit Elements (e.g., resistors, capacitors, sources, diodes, transistors, integrated circuits)	Circuits	5	18	23
Fundamental Properties of Polymers, Metals and Ceramics (e.g., strength in compression, tension and bending; elasticity/plasticity; failure mechanisms; phase diagrams; transition temperatures; surface roughness; hydrophobicity)	Materials Science	18	6	24
Mechanical Properties of Biological Tissues (e.g., elastic; viscoelastic, hysteresis, creep, stress relaxation)	Mechanics	16	9	25
Mass, Momentum and Energy Balances	Thermo-dynamics	14	12	26
Regression Analysis (e.g., method of least-squares; root mean squared error; correlations; confidence intervals)	Statistics	12	14	26

(*continued*)

Table 1.1 (Continued)

Concept	Domain	A	I	Sum
Pressure-Flow Relations in Tubes and Networks (e.g., flow rate = [change in pressure]/resistance; [Hagen-]Poiseiulle relation; Starling resistor)	Transport Phenomena	10	19	29
DC and AC circuit analyses (e.g., Ohm's and Kirchoff's laws)	Circuits	4	29	33
Forces and pressures in fluids (e.g., shear, normal, surface tension, gravitational, buoyant – Archimedes' principle; Reynolds number)	Transport Phenomena	7	26	33
Data Acquisition (e.g., sampling rates and analog-digital conversion; Nyquist criterion; aliasing)	Signal Analysis	23	11	34
Competency with (at least) One Programming Environment (e.g., Matlab, Mathematica, C, C++, FORTRAN)	General Engineering	11	23	34

From the second round of the survey, we have learned that industry and academia generally believe that some specialization is valuable beyond this core. They do not recommend that students simply take a variety of courses with no focus. Industry participants deviated in two principal ways from academics. First, they want students to have more exposure to software tools for doing design, which was also the case in an earlier survey [18]. Second, they rated concepts and skills in bioinformatics considerably higher than did academics, although neither group gave it the highest priority.

Europe and Asia have lagged behind the US in creating undergraduate programs in bioengineering. This has a great deal to do with the relation between medical school and undergraduate engineering. It was possible for the US to develop bioengineering programs with no requirement that their graduates enter industry, because medical school is a graduate degree in the US and provided an outlet for these students. For many years, the most common next step for bioengineering undergraduates from some programs was medical school, although this has changed as more job opportunities have become available. In the rest of the world, medicine is an undergraduate degree, so the graduating biomedical engineer does not have the option of entering this field. Thus, the European and Asian programs only developed gradually as an industrial market for bioengineers emerged. There has not been a comprehensive review of the programs across the world, but the authors' impression is that many of those programs are more strongly grounded in one of the more traditional engineering disciplines, and may not have the breadth or extensive biology content that characterizes the US programs.

Graduate Curriculum. The graduate curriculum in bioengineering has not been studied as extensively as the undergraduate curriculum. There have long been more graduate programs than undergraduate programs – about twice as many through the 1990s – but the differential is decreasing as more undergraduate programs reach maturity. There are currently just over 90 programs that offer graduate degrees, generally both MS and PhD, and approximately 80 that offer undergraduate degrees. As at the undergraduate level, each university has developed its graduate program rather independently. This is reasonable, particularly at the graduate level, because the subdisciplines of bioengineering require different advanced knowledge and skills. Even within a university, there may be few specific courses that are required of all bioengineering graduate students, so there is not only variation among programs, but flexibility within a program. The variation is likely to remain greater at the graduate level partly because of the needs of different subdisciplines, but also because there is no national accreditation process specifically for graduate programs, and thus no pressure for these programs to meet specific standards. While bioengineering graduate programs vary, they generally require additional courses in engineering, mathematics, and life sciences. A recent informal survey indicated that many bioengineering graduate students were expected to take courses in partial differential equations, which is generally not part of an undergraduate curriculum, but that this was not universally true. Courses in entrepreneurship and business appear to be more common than in the past. Some universities have specific bioengineering entrepreneurship tracks (e.g., Marquette, Stanford, MIT). Of course, at the PhD level, research is the key element of a graduate education. Research is sometimes a significant component of a master's degree as well, although a master's degree based solely on further coursework also seems to be of value to industry, and the job placement of both MS and PhD students is strong.

The Biological Engineering Curriculum

Several years ago, Johnson and Phillips wrote an article that set out some guiding principles for biological engineering programs [19]. They argued that biological engineering as an undergraduate program should be broad-based, and set out a desirable set of fundamental engineering concepts for biological engineers to know. The concepts were only briefly defined, and included "(1) effort and flow, (2) balances, (3) analogy, equivalence and conversion, (4) simplicity, parsimony, and incrementality (start simple and add complexity), (5) approximation, (6) calculation, (7) positive entropy (tending toward disorder), and (8) reversible and irreversible processes." These concepts are sufficiently general that they would be recognizable to almost all engineers as a basis for their work, but enumerating them is still valuable. Beyond these fundamentals, they argued for courses combining engineering and biology. Johnson and Phillips [19] also identified a list of fundamental biological concepts, and believed that biological engineers should study topics ranging from the sub-cellular to ecological. To cover the fundamental engineering, the necessary biology, and the integration of the two, they noted that a US Department of Agriculture-funded effort had identified five courses that could become the core for biological engineers:(1) Biology for Engineers, (2) Biological Responses to Environmental Stimuli, (3) Transport Processes, (4) Engineering Properties of Biological Materials, and (5) Biological Systems Control. These ideas were suggestions, but did stimulate curricular development of the field.

The question of whether biological engineering has reached a point of having core curriculum in the decade since Johnson and Phillips wrote their article was recently analyzed [8].

As noted above, the biology content of biological engineering and biomedical engineering/bioengineering programs is different, with a greater focus on biochemistry and microbiology in the biological engineering programs, and a greater focus on physiology in bioengineering. In terms of other sciences, biological engineering programs at this time tend to include more required organic chemistry. Young [8] also evaluated 10 engineering topics across biological engineering programs, but did not analyze the curriculum for some topics that might have been covered by a large number of programs, such as control theory and materials science. Nevertheless, within the 10 topics surveyed, the engineering content of biological engineering programs that have arisen from agricultural and chemical engineering backgrounds are similar in their frequent requirement for thermodynamics and fluid mechanics. Another similarity is that about 40% of each type of program requires further knowledge of transport phenomena beyond fluids. However, there are substantial differences between the two types of biological engineering programs in other areas. More than half the agriculturally-based programs require mechanics, circuits, instrumentation, and statistics (which is true of biomedical engineering programs as well), while far less than half the chemical engineering-based biological engineering programs require any of these topics. The only course required more frequently by the chemically-based programs is, not surprisingly, biochemical engineering. It must be remembered that the biological engineering programs arising from chemical engineering are newer, and whether they will remain a specialty of chemical engineering or evolve to the broader-based biological engineering envisioned by Johnson and Phillips is not known. At present, however, a student should investigate the curriculum of biological engineering to determine whether that curriculum offers the emphasis matching his or her goals.

Core Competencies

Students in high school who are considering educational paths and careers in bioengineering and biotechnology often think that if they are good at math and science and are interested in design or practical uses of that knowledge, as opposed to development of fundamental new knowledge, they should be engineers. In going into engineering, they often hope to avoid writing, public speaking, and social sciences, which may not be their strengths. We need to dispel this idea at the outset. Technical expertise is necessary, but not sufficient for success in virtually all technical careers. It may be that engineers once worked primarily in isolation, doing calculations with their slide rules, and this is the popular conception of engineering that persists, but it was probably always a false picture, fitting only a minority of engineers. Engineers have always worked to solve problems in society, and need to be aware of how their designs will be used in context. This may be particularly true of bioengineers, but applies to all engineers [20]. A broad range of professional skills beyond the technical ones are critically important to design products, processes, and systems with an understanding of societal needs, push a project through the corporate environment, obtain regulatory approvals, and successfully introduce the product in the marketplace. They include teamwork, communication skills (oral, written, mathematical, and graphical), a facility for and grounding in ethical decision-making, a sense of personal responsibility, an ability to manage finances and people, and the ability to adapt to new situations and to engage in life-long learning.

There is widespread recognition of the importance of the core competencies. The Accreditation Board for Engineering and Technology (ABET) holds engineering programs to a list of "outcomes" that their students are expected to be able to meet. The current list of ABET outcomes is shown in Table 1.2. Note that the italicized outcomes, d, f-j, and part of c are all about professional skills, not technical ones. This is Criterion 3, only one of several criteria on which programs are evaluated, but a very important one.

ABET does not provide any detail about the subtopics that should comprise these outcomes, or make recommendations on how they should be achieved in the curriculum. These important details are left up to academic programs to sort out

Table 1.2 Program outcomes from ABET Criterion 3 for engineering programs [23]

Engineering programs must demonstrate that their students attain:

(a) an ability to apply knowledge of mathematics, science, and engineering
(b) an ability to design and conduct experiments, as well as to analyze and interpret data
(c) an ability to design a system, component, or process to meet desired needs within realistic constraints such as *economic*, environmental, *social, political, ethical*, health and safety, manufacturability, and *sustainability*
(d) *an ability to function on multi-disciplinary teams*
(e) an ability to identify, formulate, and solve engineering problems
(f) *an understanding of professional and ethical responsibility*
(g) *an ability to communicate effectively*
(h) *the broad education necessary to understand the impact of engineering solutions in a global, economic, environmental, and societal context*
(i) *a recognition of the need for, and an ability to engage in life-long learning*
(j) *a knowledge of contemporary issues*
(k) an ability to use the techniques, skills, and modern engineering tools necessary for engineering practice.

for themselves. This is good, as it guarantees that academic programs have to grapple with these questions, prevents too much homogeneity, and encourages curricular innovation. However, it can be a little daunting. At least two organizations have attempted to delineate an entire taxonomy that provides more guidance than the ABET list. The first organization to do this was CDIO (Conceive Design Implement Operate), a consortium of universities initiated by work in the Department of Aeronautics and Astronautics at MIT. Their effort, which involved surveys of many industry and academic professionals, is called the CDIO Syllabus [21]. The VaNTH ERC has created another taxonomy of these skills that is similar, but was intended to apply more directly to bioengineering. It is available at the VaNTH curriculum website [22].

The Future

Given the need to learn and apply engineering and biology concepts, and also learn professional skills, biomedical or biological engineering may sound daunting as a career choice. However, there are many opportunities that make this choice worthwhile. In addition, the bio-based programs are not rigidly structured, and offer many different areas for specialization. If a student starts by being somewhat bewildered by the range of possibilities, it is beneficial to note that specialization does not typically occur until he or she has some grounding in, and knowledge of, the core biology and engineering principles. There is a tremendous amount to learn, but the integration of these fields is the source of the excitement.

Students should be motivated by the importance of engineering in society. Dr. Joseph Bordogna, former Deputy Director of the National Science Foundation, in a lecture to the American Society for Engineering Education on the challenges of engineering education, defined the engineer by the terms in Figure 1.2, most of which emphasize these important societal roles. Medical advances in devices and drugs, as well as advances in the quality and safety of the food supply, rely on engineers to translate fundamental discoveries into products and processes that benefit large numbers of people. In the decades to come, more engineers will be called upon to use their skills in the developing world.

In summary, biomedical and biological engineering are growing areas of employment. Our understanding of the biological world will continue to increase, allowing the possibility for many innovations. More engineers of all types are finding

Characteristics of the Engineer

- *Holistic Designer*
- *Acute Maker*
- *Trusted Innovator*
- *Harm Avoider*
- *Change Agent*
- *Master Integrator*
- *Enterprise Enabler*
- *Knowledge Handler*
- *Technology Steward*

Figure 1.2 The engineer, modified from a lecture at ASEE in 2003 by Dr. Joseph Bordogna

Source: Bordogna, J. U.S. Engineering: Enabling the Nation's Capacity to Perform. ASEE Annual Meeting, June 25, 2003. www.nsf.gov/news/speeches/bordogna/03/jb030625asee.jsp

that they need to know at least some biology, but it is the biomedical and biological engineers who will have the knowledge and skills to be on the forefront of new knowledge and applications.

Additional Resources

The Biomedical Engineering Society – Planning a Career in Biomedical Engineering
www.bmes.org/careers.asp
Whitaker Foundation – Overview of the Field (Biomedical Engineering)
www.bluestream.wustl.edu/WhitakerArchives/glance
The Biotechnology Institute – Careers in Biotechnology
www.biotechinstitute.org/
Institute of Biological Engineering – Academic Programs
www.ibeweb.org/academic/index.cgi
US Department of Labor – Bureau of Labor Statistics – Engineers
www.bls.gov/oco/ocos027.htm
Whitaker Foundation BME Education Summit White Papers
Annals of Biomedical Engineering, Vol. 34, number 2, February, 2006

Acknowledgments We thank Joan A.W. Linsenmeier for useful feedback. This work was supported primarily by the Engineering Research Centers program of the National Science Foundation under grant EEC-9876363.

References

1. Linsenmeier, R.A., "What makes a biomedical engineer?" *IEEE Eng Med Biol Mag*, 2003. 22(4): 32–8.
2. NIH. *Bioengineering Definition*. 1997 [cited 9/17/2007]; Available from: www.becon.nih.gov/bioengineering_definition.htm.
3. Katona, P., "Biomedical engineering and the Whitaker foundation: a thirty-year partnership." *Ann Biomed Eng*, 2006. 34(6): 904–16.
4. AIMBE. *Hall of Fame*. 2007 [cited 9/17/2007]; Available from: www.aimbe.org/content/index.php?pid=127.
5. TechTarget. *Biotechnology*. 2007 [cited 9/17/2007]; Available from: www.whatis.techtarget.com/definition/0,,sid9_gci1109187,00.html.
6. NSF. *Biochemical Engineering and Biotechnology*. 2006 [cited 9/17/2007]; Available from: www.nsf.gov/funding/pgm_summ.jsp?pims_id=13368.
7. BIO. *Biotechnology Industry Facts*. 2007 [cited 7/15/2007]; Available from: www.bio.org/speeches/pubs/er/statistics.asp.
8. Young, R.E., "Comparisons of 'Bio'-type Engineering Undergraduate Curricula from Agricultural, Medical and Chemical Origins." *Int J Eng Edu*, 2006. 22(1): 14–22.
9. Young, R.E., "The 'Bio'-Type Engineering Name Game." *Int J Eng Edu*, 2006. 22(1): 23–27.
10. Johnson, A.T., "The making of a new discipline." *Int J Eng Edu*, 2006. 22(1): 3–8.
11. U.S. Department of Labor, B.o.L.S. *Occupational Outlook Handbook, Engineers* 2006–07 Edition [cited 9/17/2007]; Available from: www.bls.gov/oco/ocos027.htm.

12. AIMBE. *Academic Survey Results.* [cited May 26, 2007]; Available from: www.aimbe.org/content/index.php?pid=201.
13. Lysaght, M.J. and A.L. Hazlehurst, "Tissue engineering: the end of the beginning." *Tissue Eng,* 2004. 10(1–2): 309–20.
14. Ideker, T., L.R. Winslow, and D.A. Lauffenburger, "Bioengineering and systems biology." *Ann Biomed Eng,* 2006. 34(7): 1226–33.
15. Chan, W.C., "Bionanotechnology progress and advances." *Biol Blood Marrow Transplant,* 2006. 12(Suppl 1): 87–91.
16. Clark, J., et al., "Design and analysis of nanoscale bioassemblies." *Biotechniques,* 2004. 36(6): 998–1001.
17. Linsenmeier, R.A. and D.W. Gatchell. *Core Elements of an Undergraduate Biomedical Engineering Curriculum – State of the Art and Recommendations.* in *9th International Conference on Engineering Education.* 2006. San Juan, Puerto Rico.
18. King, P. and R. Fries, "Designing Biomedical Engineering Design Courses." *Int J Eng Edu,* 2003. 19: 346–353.
19. Johnson, A.T. and W.M. Phillips, "Philosophical Foundations of Biological Engineering." *J Eng Edu,* 1995. 84: 311–320.
20. Billington, D.P. and D.P.J. Billington, *Power, Speed, and Form: Engineers and the Making of the Twentieth Century.* 2006, Princeton, New Jersey: Princeton University Press.
21. CDIO. *CDIO Syllabus.* [cited 9/17/2007]; Available from: www.cdio.org/cdio_syllabus_rept/index.html.
22. VaNTH. *Domain Taxonomies.* 2007 [cited 9/17/2007]; Available from: www.vanth.org/curriculum/curr_taxon.asp.
23. ABET. *Accreditation criteria.* 2006 [cited 9/15/2007]; Available from: www.abet.org/Linked%20Documents-UPDATE/Criteria%20and%20PP/E001%2007–08%20EAC%20Criteria%2011–15–06.pdf.

2 Bioengineering and Biotechnology: A European Perspective

Joachim H. Nagel, DSc

International Federation of Medical and Biological Engineering; International Union for Physical and Engineering Sciences in Medicine; Department of Biomedical Engineering, University of Stuttgart, Stuttgart, Germany

Globalization has amplified the international image of bioengineering and biotechnology. Though the disciplines are almost the same all over the world, there are, of course, regional differences in opportunities and educational systems. This chapter outlines European perspectives related to planning a career in bioengineering or biotechnology.

Research and Industry

Many careers in science and engineering are built on the framework of competitive research and thriving industry. Consequently, it is important to understand the status of research, research funding, and the biomedical industry in judging the prospects for successful careers in bioengineering and biotechnology. This section outlines the current state of research and the biomedical technology industry in Europe – it includes parts of a position paper [1] produced by the European Alliance for Medical and Biological Engineering and Science (EAMBES) in 2005.

The European Research Area (ERA)

According to the Lisbon Strategy, (a development plan put forward by the European Council in 2000), Europe aims to become *the most competitive and dynamic knowledge-based economy in the world by* 2010, *capable of sustainable economic growth with more and better jobs and greater social cohesion* [2]. This plan was further stressed through the Barcelona objective in 2002, which stated that 3% *of GDP* [*Gross Domestic Product*] *should be spent on research and development by* 2010. Two thirds of this investment was to come from business, with the bulk of

G. Madhavan et al. (eds.), *Career Development in Bioengineering and Biotechnology*,
DOI: 10.1007/978-0-387-76495-5_2, © Springer Science+Business Media, LLC 2008

the remaining 1% coming from the public sector. In 2004, average research spending in the EU was at 1.86% of the GDP, against 2.66% in the US and 3.18% in Japan, with 54.8% coming from industry, compared to the US, with 63.7%, and Japan, with 74.8% [3]. It is obvious that much remains to be done in Europe to achieve the ambitious goals of the Lisbon Strategy.

Though the EU, represented by the European Commission, is seeking the lead in European research funding by establishing well-coordinated research programs and priorities, including a significant volume of public research investment at the European level, current financial contribution is just a meager 4%–96% of research funding is still provided by the national research agencies.

The EU's primary instruments for funding research are the so-called Framework Programs (FP) for research and technical development. The FPs are research plans designed for a limited time span, usually five years, to support research in selected priority areas. The duration of the current FP7, however, has been extended to seven years. Funding is normally not provided to individual institutions, but rather to research networks incorporating participants from a number of countries, including, but not limited to, EU member states. This makes the ERA attractive to researchers worldwide. The main intent of such funding is to improve the European research infrastructure. As of 2006, research spending by the European Commission was €5.3 billion – 4.3% of the total EU budget. Increases in research funding have been pronounced, with a 41% budget growth from 2004 to 2007. By 2013, the final year of FP7, the EU research budget is scheduled to reach €11 billion.

Overall, opportunities for bioengineering and biotechnology in FP7 are ample within the thematic areas of health (€6.1 billion over the seven years), food, agriculture and biotechnology (€1.9 billion), information and communication technologies (€9.1 billion), nanotechnology (€3.5 billion), security (€1.3 billion), and, due to the ever increasing socio-economic impact of healthcare and health technologies, in the area of socio-economic sciences and humanities (€0.6 billion). While the share of research funding for bioengineering and biotechnology within the FPs can be considered as roughly representative of the national funding patterns in the EU member states, countries with a large biomedical industry, such as Germany, the United Kingdom, and France, provide a much higher rate of funding for bioengineering and biotechnology.

Competitiveness in Research and Innovation

A number of different indicators, such as the number of researchers per thousand in the labor force, share of scientific publications and citations, intellectual property rights, share of patents, and share of high tech exports, are being used to evaluate competitiveness in research and innovation.

In terms of the number of scientific publications, Europe is in the lead with 38.3%, compared with 31.1% for the US, and 9.6% for Japan [3]. The number of citations is commonly regarded as a good indicator for the quality and impact of research. Here, however, Europe is behind the US in most disciplines: about 26%

more references are from US researchers, who hold almost 50% of the world's citations. Japan's citation share over the same period (1997–2001) was 8.4% [4]. In considering these numbers, it should be kept in mind that many researchers outside the US are still publishing in their native languages. Moreover, there is a recognizable tendency by US authors to cite fellow researchers in their own country. Finally, the number of researchers per thousand labor force in the EU is low, with only 5.5, as compared to 9.1 in the US and 10.1 in Japan [3].

A field-by-field analysis shows that the citation gap is generally wider in areas of basic research, where an increase in knowledge is likely to have a particularly marked effect on competitiveness. The gap is noticeable in medicine, and is particularly marked in basic life sciences, pre-clinical medicine, and health sciences. It is relatively small in fields such as the physical sciences, mathematics, and engineering.

Among patents, the triadic patents, which are those that have been approved by the European, Japanese, and US patent offices, are an important indicator of research and development. In this category, the US leads with a share of 34.3%. Europe follows with a share of 31.5%, and Japan comes in third at 26.9%[3]. In terms of triadic patents per million population, however, Japan ranks first with 92.6, followed by the US (53.1) and Europe (30.5).

Germany had the largest share of exports worldwide until it was surpassed by China in 2007. In spite of this, average European exports of all high-tech goods, at a 16.7% share of total world high-tech exports, are behind the US, which is at 19.5%. Japan holds a 10.6% share. As for the medical technologies, German exports, which were worth €14 billion in 2006, were more than 30% higher than its imports, and Ireland's exports, though only half the value of German exports, stand at even more than four times the value of its imports [5]. As a comparison, US exports of the medical device and diagnostics industry in 2003 were at $22.5 billion, approximating to €16.1 billion [6]. Imports were at $22 billion or €15.7 billion.

In Europe, an average of 8.7% of the GDP is spent on healthcare. Of this figure, 6.3% (i.e., 0.55% of GDP) goes to medical technology. In the US, expenditure on healthcare is about 15.3% of GDP, with 5.5% (i.e., 0.84% of GDP) spent on medical technology [5].

The Biomedical Technology Industry

Health related industries play a major role in the EU economy. There are approximately 11,000 biomedical technology companies in Europe; over 80% are small and mid-sized. Overall, the European medical technology industry employs about 435,000 people [5]. With 110,000, or 25.3%, Germany has the largest percentage of total European employment in medical technology, while the number of medical technology employees per capita is highest in Ireland, at 0.62%, versus 0.13% in Germany. With a one third share of the world market (€187 billion), the value of the EU medical technology market, (€63.6 billion in 2005), is the second largest after the US (42%, €79.4 billion). For reference, the Japanese market represents about 10% (€18.8 billion) of the global medical technology market, the Chinese market about 2% (€3.1 billion), and the Brazilian market about 1% (€2.6 billion). The European medical technology market is growing at an annual rate of 5% to 6% [5].

Germany has the largest individual medical technology market among European countries, representing nearly one-third of the total. The five biggest countries (Germany, France, Italy, the United Kingdom, and Spain) represent 77% of the market, while Germany and France together constitute nearly half.

The European medical technology industry invests on average 6.35% of its sales in research and development, against 12.9% for the US, and 5.8% for Japan. This means Europe invests less than half US in future innovation compared to the US competitors [6]. Although EU medical device production has recorded growth rates that are well above the manufacturing sector averages in recent years [7], the EU healthcare industry is losing in competitiveness compared to its major competitors, the US and Asia. This is mainly due to mergers, takeovers, and buyouts, which are usually followed by a progressive move of all research and development operations to the overseas headquarters, leaving the EU out of the innovation loops.

Amongst all differences in competitiveness between the US and Europe, the most prominent is the relative lack of willingness in Europe to take business risks for fear of failure. In a recent survey among the members of Eucomed, the European association of the medical technology industry, 96% of the respondents acknowledged this fear [8]. Two main factors seem to drive this gap: a cultural aversion to risk, and the lack of venture capital in Europe.

Human Resource Development

Human resource development is an important prerequisite for the development and widespread introduction of new technologies in the health sector. Central to its support is the need for public awareness and recognition of the importance of science coupled with the need to improve the image of scientists and researchers within society. Further, investment in necessary human resources must be geared toward keeping promising human capital in Europe, and attracting world-class researchers from abroad. In an era where many basic sciences are facing declining student numbers because of loss of appeal, the multi-faceted aspects of bioengineering and biotechnology make it well suited to attracting top-quality talent, and to keeping this creative intellectual potential firmly linked to an important core science for Europe's transition to a knowledge society. Academic programs incorporating biomedical sciences are experiencing an increase in the number of students. In particular, it appears that bioengineering and biotechnology is an attractive field for women. At present, programs with a bioengineering component have considerably higher percentages of female students than other engineering programs.

Although Europe produces more new science and technology PhDs per capita than the US and Japan, the number of full-time researchers is significantly lower in comparison. This is because a large number of European PhDs are not employed in research, or leave the EU research system to work abroad. Most researchers who go abroad work in the US, which offers appreciably higher salaries to attract top scientific talent [9]. In addition, there are considerable cultural and political roadblocks for young scientists who want to return home after a post doctoral period spent in the US [10]. However, tighter immigration rules in the wake of the 2001 terrorist attacks in New York have put increasing restrictions on the mobility and recruitment of foreign science and engineering talent to the US [11,12], whereas new policy schemes by the EU Commission are designed to facilitate repatriation, and to attract and maintain promising human capital through legal initiatives to improve the entry conditions of foreign researchers into the EU [13]. An effort to enhance student and researcher mobility through out the ERA is creating a multicultural space for science and research, with attractive research training and working conditions for young researchers.

Education

Until now, only a few pure bioengineering and biotechnology programs have been available in European universities. In contrast to polytechnic schools, which offer a larger number of bioengineering (specifically, clinical engineering) and biotechnology programs, most universities provide bioengineering and biotechnology education only as part of their traditional engineering and science programs. Undergraduate bioengineering programs as a first step toward medical school are almost completely unknown. Nevertheless, there is widespread recognition of the need for high quality bioengineering education and training throughout Europe. Establishment of independent bachelors/masters programs in bioengineering and biotechnology is progressing quickly, and a number of advanced schemes are being developed [14].

The rapid development of the EU over the past decade raised the issue of mobility, and the related problems of mutual recognition of academic degrees. While according to European law, all citizens of the EU are free to settle and work in any of the member countries, there are still some restrictions due to the recognition of professional qualifications. As a consequence, the Bologna Declaration of 1999 was designed to help achieve Europe-wide harmonization and quality control of higher education [15] through the creation of a European Higher Education Area (EHEA). The Declaration was originally signed by 29 European countries, including 14 non-members of the European Union, which by 2007 has grown to incorporate 27 countries. Participation in the implementation of the EHEA, the so-called Bologna Process [16], grew rapidly and now includes 45 European countries. For most of them, the Bologna Process has caused major, in some cases even fundamental, changes in the national systems for higher education.

The Declaration asks for the promotion of European cooperation in quality assurance, with a view to developing comparable criteria and methodologies, and the advancement of the necessary European dimensions in higher education, particularly with regards to curricular development, inter-institutional cooperation, mobility schemes, and integrated programs of study, training, and research. A system of credits, the European Credit Transfer System (ECTS), is being established as a proper means of recognizing and facilitating student mobility [17].

The declaration involves six actions related to employability, mobility, compatibility, and quality assurance to be completed by 2010:

- adopt a system of easily readable and comparable degrees;
- adopt a system with two main cycles (undergraduate/graduate), which was in 2005 expanded to include doctoral degrees as a third cycle;
- establish a system of credits;
- promote mobility by overcoming obstacles;
- promote European cooperation in quality assurance;
- promote European dimensions in higher education.

The aim of the process is to make the higher education systems in Europe converge toward more transparency, where the different national systems would use a common framework based on three cycles, with the bachelors, masters, and doctorate degrees replacing the traditional national degrees, such as the "diploma" and degrees.

As a unique opportunity to promote bioengineering and biotechnology, the Bologna movement has triggered harmonization of the educational programs, specification of minimum qualifications, establishment of educational quality control criteria, and prioritization of life-long learning skills, allowing mobility in education and employment. Implementing the EHEA for bioengineering and biotechnology is also furthering the competitiveness of European education, industries, and healthcare systems. A survey of the current status of the EHEA for bioengineering and biotechnology demonstrates that its implementation is steadily progressing, and that it will substantially benefit the quality of education and training [18].

European harmonization of higher education programs, a prerequisite for student and teacher mobility, is an important political goal of the EU. All academic disciplines will be harmonized sooner or later, although most of them have no inherent reason to do so. Outcomes, course contents, and specializations in the traditional engineering disciplines are well established throughout Europe (and even globally), such that finding a suitable higher education program is straightforward between any of the European countries. Mutual acceptance of degrees, as well as student/teacher mobility between universities and countries, also pose no major challenges. However, given the rapid development of bioengineering and biotechnology, and particularly the initialization of many new specialty areas, no single European university can keep up on a broad scale. After all, most European bioengineering departments are small, employing no more than three to six faculty members who can concentrate on only a rather limited set of specialties.

Since students should be able to select and study the specialties of their choice within bioengineering and biotechnology, no matter where they come from, and what the specialties of their home university, the need for harmonized programs eliminating the obstacles for unimpeded mobility between universities has become obvious. As a response, the European project *Biomedical Engineering Preparing for the Higher Education Area* (*BIOMEDEA*) was established in 2004 with the participation of more than 80 universities, universities of applied sciences, and polytechnic schools. Additionally, guidelines for the harmonization of bioengineering programs have been developed [www.biomedea.org]. As a consequence of these developments, mobility between academic institutions has grown tremendously. This mobility is funded by various European student and faculty exchange programs, such as Erasmus and the Marie-Curie Actions [www.europa.eu]. An example of the type of impact these developments have had can be seen in the education of German bioengineering and biotechnology students, who generally spend six to twelve months in a study-abroad program.

In addition to the Bologna process, innovative teaching strategies and methods, as well as new educational programs, are emerging throughout Europe, filling the gap between engineering and life sciences with a variety of new specializations. It is no surprise that *Nature* magazine stated in its September 2003 *'Careers and Recruitment'* section [19] that "Europe chips in for training. The US may have more coordinated funding, but Europe is taking the lead in training bioengineers." This development may be due to the diversity of the European education systems, which contrasts with the homogeneous higher education system of the US, and the extra effort that European educators have taken to harmonize their programs. It should be noted in this context that harmonization within the EHEA is not to be confused with standardization. Each university in every country is allowed, even encouraged, to keep its special character and programs, as long as the criteria for outcome qualifications are met and easy student/teacher exchange is made possible. In addition to European cooperation between universities, there have been strong interactions between Europe and the US, including European participation in the Whitaker Summits[1] on bioengineering education, where new methods, best practices, and contents have been discussed and accepted in transatlantic agreement. The education environments are different, however. In the US, the universities are largely autonomous, unlike most European countries, where higher education institutions are under direct governmental control.

Sadly, European bioengineering and biotechnology have not received independent funding, such as that of the US based Whitaker Foundation, for the initiation of new bioengineering programs and research. Also, language differences form a major barrier. With the exception of a few countries, such as Sweden and The Netherlands, which offer courses in English despite the fact that it is not

[1] The Whitaker Foundation, founded in 1975, was an organization which promoted biomedical engineering education and research. To have the maximum impact, the foundation decided to spend its financial resources over a limited period and closed in 2006, after contributing more than US$700 million to universities and medical schools. The foundation helped create 30 biomedical engineering programs at various universities in the United States and organized two Biomedical Engineering Educational Summits to help universities design and modify biomedical engineering programs to meet future needs.

their official language, most EU countries use their own languages for teaching. Thus, graduate and undergraduate student exchanges between, for example, Spain and Finland, or Denmark and France can be difficult. Instructors have the same problem.

Present Situation and Expectations for the Future

Information on post-Bologna developments involving bioengineering education and training in 28 countries has been gathered in the International Federation of Medical and Biological Engineering (IFMBE)/EAMBES White Paper on the status of bioengineering and biotechnology in Europe [18]. The Bologna Declaration envisages two main educational cycles, undergraduate and graduate. The first cycle lasts a minimum of three years, and leads to the award of a degree that is, in the words of the Declaration, "relevant to the European labor market as an appropriate level of qualification" [15]. In contrast to the US, this first level of professional qualification is not yet accepted by employers in most European countries. Thus, more than 90% of European university graduates are enrolling in the second cycle, though some countries intend to limit access to the second cycle to a certain percentage of first cycle graduates. The successful completion of the first cycle degree is required for access to the second cycle degree that leads to the masters and/or doctorate degree. The implementation of the Declaration has led to a situation where the minimum of three years for the first cycle is becoming the standard duration.

Bioengineering education is available in nearly all the countries covered by the White Paper. The programs are based on two models, one in which bioengineering is a component of another major, and the second in which the degrees are nominally in bioengineering.

There are many undergraduate degree programs in conventional engineering subjects – particularly electrical, mechanical and chemical – which contain bioengineering options or electives. The "bio" or "medical" content appears not to be a route to practice as a professional bioengineer, but instead has the educational aim of providing examples of the application of conventional engineering that forms the majority of the degree content in an interesting and demanding context. Biomedical applications appear to be particularly popular in electrical engineering, with an emphasis on biomedical electronics and instrumentation, as well as signal and image analysis. The situation with post graduate degrees is less clear, but there are some graduate degrees in electrical and mechanical engineering that are awarded with a major in bioengineering.

First cycle programs that lead to a bioengineering degree are either standalone programs, or the first stage of a two-cycle degree. Both types are generally three years in duration. The standalone programs usually lead to technical/technician level qualification. As with the mixed undergraduate courses, there is a strong emphasis on electronics and instrumentation, but there is also an industrial bioengineering qualification in some countries. The situation in Ireland is rather different. Here it may take four years to receive the first degree, which is at a professional rather than technical level. Some graduates may still find employment at a technical level. Some universities provide two-cycle (3+2) fully integrated bioengineering programs; that is, instruction in the life sciences is integrated into the coursework from the beginning.

The structure of second cycle bioengineering degrees is particularly variable, and even within a country there are often major differences. There are, however, three common models:

The second cycle follows the first cycle as an integrated course leading to a single degree. The second cycle components last either one or two years, commonly producing a five-year degree, although four years is more usual in England, where the resulting qualification is generally called an undergraduate masters degree.

A wholly bioengineering postgraduate degree, with completion of a first cycle degree as an essential entry requirement. The nature of the first cycle degree is, once again, variable. In many cases, the first cycle degree must be in engineering or a physical science, but there are also degree programs that will accept those with life science, medical, or paramedical degrees. The degrees are of one- or two-year duration, and generally contain both instructional and research components, although the balance of components appears to extend from virtually all instruction to virtually all research. There is a lack of common degree names, although Master of Science is widespread.

Third cycle bioengineering degrees. The Bologna Declaration originally envisioned only two educational cycles, the second cycle being a masters or doctoral degree. In practice, students can, and often do, progress from a masters degree to a separate doctoral degree, usually a PhD, which represents a third rather than a second cycle degree. Entry conditions often, but not always, require the candidate to have a second cycle degree. The PhD in Europe has traditionally been rather like a research apprenticeship, being almost entirely research based. There is a growing tendency for the degree to contain instructional material, often with credit for prior material in a preceding masters degree. The usual minimum duration of study is three years.

Provision and Accreditation of Training

There appears to be little provision in Europe for training those entering employment in a hospital setting. There are, however, training schemes reported for Denmark, the United Kingdom, the Netherlands, and Ireland. In Denmark, there is a three-year integrated postgraduate educational and training program that leads to certification as a clinical engineer. Similarly, there are two-year programs for clinical technicians.

In the United Kingdom, the trainees are employed in the National Health Service at an appropriate training grade. The training scheme normally lasts for 6 years, being divided into 3 training periods. Basic training that combines both training and education is normally the entry into the profession. This basic training usually lasts two years, reduced to at least 15 months for trainees with an MSc accredited by the national society, the Institute of Physics and Engineering in Medicine (IPEM). Successful completion of the basic training scheme leads to the award of a Diploma of the national society and an MSc. Basic training is followed by a four year program of advanced training and responsibility. Training takes place in centers that have been accredited by IPEM, which also has a syllabus of the competencies to be developed by the trainee.

Registration for Bioengineers

Registration for bioengineers is only possible in a few European countries. There is a voluntary register organized and administered by the national bioengineering societies in Ireland, Norway, and Sweden. In the United Kingdom, voluntary registration organized by the national bioengineering society, the IPEM, has been replaced with compulsory registration. All those who interact with patients, either directly or indirectly, must now be registered with the Health Professions Council as clinical scientists. Applicants for registration must show that they have achieved the competencies required after four years participation in the IPEM training scheme, although the competencies may also be obtained by alternative, non-IPEM routes. Ongoing professional development will be a requirement for continuing registration.

Employment, Career Development and Mobility

Job opportunities for bioengineers and biotechnologists are vibrant in Europe, with more than double the average predicted rate of increase compared to other engineering fields [20]. In addition to job offers from the bioengineering, medical device, and biotechnology industries, graduates with a bioengineering or biotechnology degree are also sought after for innovative career opportunities, primarily as a consequence of interdisciplinary education.

About 80% of all graduates in bioengineering develop careers in sales and marketing, meaning they do not necessarily need the research oriented education which is the forte of most universities. Thus, we see a clear split. Graduates from practice-oriented institutions, which include universities of applied sciences and polytechnic schools, move primarily towards sales, marketing, and clinical engineering jobs. University graduates, on the other hand, move into bioengineering research jobs at the universities, in research institutions, and in industry. This educational incongruence may partially disappear in the near future with the cessation of the traditional "diploma" degrees and the introduction of the new bachelors and masters degrees mandated by the Bologna Process. The new bachelors degrees in particular are becoming more practice-oriented, since they must be suitable as a first professional qualification. While more than 90% of students with a bachelors degree currently go on to receive a masters degree, or even head directly toward a doctoral degree, it remains to be seen how the situation will change industry starts accepting the bachelors degree as a professional qualification and some governments begin limiting access to graduate degrees.

Availability of higher education bioengineering-related programs for students other than the traditional full-time student, such as those already holding a job, or looking to gain continuing education credit, is not wide-spread in most European countries, except for clinical engineering. An established link between scientific education and professional formation exists only in the United Kingdom, Finland, and France.

Today, more than 200 universities, polytechnic schools, academies, and other institutions in Europe offer educational programs in bioengineering and biotechnology at all academic levels. Coordination of contents and required outcome qualifications is in most cases based on the *Criteria and Guidelines for the Accreditation of Biomedical Engineering Programs in Europe* [21], and *European Protocol for the Training of Clinical Engineers* from the BIOMEDEA project [22]. With high mobility, no or very low tuition fees in the European higher education systems, and ample funding for exchange programs, there is a high quality education in bioengineering and biotechnology available for everyone, anywhere in Europe.

References

1. Siebes, M. et al., Engineering for Health: A partner in the development of a knowledge-based society for the benefit of European healthcare. 2005; Available from: www.eambes.org/docs/MBES-position-paper-final.pdf
2. The Lisbon Strategy, ftp://ftp.cordis.europa.eu/pub/indicators/docs/3rd_report_snaps0.pdf
3. EU-funded research, www.ec.europa.eu/research/fp7/pdf/fp7_press_launch.pdf
4. King DA, "The scientific impact of nations." *Nature* 430: 311–316, 15 July 2004
5. Eucomed Industry Brief 2007, www.eucomed.org/publications.aspx
6. The medical technology industry at a glance, AdvaMed Chartbook, 2004
7. Medical Devices – Competitiveness and impact on public health expenditure, Independent Study by CERM, Rome, for the EU DG Enterprise, July 2005
8. Competitiveness and Innovativeness of the European Medical Technology Industry, 2007, www.eucomed.org/sitecore/shell/Controls/Rich%20Text%20Editor/~/media/pdf/tl/2007/portal/publications/compsurvey.ashx
9. Third European Report on Science and Technology Indicators 2003 "Towards a Knowledge-based Economy." European Commission DG for Research, EUR 20025 EN
10. S. Goodman, "Getting mobile in Europe." *Nature* 427:868–869, 2004
11. G. Brumfiel, "Security restrictions lead foreign students to snub U.S. universities." *Nature*, 431: 231, 2004
12. Policy Implications of International Graduate Students and Postdoctoral Scholars in the United States, National Academies report, May 10, 2005, www.books.nap.edu/catalog/11289.html
13. Taking Action to Stop EU Brain Drain, EC Press release, IP/03/1051 July 18, 2003
14. Nagel, JH, Slaaf, DW and Barbenel, J, "Medical and Biological Engineering and Science in the European Higher Education Area." *IEEE Engineering in Medicine and Biology Magazine*, Vol. 26, No. 3, 2007, pages 18–25.
15. The Bologna Declaration of 19 June 1999, www.bologna-berlin2003.de/pdf/bologna_declaration.pdf
16. From Bologna to Bergen, www.bologna-bergen2005.no/
17. ECTS User's Guide, www.hrk.de/de/download/dateien/ECTSUsersGuide(1).pdf
18. J.H. Nagel (Ed), Biomedical Engineering Education in Europe – Status Reports, www.biomedea.org/Status%20Reports%20on%20BME%20in%20Europe.pdf

19. Jox, R, Europe chips in for training, *Nature*, vol. 425, p326, 2003.
20. Gewin, V, Biomedicine meets engineering, *Nature*, vol. 425, 324–325, 2003.
21. Criteria and Guidelines for the Accreditation of Biomedical Engineering Programs in Europe, www.biomedea.org/Documents/Criteria%20for%20Accreditation%20Biomedea.pdf
22. European Protocol for the Training of Clinical Engineers, www.biomedea.org/Documents/European%20CE%20Protocol%20Stuttgart.pdf

3

Bioengineering and Biotechnology: An Asia-Pacific Perspective

Makoto Kikuchi, PhD

Department of Medical Engineering, National Defense Medical College, Saitama, Japan

James C.H. Goh, PhD

Department of Orthopaedic Surgery and Division of Bioengineering, National University of Singapore, Singapore

The world's highest priority in the 21st century is healthcare. In economically developed Asian-Pacific societies, there is a well-established perception that improvements in healthcare and quality of life depend on advancements in bioengineering and biotechnology.

Many countries in the Asia-Pacific region have strong human resources, rich clinical expertise and, of course, a large population – all valuable for translational medical research. By the second half of the 21st century, countries such as China and India, with their large populations, will also have a high percentage of aged individuals. Therefore, bioengineering and biotechnology developments directly related to medical care, healthcare, and welfare are considered crucial. In addition to demands placed by globalization on the development of medical technologies, there are also increasing expectations for the creation of new bioengineering and biotechnology products stemming from breakthroughs in life sciences.

For example, in Japan, the Second Period Research Strategy of the AIST (National Institute of Advanced Industrial Science and Technology), established in April 2006, has indicated that developments in science and technology focusing on humans will become a mainstream trend. Further, a strategic target for research in bioengineering and biotechnology has been set in the field of health. Recently, several governments in the Asia-Pacific region have substantially increased their support for research and training in bioengineering and biotechnology. Although absolute funding levels are lower than in the US, the rates of increase in many

G. Madhavan et al. (eds.), *Career Development in Bioengineering and Biotechnology*, DOI: 10.1007/978-0-387-76495-5_3, © Springer Science+Business Media, LLC 2008

countries in the Asia-Pacific region, such as Japan, Korea, Taiwan-China, and Singapore, are higher than those in the US and European countries.

In this chapter we focus on an overview of the recent status and developments in bioengineering and biotechnology in Asia, using Japan and Singapore as specific examples.

Bioengineering and Biotechnology in Japan

Education and Training

Generally speaking, Asia-Pacific countries have lagged behind the US in creating a well defined curriculum of bioengineering education for universities and colleges. This has a great deal to do with the relation between medical schools and undergraduate engineering education. In the US medical school is considered a graduate-level professional degree that can provide many career opportunities. In Japan, however, medicine is an undergraduate degree, so a person graduating with an undergraduate degree in bioengineering typically does not have the option of returning for another undergraduate degree to enter medicine. The Asia-Pacific programs therefore developed only gradually as the industrial market for bioengineers emerged. In Japan especially, many bioengineering programs are strongly grounded in one of the more traditional engineering disciplines, and do not have the breadth or extensive bio-content that characterizes US programs.

Collaborative, inter-disciplinary research in Japan is regarded as very important in the life sciences, and is expected to help further develop related human resources. With advancing life sciences research, and a solid commercial market in Japan, bioengineers are in great demand. This is all the more so due to the revision of the Pharmaceutical Affairs Law, in April of 2005, which mandated placing a person with expertise in bioengineering in charge of supervision and management in each medical device or pharmaceutical firm engaged in manufacturing and sales. Educating bioengineers has clearly emerged as a critical issue.

As mentioned above, in spite of high demand and increasingly diverse global trends, the status of bioengineering faculties and graduate schools in the field of bioengineering is relatively poor in Japan compared to the US, Europe, and other Asia-Pacific countries such as Singapore and Taiwan. However, a major movement is emerging to establish educational organizations related to bioengineering at the Japanese National University Corporation, and in public and private colleges and universities as well. For example, once support for Centers of Excellence in bioengineering was approved by the Japanese Ministry of Education, Culture, Sports, Science, and Technology, bioengineering research and educational institutions achieved additional support. Unfortunately, the educational aim of the undergraduate bioengineering courses currently being taught in some universities is not clear. But graduate schools are making changes so as to excel in advanced creative research across the board in the international arena, particularly in micro and nanotechnology areas. These schools will be aiming to gain the acceptance of students who have backgrounds in a variety of fields, including engineering, medicine, dentistry, science, pharmacology, agriculture, economics, and law.

There is also a growing need for clinical engineers in Japan. Clinical engineers are classified as those who provide support for the use of medical devices by healthcare professionals. Hospitals are the largest buyers of medical devices, with the purchase processes usually managed by clinical engineers. In Japan, the "Clinical Engineering Law" was established in May 1987, and in the following year the "National Examination of Clinical Engineering Technologists" was initiated, based on the provisions in Article 17 of the Law for Clinical Engineering Technologists. The Japan Association for Advancement of Medical Equipment (JAAME) was appointed by the Ministry of Health, Labor, and Welfare as the designated testing agency. It, executes a program related to the administration of the national examination of clinical engineering technologists. As of March 2007, after the 20th clinical engineering national examination, the number of clinical engineering technologists in Japan reached 23,101–a small number, compared with other governmental medical jobs. For instance, as of 2005, when we compare the number of medical staff personnel per 100 beds in general hospitals, the number of medical engineering technologists was only 0.7, whereas the numbers of doctors, nurses, and clinical laboratory technicians were 12.5, 38.4 (nursing assistants 10.6), and 3.2, respectively. (Note that in Japan, the English expression, "clinical engineering technologist" is now used in official documents.)

According to the revised Medical Service Law established in April 2007, the placement of a "manager for medical device safety" became obligatory in all Japanese medical facilities. Thus, the importance of clinical engineering technologists is growing. At the dawn of education in bioengineering, about 30 years ago in Japan, emphasis was mainly placed on training "dialysis technologists" and technologists in manufacturing companies of medical devices. Subsequently, however, there has been a great change in the training of clinical engineering technologists. In 1988, there were five facilities training clinical engineering technologists, with a total of 370 students. By April 2007, there were 2,815 students (about a 7.5-fold increase) in 62 training facilities (12.4-fold increase). The increase in the number of training facilities is attributable to not only a trend that clinical engineering technologists are increasingly required, with the changes in medical technology and incidences of diseases and/or causes of deaths, but also to the growing numbers of students, who are considering their future professions and have expectations toward job with national credentials, as well as the innovation that has flourished under government-promoted deregulation. Unfortunately, the establishment of a strategy to develop national credentials for clinical engineering technologists has lagged behind other medical job categories. The situation has been made more complex because, as an urgent demand for clinical engineering technologists was expected, a number of training programs were legally established, including programs students could take after high school graduation.

The "clinical engineering technologist" training program in Japan, a 3-year technical training program, was established as an expanded version of a

2-year bioengineering program. In 1994, a 4-year college program and a non-degree graduate program (a 2-year night class equivalent to a 1-year accredited program) were established. A college program to enhance training and research activities for leaders was subsequently established. In some institutions this was accomplished by simply reorganizing the engineering faculty. Ultimately, the value of university training for clinical engineers can sometimes be difficult to deduce.

At present, if students fulfill the requirements assigned by the Ministry of Education, Culture, Sports, Science, and Technology in a 4-year training program (for a total of more than 3400 lecture hours), they can receive an advanced specialist degree. Until recently, after the students fulfilled the 3-year program qualifications for a specialist, they had been examined for admission to graduate school individually by the judgment of each graduate school. Nowadays, they are entitled to go on to a graduate school as 4-year college graduates as long as they hold the "advanced specialist" degree. With the support and backing of the newly renovated educational system, at present, only three institutions nationwide, totaling 120 students, receive training in a 4-year specialized program. The percentage of students in clinical engineering technologist programs is only 4.1%, in contrast to those in training to become physical therapists, for example, at 37.4% (75 schools with 4402 students). Training in a 4-year college program should produce highly qualified clinical engineering technologists in the not too distant future.

Research Trends

Since the beginning of the 21st century, the Japanese government has tried to stimulate the economy and new business by emphasizing specific fields, such as information and communications technology, nanotechnology, novel materials, tissue engineering, molecular biology, and environmental science. Bioengineering has diversified markedly, and Japanese research has expanded to involve neural and tissue engineering, although electrical and mechanical engineering applications still make up the major part of the medical device industry. In the last few years, Japanese bioengineering has been at the forefront of cardiovascular research, biorobotics, and rehabilitation engineering for elderly care. The Ministry of Economics, Trade, and Industry has focused on an experimental house called "The Welfare Techno-House." Sixteen such houses have been built around the country, and research and development is producing minimally invasive or wearable technologies specifically for elderly care.

Japan is also a world leader in minimally invasive instrumentation technologies and molecular imaging, and is currently expanding efforts in the area of advanced endoscopic methods for diagnostics and therapeutics, with interface ability into drug, cell, and gene delivery systems. With these kind of advancements in the offing, the exciting coordination between medicine, engineering, and business will create more challenging intellectual and professional opportunities for bioengineers and biotechnologists.

Bioengineering and Biotechnology in Singapore

Education and Training

Significant growth in bioengineering and biotechnology is projected in Singapore, making investment in the development of human capital crucial. As a consequence, Singapore's Education Ministry has revamped the school curriculum to include the life sciences in addition to mathematics, the physical sciences, and engineering. In February 2001, the Biomedical Sciences Manpower Advisory Committee (BMSMAC) was established to identify the key manpower needs for the manufacturing industry, and to propose solutions for these needs. Subsequently, the National Science Scholarships were launched in July 2001 to provide financial support and advanced educational training to strengthen the manpower capabilities of the bioengineering and biotechnology industry. The BMSMAC works closely with Singapore's post-graduate institutions to ensure that bioengineering and biotechnology program are current and able to meet the industry's manpower needs. In recent years, the National University of Singapore (NUS) and the Nanyang Technological University (NTU) have established new departments in bioengineering and life sciences to offer new undergraduate degree programs. There are also new specializations emerging in traditional engineering departments, for example, biomechanical engineering in the mechanical engineering department, and biopharmaceutical engineering within the chemical engineering department. Academic departments are going through a makeover to ensure that their programs remain relevant and able to train manpower for the bioengineering and biotechnology industry. The typical academic coursework in Singapore universities provides the students with strong training in engineering and life sciences. Laboratory rotations are mandatory for students to allow them to identify opportunities to train under top researchers, and to allow them to thrive in a community of motivated student peers.

Many innovative research and academic partnerships are also currently under formation, including the notable collaboration between the NUS, NTU, and the Massachusetts Institute of Technology (MIT) in the US, commonly called the "Singapore-MIT Alliance." This partnership is between the computational and systems biology program at MIT, the biology, bioengineering, and biotechnology programs at NUS, NTU, and the A*STAR Research Institutes. With research funding and innovative partnerships, Singapore expects to be home to hundreds of key researchers at the doctoral level, as well as to many engineers and scientists in the bioengineering and biotechnology industry.

Industry and Research Trends

Bioengineering and biotechnology activities in Singapore started more than two decades ago. However, these remained largely academic and confined to pockets of researchers in the universities and research institutes. The big boost came at the turn of the millennium, when the Singapore government announced the development of a Biomedical Sciences Cluster as one of four key pillars of the Singapore economy. The cluster included pharmaceuticals, medical device technology, biotechnology, and healthcare services. This initiative was spearheaded by the Singapore's Economic Development Board; its investment arm, Bio*One Capital; and the Agency for Science, Technology, and Research's (A*STAR) Biomedical Research Council. These three organizations were charged with developing Singapore's industrial, intellectual, and human capital to support this initiative.

The Agency of Science Technology and Research (A*STAR) was tasked to build strong capabilities in basic research, which included the development of physical research infrastructure. In 2003, a dedicated $300 million R&D complex known as Biopolis was opened. The 185,000 m^2 complex is home to five biomedical public research institutions and laboratories. The five research institutes are the Institute of Molecular and Cell Biology, the Institute of Bioengineering and Nanotechnology, the Singapore Genome Institute, the BioInformatics Institute, and the Bioprocessing Technology Institute. The Biopolis complex is expected to house up to 30,000 researchers and 1,000 residents. It will also cater to pharmaceutical and biotechnology companies, allowing them to access the scientific facilities and services, thus facilitating cross-disciplinary research and public-private research collaborations. In tandem with infrastructural development, the Singapore government has injected fresh research funding into basic and translational research. Above and beyond these investments in June 2002, the Ministry for Trade and Industry had announced a S$1 billion fund to attract world class companies to carry out their commercial R&D activities in Singapore. EDB, through its investment arm (EDB Investments Pte Ltd.), also allocated S$ 1 billion to a dedicated Biomedical Sciences Investment Fund (BMSIF) to support the development of the Biomedical Sciences industry. Currently, Singapore hosts six of the top ten pharmaceutical conglomerates, and there is a growing base of medical technology companies within the 183-hectare Tuas Biomedical Park. Overall, the bioengineering manufacturing industry is enjoying strong growth in Singapore, with a significant 30.2% manufacturing output growth to S$23 billion in 2006. This growth is expected to continue as global bioengineering companies set up or expand their manufacturing activities.

In a further intensification of R&D efforts, the Singapore government established the National Research Foundation (NRF) in 2006 to coordinate the research activities of different agencies within the larger national framework, and to fund strategic initiatives. The three areas of research focus are Biomedical Sciences, Environmental and Water Technologies, and Interactive and Digital Media. An initial funding of S$5 billion over 5 years was approved for NRF to achieve its strategies. The goal in R&D expenditure is to reach 3% of GDP by 2010. The NRF is targeting the three areas to double their total number of jobs to 80,000, with value added of S$27 billion a year by 2015.

Clinical research and clinical trials management are also on the increase within the bioengineering and biotechnology framework. Singapore is the Asia Pacific Economic Cooperation (APEC) Coordinating Centre for Good Clinical Practice, and has implemented many initiatives, such as the training of clinical research personnel and the creation of a conducive environment for multi-site clinical trials in the region. A number of leading global pharmaceutical companies, including AstraZeneca, Aventis, Eli Lilly, GlaxoSmithKline, Merck & Co, Novartis, Novo Nordisk, Sanofi-Synthelabo, and Schering-Plough, have set up their regional clinical trial centers in Singapore. To complement the push in bioengineering and biotechnology research, a national Bioethics Advisory Committee has been recently established to address ethical, social, and legal aspects of research, and to provide guidance on the development of national policies and regulations. The principle is to protect the welfare and rights of individuals, while allowing research to develop and realize its full potential for benefit of mankind. A regulatory framework has also been established to oversee research with intellectual property protection.

Concluding Remarks

As mentioned above, bioengineering and biotechnology have been experiencing significant transformations since the dawn of the 21st century. In addition to conventional technologies, biotechnology and nanotechnology are being fused anew into health technologies. The advanced healthcare that our citizens are benefiting from every day is made possible, in part, due to significant efforts of bioengineers and biotechnologists. Medical technologies have, in the past, shown progress in ten-year stages. For example, computerization and systemization of measuring and monitoring equipment for biophenomena were implemented in the 1960s. Various medical imaging technologies were born in the 1970s. Minimally invasive diagnostics and treatment technologies appeared in the 1980s, and robotic surgery was realized in the 1990s. An extremely interesting fact is that the many medical systems also underwent radical transformations synchronously with these stages. In the 1960s, it was "scaling-up of the volume of health care" by establishing more medical facilities. In the 1970s, it was "improving the quality of health care." In the 1980s, it was "the issue of medical costs." And finally in the 1990s, it was "the balance between the cost and quality of good health care." We now see that the nature of healthcare has actually been affected directly by the progress in, and supply of, bioengineering and biotechnology. This fact indicates the necessity of strong coordination between technology development and the social system, suggesting that developments in bioengineering and biotechnology are not merely limited to the realm of scientific research. A stable source of advanced healthcare represents the needs of the public and society. As the future continues to be bright, bioengineering and biotechnology will be key in effectively forming translational research partnerships to further new knowledge and health.

References

1. Makoto Kikuchi, "Present State of Biomedical Measurement Technologies and Future Prospects," *AIST TODAY*, English version, pp. 4–7, 2007
2. Toshio Tamura, (guest editor), "Recent Technological Developments in Japan," *IEEE EMB Magazine*, Vol. 24, No.4, 2005

3. Japan Association for Clinical Engineering Technologists; www.jacet.or.jp.
4. Japan Association of Educational Facilities of Clinical Engineering Technologists: JAEFCET); www.jaefcet.org. Biomedical Sciences in Singapore; www.biomed-singapore.com/.
5. Singapore's National Research Foundation; www.nrf.gov.sg/.
6. Biotechnology in Singapore. Goh R, Biotechnology Team, Western Australian Department of Industry and Resources, February 2005; www.doir.wa.gov.au/documents/business andindustry/BIOTECHNOLOGY_IN_SINGAPORE.pdf.
7. National University of Singapore; www.nus.edu.sg/.
8. Nanyang Technological University; www.ntu.edu.sg/.

4

Employment Outlook and Motivation for Career Preparation

John D. Enderle, PhD

Department of Biomedical Engineering, School of Engineering, University of Connecticut, Storrs, Connecticut, USA

Bioengineering has emerged as a professional discipline largely in answer to the increasing dependence of modern health care on complex diagnostic, screening, monitoring, prosthetics, surgical instrumentation, and biotechnology. Bioengineering education first evolved as a bridge discipline that united engineering expertise with medical or biological science. Today, bioengineering education integrates engineering and life sciences [1].

A vast number of bioengineering specialties have arisen, ranging from theoretical, non-experimental undertakings to state-of-the-art applications, and can encompass research, development, implementation, operations, and management. Bioengineers may be involved in the design and development of a broad spectrum of medical and biological products, cutting-edge research at the molecular or systems level, computer modeling, or telecommunications.

Bioengineers are employed in hospitals, industry, government, universities, and medical schools, and are essential to the success of health care systems throughout the world. Furthermore, industry around the world is increasingly embracing bioengineering and science as both a leading and lucrative investment area, with the potential to drive huge technological advances. Bioengineering graduates will find a strong and expanding job market with many diverse opportunities.

This chapter first introduces trends in bioengineering educational programs. A description of future opportunities is described, followed by a survey of employment and a career outlook, with some thoughts on finding a job in the realm of bioengineering.

Trends in Bioengineering Education

Advances in physics and chemistry characterized the major accomplishments of the 20th century, which dominated most aspects of bioengineering education. The 21st century, on the other hand, is being characterized by advances in biotechnology.

G. Madhavan et al. (eds.), *Career Development in Bioengineering and Biotechnology*,
DOI: 10.1007/978-0-387-76495-5_4, © Springer Science+Business Media, LLC 2008

This has supported the evolution of bioengineering education from a bridge to a seamless integration of engineering and the life sciences [1]–[2].

Bioengineering graduate programs have existed since the early 1960s. From these graduate programs evolved undergraduate degree programs. Accredited undergraduate bioengineering programs, first created in 1972 at Rensselaer Polytechnic Institute and Duke University, grew to 19 by 1995. In 2003, there were 74 unaccredited undergraduate bioengineering programs [3], and 28 accredited programs. In 2006, there were 44 accredited programs. It is expected that the remaining programs from 2003, and many additional programs, will become accredited in the next few years, delayed because of the requirements for accreditation [3]. Around the world, bioengineering education activities are similar to the experience in the US. For example, in Europe there is an ongoing major harmonization effort in accrediting bioengineering programs through the BIOMEDEA project [4].

One of the major reasons for the growth of undergraduate bioengineering programs in the US is the funding provided by the Whitaker Foundation, which provided $710 million over a thirty year period, beginning in 1976, to support bioengineering [2]. The Whitaker Foundation, which closed on June 30, 2006, has left a lasting imprint on the field of bioengineering.

Paralleling the growth of new programs, there has been an increase in the number of bioengineering undergraduate students, from approximately 5,000 enrolled in the US bioengineering programs in 1995, to over 14,000 in 2004 [2]. Increases in the number of bioengineering graduate students are expected in the future.

Bioengineering is an unusual field in that, unlike other fields of engineering, 56% of undergraduates went on to further education (medical, dental, and engineering graduate programs) during 2001–2004 [5]. Thus, the impact of the almost three-fold increase in the number of bachelor-level graduates over the past decade is smaller than what would have been in other engineering fields because less than half go directly into industry. An indication of the growing popularity of bioengineering programs is that while the number of students pursuing bioengineering is only approximately 4% of the total engineering enrollment in the US, those universities that do offer an undergraduate bioengineering degree program typically find these programs to have the largest enrollments.

The ways in which bioengineering programs offer a curriculum are changing, especially with ABET allowing programs to define themselves according to the needs of their constituencies [3]. Advances in life sciences, for example, have caused many bioengineering programs to increase life science content. However, bioengineering programs are also constrained by the fact that accreditation requirements require a minimum of 1 1/2 years of engineering topics. Novel programs are now introducing bioengineering courses that integrate both the life sciences and engineering in the same classes, rather than offering a curriculum of engineering classes that are separate from those of the life sciences [1].

Bioengineering Industry: Past, Present and Future

When did the bioengineering industry begin? If one is flexible, the first bioengineering companies began with the dawn of history [6]–[7]. However, one could argue that before a professional society is established, the field does not exist. For bioengineering, that began with the creation of the professional group on Medical Electronics as part of the Institute of Radio Engineers in 1952 [8]. It might be surprising to realize, then, that the bioengineering industry was started by non-bioengineers and scientists.

The bioengineering industry is conventionally thought of as beginning in the 1950s and 60s, with the development of the heart-lung machine, cardiac pacemakers, artificial heart, and artificial joint replacement. In the 1970s, medical imaging became more prominent, with the use of computerized axial tomography (CAT), computerized tomography (CT), ultrasound, and nuclear magnetic resonance (NMR) imaging. The 1980s saw the further development of imaging devices, endoscopy, and lasers in medicine. It was during this time period that bioengineers began to be hired by bioengineering companies and the federal government. During the 1990s, the bioengineering industry incorporated robots in their product line, as well as implantable devices and the use of computers. In parallel, the human genome project, one of the most ambitious engineering projects of the decade, was begun in 1991, and completed—at least in rough draft form—in 2003. It is no coincidence that this decade also marked a major shift in bioengineering from organ-system level (such as cardiac prosthetics) toward the cellular and molecular level (for example, angiogenesis). In 1995, the fastest growing segment of the US economy was the bioengineering industry, with sales of $40 billion annually [8]. One drawback of the bioengineering industry is that it is very research intensive as compared with other industries, with research growth from 5.4% 1990 to 11.4% in 2002 [7]. In fact, many small startup companies spend almost their entire budgets on research and development [7]. Today, bioengineering companies generate over $77 billion in sales annually at a rate of 6% growth, twice the United States' gross domestic product growth rate [7].

Opportunities for a bioengineering faculty career in academia are excellent. The expanding number of new bioengineering departments is not expected to end, as approximately 250 engineering schools and colleges are without a bioengineering department and will probably be adding one. The bioengineering industry is currently moving in two general directions: medical devices and biotechnology. A certain blending of these technologies will occur as medical devices are miniaturized and used at the cellular and sub-cellular level [9].

Medical Device Industry

It is clear that the medical device industry will continue to create new diagnostic devices, as well as find therapeutic and assistive solutions for chronic conditions in an ever-aging population. Advances in biosensors, computers, analysis methods and algorithm development, materials, telemedicine, miniaturization technologies, and the use of organic materials with traditional medical device materials will drive the medical device industry in the future. In addition, better understanding of human physiology from gene to organ level will promote more accurate diagnostic and therapeutic devices. Organ replacement, such as the artificial heart, where donors are too few in numbers, will continue to evolve and mature.

Miniaturization of components and devices will allow short-term placement of diagnostic devices into the body, such as the 'swallowable' pill endoscope [8], and permanently implanted therapeutic and diagnostic devices that can monitor the health of an individual on a continual basis [10]–[11]. Repair and restoration of cells, tissues, and organs are important areas of research and development using biomaterials and tissue engineering [12]. Telemedicine, biosensors, and wearable technology are offering exciting opportunities that have evolved to the point where commercial products are being developed for clinical applications [13]–[14]. Assistive technology is becoming much more important with the rapidly expanding aging population. Major strides in research and treatment of balance disorders in the elderly are occurring; it should be noted that only automotive injuries are more costly than those associated with falls [14].

On the horizon is the hope that nanotechnology will underpin the development of therapeutic devices that can provide targeted drug delivery and treatment to diseased organs, and treat diseases of the cell. The development of device interfaces to organs, tissues, and cells are also important areas of research [12]–[14]. New technologies in remote imaging detectors are currently being developed to assist in viewing cells and molecules [15]–[16].

In response to the tragic events of September 11, 2001, government spending on homeland security has increased dramatically. This has spawned the development of an entire industry. In particular, the need for rapidly detecting and preventing the spread of pathogens to combat terrorism has created opportunities for pharmaceutical, biotechnology, medical device, and diagnostic and treatment companies [17]–[19].

Clinical engineering opportunities for employment in hospitals are also important aspects of the medical device industry [20]. Clinical engineers provide support for the use of medical devices by health care professionals. In addition, hospitals are the largest buyers of medical devices, with the purchase process usually managed by the clinical engineer.

Biotechnology

Biotechnology has its roots in genomics, therapies and drug development, and organic tissues [2]. An overarching development in this field is the promise and impact of bionanotechnology and bioinformatics. Many biotechnology companies are investing their R&D efforts in drug discovery and genome data mining.

Genomics will continue to grow and provide increasingly significant opportunities related to the health care industry. Some of the work in this field involves the creation of new medical devices for DNA analysis and chip based systems. As the genomes of animals are decoded and profiled, opportunities have opened for developing better diagnostics for disease detection and treatment (via manipulation of genetic code in genetically based diseases), as well as in the design of new drugs. Genetically engineered biosensors may be used in detecting bioterrorism threats. Newly developed analysis tools may also aid in advancing genomics, such as using digital communication theory and treating DNA as digital data [21]. Unfortunately, the outcomes of the great promise of genomics may not arrive until 2015, or even at all, because of technological and ethical barriers [2].

Another exciting area of exploration is proteomics [22]. Opportunities will follow in creating high-throughput devices and algorithms to study proteins expressed in cells.

Cloning opportunities in engineered crops and livestock are moving forward, and offer great promise [23]. However, restrictions on human cloning in regulated parts of the world, due to profound ethical, moral, and religious issues, will limit advances in therapeutic cloning to those developed in unregulated parts of the world.

Tissue engineering opportunities also hold great promise; these involve creating replacement parts for tissues and organs for humans [2], [12]. Some advances are moving forward very quickly, such as engineered skin replacements, and the growth of cartilage to treat and replace damaged cartilage. Other projected advances will take a decade or more (for example, diseased heart tissue replaced by the growth of new tissue). Resolving biocompatibility issues with materials implanted in the body will offer many opportunities.

Stem cell research and therapy have the same ethical issues as cloning. If these ethical issues can be resolved, major advances are expected in using stem cells to repair and replace cells, tissues, and organs. Some states in the US are funding stem cell initiatives, and this may spur the developments and opportunities in this field.

Employment Outlook

The aging of the US population is driven by the 77 million baby boomers who will retire in 10–15 years [24]. Generation X that follows the baby boomers is about one-half the size. In general, engineering employers will have a difficult time with staffing in the future because of a shrinking pool of candidates. For bioengineering employment, a similar scenario – will occur, but an even higher demand for bioengineers is expected because of the needs of those over 65. The demand for better health care services will drive better medical devices and equipment for diagnostics, therapy, and assistive technology. These expectations will increase the demand for bioengineers.

According to the US Bureau of Labor Statistics (BLS), bioengineers are expected to have employment growth of more than 27% each year through 2014 [25]. BLS indicates that job growth across all engineering disciplines is 9% to 17% per year. In 2004, BLS reported that there were 9,700 bioengineers out of a total of 1.4 million engineers. Of the 9,700 bioengineers, 15.6% worked in pharmaceutical and medicine manufacturing industries, and 18.7% worked in scientific research and development services. Hospitals and government agencies

Table 4.1 Engineering earnings by specialty, 2005 [25]

Curriculum	Bachelor's	Master's	PhD
Bioengineering	$48,503	$59,667	—
Chemical	$53,813	$57,260	$79,591
Civil	$43,679	$48,050	$59,625
Computer	$52,464	$60,354	$69,625
Electrical/electronics and communications	$51,888	$64,416	$80,206
Environmental/environmental health	$47,384	—	—
Industrial/manufacturing	$49,567	$56,561	$85,000
Materials	$50,982	—	—
Mechanical	$50,236	$59,880	$68,299

also employ a large number of bioengineers. Average starting salaries reported in the government report for 2005 are $48,503 for a bachelor's degree, and $59,667 for a Masters degree. No figure was given for the PhD Table 4.1 shows the earnings by degree according to a 2005 survey by the National Association of Colleges and Employers [25].

AIMBE Employment Survey

Since the 2001–2002 academic year, the American Institute for Medical and Biological Engineering (AIMBE) Academic Council has collected data on bachelors, masters, and PhD graduating bioengineering students from US institutions [5], [26]. These four years of data are the most accurate reflection of bioengineering graduation rates and the careers selected. The participation in the survey is voluntary and carried out via email in the fall of each year.

In the latest survey, 88 programs were contacted to complete the survey, with 55 respondents (63% response rate). Of those programs that responded, 6 reported no graduates, and 5 indicated the data was too difficult to obtain. The fact that 6 programs had no graduates is not a surprise because of the number of new programs. Table 4.2 provides the number of bioengineering graduates during each academic year from 2001–2005.

As reported at the website [26],

> Each graduate was placed into one of four broad categories: pursued further education, obtained a job, seeking employment, and unknown. The "pursuing further education" category was further broken down into six detailed categories: engineering, medicine, business/management, law, post-doc, and other. The "obtained a job" category was further broken down into six categories: bioengineering in industry, consulting, hospital, government, academia, and other.

Table 4.2 Number of bioengineering graduates, 2001–2005

Academic Year	Universities Responding	Total Number of Graduates	Bachelors	Masters	PhD
2001–2002	30	808	445	223	112
2002–2003	37	1,334	906	288	138
2003–2004	44	1,475	974	318	151
2004–2005	42	1,913	1,211	485	193

Data and analysis for the academic years 2001–2004 are presented in [5]. Overall for the 2004–5 academic year, the AIMBE survey reported placement rates of: 92% bachelors, 96% masters, and 96% PhD. Of the 1,211 bachelors graduates, 45% continued their education (20% in medical school), 28% obtained a job, 6% were still seeking a job, and 20% were unknown. Of the 485 masters graduates, 40% continued their education, 43% obtained a job, 4% were still seeking employment, and 13% were unknown. Of the 193 PhD graduates, 44.5% continued their education, 44.5% obtained a job, 3% were still seeking employment, and 8% were unknown.

In general, the number of graduates at all levels continued to increase as a function of time. This trend is due to the number of engineering programs that have started a bioengineering program (or option), with 65 bioengineering members of the Academic Council in 2002, to 88 in 2005. The numbers also reflect the popularity of bioengineering.

Advice for Finding a Job

There are certain skills and attributes that are required regardless of the type of employment one is seeking. A large number of studies and employer surveys focus on what are called "transferable" skills [27]. These include:

- Communication
- Decision making
- Dependability
- Leadership
- Life-long learning
- Positive attitude
- Cooperativeness
- Problem-solving
- Self confidence
- Teamwork
- Technical competency

Interestingly, these skills have been sought by employers without much change for the last 20–30 years. Moreover, many of these skills are those required for all graduates by an ABET engineering accredited program [3].

Students can develop these skills in a variety of ways. For instance, a robust senior design course cultivates these attributes while completing the design project. As part of the course requirements, undergraduates hone their verbal and written communication skills with project work, reports, and oral presentations. Each student in a team typically is responsible for an aspect of the project. Work responsibility for this aspect often develops leadership skills. Since many of the skills required to successfully complete the project are not taught in the curriculum, students need to learn new concepts and skills, much as they will while working in industry (life-long learning). These skills should be highlighted in one's resume and discussed during interviews to set one apart from other candidates. A student should make certain to research the employer's needs and then match his or her skill set to that of the employer's.

In addition, the NSF Research Experiences for Undergraduates (REU) grant program provides funding for students to work on focused research during the summer. Apart from the research experience gained by the student, these experiences also develop the skills desired by employers. Traditional internships and co-ops* (that is,) are also invaluable activities for honing these skills, as well as providing industry contacts for future employment. Work in faculty labs during the academic year also provides important experiences that employers are looking for.

Another website lists important personality traits that are highly desired in employees [28]:

1. Honesty/integrity
2. Teamwork skills (works well with others)
3. Interpersonal skills (relates well to others)
4. Motivation/initiative
5. Strong work ethic
6. Analytical skills
7. Flexibility/adaptability
8. Computer skills
9. Organizational skills
10. Detail oriented
11. Leadership skills
12. Self-confidence
13. Friendly/outgoing personality
14. Tactfulness
15. Well-mannered/polite
16. Grade point average/percentage (3.0 or better; 80% or more)
17. Creativity
18. Entrepreneurial skills/risk-taker
19. Sense of humor

* "Co-ops" (short for Cooperative education programs) are programs that partner students with industry. This allows students to gain real world work experience while they are earning their college degree.

20. Read up for the interview
21. Interacting easily with others
22. Working effectively with team members
23. Good work ethic
24. Listening intelligently
25. Thinking through problems
26. Critical thinking
27. Communication skills (verbal & written)

Concluding Remarks

Opportunities in bioengineering education and employment are the best they have ever been. The enormous growth of new bioengineering programs will continue in the future. This growth, initially spurred by the Whitaker Foundation, is now driven by the huge potential inspired by advances of biotechnology and the nanotechnology-inspired medical devices. Advances in life sciences will continue to reshape the bioengineering program of study with an integration of the life sciences with engineering. Opportunities for employment in the bioengineering industry are excellent, with predicted growth at the highest level compared with all other engineering fields.

References

1. Katona, PG, "Biomedical Engineering and the Whitaker Foundation: A Thirty-Year Partnership." *Annals of Biomedical Engineering*, Vol. 34, No. 6, 2006: 904–916.
2. Antón, PS, Silberglitt, R, and Schneider, J, The Global Technology Revolution: Bio/Nano/Materials Trends and their Synergies with Information Technology by 2015. *RAND*, 2001.
3. Enderle, JD, Gassert, J, Blanchard, SM, King, P, Beasley, D, Hale Jr., P and Aldridge, D, "The ABCs of Preparing for ABET," *IEEE EMB Magazine*, Vol. 22, No. 4, 122–132, 2003
4. Nagel, JH, Slaaf, DW, and Barbenel, J, "Medical and Biological Engineering and Science in the European Higher Education Area." *IEEE EMB Magazine*, Vol. 26, No. 3, 2007, pp. 18–25.
5. Schreiner, S, "Placement of Bioengineering and Biomedical Engineering Graduates." *Proceedings of the 2005 American Society for Engineering Education Annual Conference and Exposition*, Portland, Oregon, 2005.
6. Enderle, JD, Blanchard, SM and Bronzino, JD, *Introduction to Biomedical Engineering (Second Edition)*, Chapter 1, Elsevier, Amsterdam, 2005, 1118 p.
7. Panescu, D, "Medical Device Industry." In *Wiley Encyclopedia of Biomedical Engineering* (Metin Akay, ed.) Hoboken: John Wiley & Sons, Inc., 2006.
8. Nebeker, F, "Golden Accomplishments in Biomedical Engineering," In *Charting the Milestones of Biomedical Engineering* (John Enderle, ed.), IEEE, Piscataway, New Jersey, 2002.
9. Gough, D, *Employment and Career Prospects for Bioengineers.* Nature Publishing Group, www.biotechn.nature.com, 2002.
10. Weiland, J, Humayun, M, (guest editors), "Biomimetic Systems," *IEEE EMB Magazine*, Vol. 24, No. 5, 2005.
11. Hughes, M, (guest editor), "Micro- and Nanoelectrokinetics in Medicine," *IEEE EMB Magazine*, Vol. 22, No. 6, 2003.

12. Laurencin, C, Katti, D, (guest editors), "Tissue Engineering," *IEEE EMB Magazine*, Vol. 22, No. 5, 2003.
13. Bonato, P, (guest editor), "Impact of Wearable Technology," *IEEE EMB Magazine*, Vol. 22, No. 3, 2003.
14. Kern, S, Jaron, D, (guest editors), "Balancing Healthcare Technology, Economics, and Policy," *IEEE EMB Magazine*, Vol. 22, No. 1, 2003.
15. Loughlin, P, Redfern, M, (guest editors), "Analysis and Modeling of Human Postural Control," *IEEE EMB Magazine*, Vol. 22, No. 3, 2003.
16. Principe, J, Sanchez, J, *Brain-Machine Interaction*, Morgan & Claypool Publishers, San Rafael, California, 2007.
17. Thakor, N, (guest editor), "Clinical Neuroengineering I," *IEEE EMB Magazine*, Vol. 25, No. 4, 2006.
18. Thakor, N, (guest editor), "Clinical Neuroengineering II," *IEEE EMB Magazine*, Vol. 25, No. 5, 2006.
19. Li, K, (guest editor), "Molecular Imaging I," *IEEE EMB Magazine*, Vol. 23, No. 4, 2004.
20. Li, K, (guest editor), "Molecular Imaging II," *IEEE EMB Magazine*, Vol. 23, No. 5, 2004.
21. McIsaac, JH, (Editor), *Hospital Preparation for Bioterror: A Medical and Biomedical Systems Approach*, Elsevier, Amsterdam, 2006.
22. Painter, F, (guest editor), "Clinical Engineering," *IEEE EMB Magazine*, Vol. 23, No. 3, 2004.
23. Senhadji, L, Siebes, M, vander Sloten, J, and Saranummi, N, (guest editors), "Recent Technological Developments in Europe," *IEEE EMB Magazine*, Vol. 26, No. 3, 2007.
24. Tamura, T, (guest editor), "Recent Technological Developments in Japan," *IEEE EMB Magazine*, Vol. 24, No. 4, 2005.
25. May, E, (guest editor), "Communication Theory and Molecular Biology," *IEEE EMB Magazine*, Vol. 25, No. 1, 2006.
26. Hamady, M, Cheung, THT, Resing, K, Cios, KJ, and Knight, R, (guest editors), "Challenges and Issues in Proteomics," *IEEE EMB Magazine*, Vol. 24, No. 3, 2005.
27. Yang, X, (guest editor), "The Cloning Debate," *IEEE EMB Magazine*, Vol. 23, No. 2, 2004.
28. 2007 Engineering Outlook. www.BioMedicalEngineer.com.
29. Kun, L, Laxminarayan, S, (guest editors), "Combating Bioterrorism with Biomedical Engineering," *IEEE EMB Magazine*, Vol. 21, No. 5, 2002.
30. Kun, L, Laxminarayan, S, (guest editors), "Homeland Security," *IEEE EMB Magazine*, Vol. 23, No. 1, 2004.
31. Bureau of Labor Statistics, U.S. Department of Labor, Occupational Outlook Handbook, 2006–07 Edition, Engineers, on the Internet at www.bls.gov/oco/ocos027.htm
32. www.aimbe.org/content/index.php?pid=201
33. www.shu.ac.uk/research/cre/publications/gws/gwsskills.html.
34. www.studentservices.engr.wisc.edu/international/employers.html

Part II

Traditional Careers in Bioengineering and Biotechnology

5

Academic Research and Teaching

Nitish V. Thakor, PhD

Department of Biomedical Engineering, The Whitaker Biomedical Engineering Institute and the Neuroengineering Training Initiative, School of Medicine, The Johns Hopkins University School of Medicine, Baltimore, Maryland, USA

Academics are beacons of knowledge through excellence in teaching, scholarly research, and service to the society, and for that they earn honors in many ways. The pinnacles of honor—won by many who have been involved in bioengineering and biotechnology—are awards such as the Nobel Prize, the US Presidential Medal of Science and Technology, and the Kyoto Prize. Additionally, various professional societies and national bodies have their own honors and recognitions, such as the Institute of Electrical and Electronics Engineers' (IEEE) Engineering in Medicine and Biology Society (EMBS) Distinguished Career Award, and the Biomedical Engineering Society (BMES) Distinguished Career Award. Another type of honor accorded to academics is election to learned academics; for example, one of the US National Academies (which include science, engineering, and medicine). The professional societies also honor the leading researchers and academicians by selection or appointment as *Fellows* of societies such as the American Institute of Medical and Biological Engineering (AIMBE), the American Society of Engineering Education (ASEE), the IEEE, and the BMES. Bioengineering faculty members are selected to serve on the bodies of major funding agencies, such as the National Institutes of Health (NIH), and the National Science Foundation (NSF). Throughout their careers, academics write dozens, hundreds, or even more than a thousand research papers. Some are very inventive and have numerous patents and startup companies credited to them. Some are prolific in writing or editing books. Successful professors are sought as invited lecturers, and keynote and plenary speakers at major national and international conventions. Perhaps most importantly, such professors are the mentors of the next generation of students. The intent of this chapter is to provide you with ideas to help you enter and succeed in academia.

G. Madhavan et al. (eds.), *Career Development in Bioengineering and Biotechnology*,
DOI: 10.1007/978-0-387-76495-5_5, © Springer Science+Business Media, LLC 2008

My Own Journey

It would be useful to present my own journey, and integrate the views of others as a case study. My own path is not unique—rather, it exemplifies the challenge that many young engineers and budding scientists face on the road to academia. Early on, as with many students in India in the 1950s and 60s, I was excited by scientific discoveries and technological innovations that were briskly unfolding worldwide. The transistor was invented by Bardeen, Brittain, and Shockley while I was in grade school, and the electronic revolution based on the integrated circuit swept in while I was in high school. The 1960s and 70s saw medical devices such as electrocardiographic (ECG) monitors and defibrillators being introduced into hospitals and intensive care units. However, the computers of the time were quite basic (there were no microprocessors or personal computers, for example), and the telecommunication revolution had not even started—cellular telephones were still decades in the future. Nevertheless, the Apollo mission to the moon and the revolutions underway in both science and technology were inspiring and motivating for young students and budding engineers and scientists.

Upon completion of high school, I was fortunate to be admitted to the highly selective Indian Institute of Technology (IIT) in Bombay (now Mumbai) through a very competitive national entrance exam. I was launched into a career in engineering. There was only one problem: by the time I reached high school, even though I was very much enamored of technology, I had decided that I wanted to be a physician. How does one reconcile these two diverse interests and passions? This predicament was what led me to discover the field of bioengineering.

I studied engineering but scoured biomedical literature. I did the usual engineering coursework, but carried out a senior design project to build an ECG machine and heart murmur

Could you tell us something about your entry into academia and your highly successful career?

"I belonged to my professional society. In the summer of 1962 I read an article in a magazine published by my professional society that the National Institutes of Health (NIH) was encouraging experienced engineers to get their PhD and become teachers in the new field of bioengineering and NIH would pay the bills. I applied and was accepted at the University of Rochester in 1963 and four years later started teaching and research at the University of Wisconsin."

—John Webster, PhD, University of Wisconsin – Madison, USA

"While still completing a PhD in cardiac neurophysiology, I was fortunate enough to be awarded a lecturing position. I was thus introduced early into aspects of academia including teaching and learning, scholarship and commercial interactions. The reason behind commencing a PhD was quite simple. I was working in industry and felt that I was not being challenged and that I would need to spend many years working in rather menial roles until I could engage in more stimulating activities. A return to the University was an easy choice."

—Nigel Lovell, PhD, University of New South Wales, Australia

"During my undergraduate years, corporate engineering positions or clinical medicine were more visible and 'understandable.' A summer opportunity through the National Science Foundation/Research Experience for Undergraduates during my undergraduate years, in the field of computational neuroscience, opened my eyes to academic research. I also realized that research could provide a fascinating career. I ultimately received a PhD in Electrical Engineering and then a change of direction as a neurobiology post-doc. Fusing these experiences together, into 'neural engineering' / 'computational neuroscience' is what helped open doors the doors for me as an academic."

—Krishna Shenoy, PhD, Stanford University, USA

detection device. It was both a tremendous challenge and tremendous fun to design and build an ECG monitor when transistors and IC chips were just coming into vogue (remember—this was in the late sixties). Ultimately, my senior design project strengthened my resolve to go on to graduate school.

The path to academia need not always follow the straightforward route from undergraduate to graduate school to postdoctoral fellowship and, eventually, to a faculty position. For example, after I received my bachelor's and master's degrees, I worked in industry for two years to gain industrial engineering and managerial experience with Philips, a multi-national corporation in India. I also wanted to start my own company, but found I was unable to get it off the ground. Throughout this period, I realized that I was deeply committed to developing novel technology and learning new and original scientific research. I then decided to work toward a doctorate at the University of Wisconsin in Madison. I approached Prof. John Webster for a graduate fellowship—his papers had helped me complete my ECG device senior project. After completing my Ph.D. dissertation at the University of Wisconsin, I happily remained in academia.

What are the key steps after defending the dissertation to beginning a post-doctoral or faculty career?

"Get the remaining journal papers from your doctoral research finished and submitted now while you have the time!"

—Cameron Riviere, PhD, Carnegie Mellon University, USA

"For a successful academic career with a strong research component, a post-doctoral fellowship is nearly essential. Start the search at least six months before you defend your doctoral dissertation. If you are not tied to a given location by family concerns, consider going to a new part of the world. First and foremost, find a post-doctoral advisor who is both well-respected and altruistic. For potential advisors who are senior faculty, the quality of jobs found by their past post-doctoral fellows is the best single indicator of your probability of success."

—John White, PhD, University of Utah, USA

What is your advice to a student trying to complete her/his dissertation and planning a career in academia?

"The mental transition from undergraduate to graduate student is a transition where the student him or herself becomes the central, active force in deciding what is to be mastered. The next transition—to a faculty position—involves realizing what one wants to discover, invent or speculate about. Finding that vision and then developing the skills and strategies to execute it are a critical aspect of a successful career in academics."

—Christopher Chen, MD, PhD, University of Pennsylvania, USA

"The student should be passionate about research, working hard, and being honest. It may sound glib, but it has not failed me yet."

—Kam Leong, PhD, Duke University, USA

"Learn about the real-world societal aspects of your research field; know the big picture. It is not enough to know how to solve a research problem; to be successful at the faculty level you need to be able to state convincingly why the problem is one worth solving. Know what is worth doing as a researcher in your field, and be able to present the case for it, verbally and in writing."

—Cameron Riviere, PhD, Carnegie Mellon University, USA

First Step: Getting Your Doctorate

The path to academic research starts with a doctoral degree. It is possible to get some industrial experience on the way, or immediately after getting your doctoral degree. However, you should be aware that it can be challenging to return to academia if you are away for an extended period. It is difficult to return because in industry, it is generally not possible to keep up with the fast pace of modern research, stay current with academic coursework, publish your work in journals, and retain the ability to enjoy a frugal and hardworking life style. However, working in industry may widen your outlook and open your horizons. This type of experience and outlook can be useful when you are trying to advise your students on their career choices, or if you are trying to develop a collaboration with industry, or spin off a start up company.

After Getting Your Doctorate

After earning your doctorate degree, you may continue to do postdoctoral research work or, less commonly these days except in the traditional engineering disciplines, look for an academic position immediately. A postdoctoral fellowship gives you another two to five years (three years being the most common) to (1) further hone your research skills, (2) master research methodologies, (3) write more scientific papers to augment your credentials, (4) learn the process and the critical art of writing research proposals, and (5) experience the academic responsibilities without the burden of managing your own lab, and without the distractions of teaching and advising students. As important as these responsibilities might be, it is quite helpful in the early years to fully commit to research and strengthen your scholarly credentials and skills for becoming an independent scientist.

It is common in most engineering fields to go directly into academic teaching and research after getting the doctoral degree. However, it is quite uncommon in biological and life sciences. Perhaps this is due to tradition, or simply the dynamics of supply and demand (how many candidates are looking for faculty positions and how many positions are actually available). The differences between disciplines may also be due to the greater need to gain and skillfully master laboratory

What would you say to a graduate student about how to prepare for an academic career?

"Choose a new, high-impact area but hedge with a side project that is lower risk. Then pursue your research passionately, and publish well. Finally, lead the way – do not wait for your advisor to suggest every next step."

—Krishna Shenoy, PhD, Stanford University, USA

"Develop a solid foundation in an area of your choice. Learn to write proposals early. Practice teaching skills."

—Dominique Durand, PhD, Case Western Reserve University, USA

"It is important for the graduate student to start as early as possible to prepare for an academic career. Of course, this requires a great deal of maturity and soul searching to determine whether a career in academia is the right path. Once this decision has been made, the graduate student must focus on a few 'must haves'. These include a number of highly influential publications in prestigious journals, excellent letters of reference from mentors and colleagues who are respected in academia. Also, it is important to pick a research direction that is futuristic, and to have a clear and viable plan of the future direction of research."

—Ali Khademhosseini, PhD, Massachusetts Institute of Technology, USA

"Today almost all research is collaborative or interdisciplinary. Frame your research goals, and focus on making yourself visible. Become part of the research community by meeting people who will be your peers and collaborators in the future. Pick the low-hanging fruit first. Do not be so focused on the long-term goals that you fail to publish the intermediate steps along the way. Make a habit of carrying projects through to the point of journal publication. It is very easy to do things halfway, maybe to the point of conference publication, and then fail to follow them through to journal publication."

—Cameron Riviere, PhD, Carnegie Mellon University, USA

Is post-doctoral experience desirable or necessary?

"The answer to this question depends very much on the field. For some academics responsible for hiring, having post-doc experience is non-negotiable for every job candidate. Indeed, the post-doc experience provides additional time to prepare for the funding, hiring, and managing complexities of modern faculty life."

—Krishna Shenoy, PhD, Stanford University, USA

experience in life sciences. Bioengineering falls somewhere in between; however, it is safe to say that these days, most faculty candidates are expected to have some amount of postdoctoral experience.

By all means, use your time as a postdoctoral candidate to learn new research techniques and publish more original work. However, this is also a precious opportunity to observe and learn from an experienced mentor about the art of juggling teaching, research, and administrative responsibilities, and the science of grantsmanship. After a solid postdoctoral experience, you can head to the halls of academia – to teach, do research, mentor, and pursue new research directions and technologies.

"A post-doc experience is not essential if your graduate training is extensive and you have good mentors to help you transition to an academic position. However, doing a post-doc allows you to get more experience and publications under your belt before taking on a hectic faculty position. That extra time and experience will give you a leg up when getting your own research career underway."

—Dawn Taylor, PhD, Case Western Reserve University, USA

"Once in a while you see someone truly brilliant who can enter academia without doing a post-doctoral stint. Still, you wonder if that person might have been even more spectacular if he or she had branched out and learn another field before building a own research program. I would very strongly encourage my students to do post-doctoral training before going into academia."

—Kam Leong, PhD, Duke University, USA

Academia Versus Industry

One of the first questions you will be asked by your friends, family, and mentor is, "Are you planning to go into industry or academia?" This is definitely a personal and value oriented choice. Only you can make this decision, which should be based on your values and career aspirations. It may help to know that the decision you make is not irrevocable.

If you want to teach, contribute to knowledge in an open intellectual atmosphere, do basic research, and disseminate your teaching and research ideas, then the answer is academia. If, on the other hand, you want to only do research (i.e., not teach and educate/mentor students), or you want to pursue translation and application of your ideas, applying those ideas to product development, manufacturing, marketing, and so on, then the answer is industry. Of course, there are many other traditional and non-traditional career categories to consider, many of which this book superbly lays forth. However, if you are considering a career in academia, you need to feel a

What would you say to a student who is undecided about an industry versus an academic career?

"It depends very much on the personality, but if undecided I would tend to encourage the person to experience industry. A prospective academic should not enter academia with ambivalent views."

—Nigel Lovell, PhD, University of New South Wales, Australia

"Follow your passion. See what you like more. That is the most important."

—Ali Khademhosseini, PhD, Massachusetts Institute of Technology, USA

"I suggest that an internship might be appropriate for them. That is why we have industry socials in our university where industry practitioners come to have dinner with our students."

—Herbert Voigt, PhD, Boston University, USA

Is it possible to come back to academia after some time in industry?

"Absolutely! It is often a good idea to gain experience from both sides of the fence."

—Nigel Lovell, PhD, University of New South Wales, Australia

special love for teaching, educating, and doing research in an open academic atmosphere.

Academic Job Search

Looking for an academic position starts in the final year of your doctoral work or your postdoctoral work. You can never be sure about when you might finish your doctoral research or when you might land a job. It is safe to say that you need about nine months to a year for the job search. You will preferably start the process in advance, in anticipation of joining at the beginning of a typical academic year, which usually begins in the fall. Thus, you need to spruce up your curriculum vitae, push your papers out the door, and start focusing on job hunting while you wind down your research in preparation for graduation.

Where and how should you look for an academic position? Just as with your research, your doctoral and postdoctoral advisors will be your closest mentors for your job search, followed by those who were on your dissertation committee. You should review your curriculum vitae, research, teaching, and professional plan with these individuals. They are also likely to be well known in academia, thus providing an indirect boost to your reputation. Additionally, they are likely to have many connections to important players in the field, and to know where the job opportunities lie. They are also likely to have a feel for which universities have programs that are recruiting and growing, improving in reputation, and generating new resources.

"Indeed, it is quite possible to return to academia after some time in industry. However, one must be mindful that it is very challenging to do so. If this is a goal, then while in industry it is advisable that you maintain an active publication track record. Stay active in the professional societies; organize workshops, conferences and serve on committees of professional societies; this will bring you into contact with academic professionals who may serve as referees when you are seeking that return to academic life."

—Anthony Guiseppi-Elie, ScD, Clemson University, USA

As a professor, do you interact with industry? What attributes or attitudes help you?

"Yes – both through collaborative grants and by way of two start-up companies of my own. Working toward realistic and achievable milestones and recognizing the important contributions of product development, manufacturability and marketing are important attributes."

—Nigel Lovell, PhD, University of New South Wales, Australia

"As a professor I have extensive interactions with industry. I serve on scientific advisory boards and I am also a consultant with several companies. The most important attribute is empathy for the agenda of company executives and the technical people who work in companies. That empathy allows one to appreciate their perspectives on research and the timelines to which this research must be responsive."

—Anthony Guiseppi-Elie, ScD, Clemson University, USA

How did you go about looking for a faculty position?

"I looked at the departments that were doing exciting work, where I thought I could impact the future direction of the school, and then introduced myself to the chair."

—Christopher Chen, MD, PhD, University of Pennsylvania, USA

"I spoke to my advisor and then followed up on his suggestion by doing a search for posted advertisements. Also, conferences and journals in one's area can be helpful."

—Sangeeta Bhatia, MD, PhD, Massachusetts Institute of Technology and Harvard Medical School, USA

"I began my search by asking individuals in my professional network who might be hiring. My training is fairly broad, so I was looking for positions in several types of academic departments. For this reason, I needed to perform extensive searches on my own. I found there was no single clearinghouse for job postings, and I eventually began routinely checking faculty postings on the web sites of each individual university and department in which I was interested."

—Dennis Barbour, MD, PhD, Washington University in St Louis, USA

Job postings are typically available on various departments' websites, at career portals (such as www.bmenet. org), and advertised in general periodicals such as *Science* and *Nature*, as well as in more specific journals and magazines such as *IEEE Spectrum* and the *IEEE Engineering in Medicine and Biology Magazine*. Some advertisements are sent directly to your department, typically to your department chairperson. A well known "secret" is that most academic departments are *always* looking for top notch candidates. They may open up a new position just for you, or be planning to open a new position soon for which they may keep you in mind. So another approach is to identify which universities and departments you might want to join and apply to them; you just never know what might happen. Similarly, do not limit yourself only to bioengineering departments. Various engineering departments; basic science departments, such as biology, physiology, biophysics, psychology, and neurosciences; and occasionally clinical departments such as radiology, neurology, physical medicine, and various surgical specialties also recruit faculty candidates in bioengineering and biotechnology. You probably have more options than you think, so plow ahead!

What were the critical steps or events that finally landed you the faculty position?

"I think that in my case, there were a few practical and psychological key steps that lead to a faculty offer. On the practical front, good contacts and a critical mass of publications in my chosen field were absolutely essential. A well formed interdisciplinary research plan with room to grow was another essential write-up to complete the package. Having a NIH funding was also a big plus in my case. I also had a couple of publicity articles in the literature.

Further, by far the most important thing is to be psychologically well prepared during the full process of job search and interviewing. In particular, a good seminar presentation, in which I could clearly explain my work, and show my vision of where the work fits in its general scientific field, and where the future of this scientific field lies, was the most critical step. It was more about showing that I had thought about my research plans a great deal and at the same time projecting a well founded conviction that I could handle them successfully."

—*Nada Boustany, PhD, Rutgers University, USA*

"Having a strong curriculum vitae with a number of first-author publications in good journals, and having specialized in an area of growth (in my case it was microfabrication and tissue engineering) are the best things you can do to prepare yourself. I spent time on my package, especially the research plan. Then, when I was interviewed, I tried to do my homework—to know the department and the faculty I would meet. I also made sure to give a strong seminar."

—*Sangeeta Bhatia, MD, PhD, Massachusetts Institute of Technology and Harvard Medical School, USA*

At this crucial stage in mapping out your career, it is very important to go to various national and international conferences. You should seek out opportunities to present, so that your work receives exposure and you obtain feedback. The conferences are great places to network, to find out about postdoctoral or faculty positions (direct advertisements or word of mouth), to meet students and postdocs from the institutions you are considering, and, most importantly, to meet faculty from these institutions. You can not only learn about openings and opportunities—but you also have a chance to make a good impression. At this dissertation wrap-up stage, you might also consider making exploratory visits to the universities, laboratories, and departments. Unless tied in with a trip to a conference, however, this can expensive and time-consuming at a time when you already have much to do.

Curriculum Vitae

First, of course, prepare your curriculum vitae (CV, in short). Your CV presents your academic credentials, awards and honors, teaching experience, and mainly your research accomplishments. These should include both conference abstracts and papers, and any journal papers. As a starting faculty, it is acceptable to add papers under review or in preparation.

In addition, you should prepare a two to three page document summarizing your current and future research. This will be read with interest. It lays out for the interested readers, some of the faculty in your area at the recruiting institution, the details of what you are doing and what you plan to do. Your essay should reflect a focused and coherent program; exciting but not overly ambitious. It should lay out your path from the doctoral/ postdoctoral work to your independent research. The readers will try to gauge what research grants your research work will result in, or how successful you will be at getting the grants (you will also be asked about this directly during your personal interviews). However, most importantly, this document lays out your vision, and the path you propose to take to be a successful and independent scientist and academician. The recruiting committee will compare your credentials with their needs, use their experience, and compare your credentials with all the candidates to produce a short list and invite those candidates for an on campus interview. If you are invited, you will have cleared the first hurdle. If you are not invited, do not be dismayed—in many,

When you evaluate a prospective faculty candidate, what do you look for in the CV?

"I always look at the publications first, to assess what the candidate has accomplished and how long he or she took to do it. Then I read the person's statement, to understand what the vision of their career is, and whether it is exciting. Lastly, I look back at the entire CV to evaluate whether I think the person's past experiences and accomplishments would feasibly accomplish what they have proposed. But we are all individuals, and so candidates who do not fit a particular mold are always considered as well."

—*Christopher Chen, MD, PhD, University of Pennsylvania, USA*

"The first thing is the publications. Are they in an area of interest? Are they of good quality? Do they show creativity and hard work? Are other people reading them (citation analysis)? Value added comes from experience in other areas, teaching, volunteer service, etc. The number of these is not as important as the quality. Usually the CV comes with a statement of research interests and plans for career development. Are these creative and far-sighted? Are they good ideas that match the needs or interests of the position available? Are they feasible research topics?"

—*Eric Young, PhD, Johns Hopkins University, USA*

"I review the CV to see if the applicant has the appropriate background for the specific needs of the department. Next, I examine the CV to determine the productivity of the applicant. I also look to see whether the applicant has presented at meetings and is involved in other scholarly activities such as reviewing papers or writing review papers. I also scan the CV to determine if the applicant worked on different research projects in different labs as undergraduate, graduate student and postdoctoral fellow. The candidate's research and teaching statement are very important as are references. When reading the research and teaching statement, I want to determine whether the candidate is moving into new areas that will distinguish her from her mentors and have an impact on the field."

—*George Truskey, PhD, Duke University, USA*

What do you value most in a CV?

"For junior candidates, I look for hints in CVs and letters that the person can think independently. First-author papers are important, of course, but I like to see independent evidence that the person was truly responsible for the work. The intellectual coherence of the research plan is a big factor for me."

—*John White, PhD, University of Utah, USA*

"Creativity and originality of the candidate's work. People who think outside the box are what you want when you build a research program."

—*Paolo Bonato, PhD, Harvard Medical School, USA*

if not most, cases, your area of specialization simply was not what the recruiting committee was looking for.

The Interview

The next step is the interview. If all has gone well, you will be invited to several universities. Your interviews will typically span a day or two. You will most likely be hosted by one or more interested faculty in the department who are eager to expand into your area of specialization or expand their research group. You will meet with the department chair and some senior faculty and administrators. This will all appear to be informal, with these discussions meant to get to know you as a person. But subtly the faculty you meet are gauging your abilities, ambitions, career goals, priorities and future directions. They will quietly find out from you what you would like to teach and how you will contribute to the undergraduate and graduate programs. Generally these interviews have a limited positive benefit, but if they do not go well, senior faculty may get the impression that you might not be a good fit for the department. Hence being cordial and professional is very desirable, while also being alert and at your best personal self.

Sometime during the day you will present your seminar. The seminar is extremely important. This is where the broader faculty audience as well as the faculty interested in your research get to hear your story from your mouth and to learn about your research in depth (remember they have already seen your CV and know your pedigree and credentials). This is where your presentation and communication skills matter. It is important during the latter part of your seminar to present future directions of your research. These seminars can be "make or break" as the faculty will quickly judge your teaching and communication skills, as well as whether your research is at the cutting edge and your vision looks worthwhile. In general, your seminar will last for an hour or so, with time for questions. However, there can be many or few questions along the way and how you handle them may be equally or more important than making your presentation.

You will have lunch and dinner with different groups of faculty. Often these groups will include faculty from other departments who may have an interest in fostering a collaboration to help strengthening their research program. Sometimes it is the other way around—these faculty members may be present to help recruit you. The lunch and the dinner are generally informal with a lot of bantering that may not directly involve you; it may not be unusual for the colleagues to catch up with each other and what is going on in the department and the university. Implicitly, this is also your chance to learn about the department and the university, the chemistry of the people involved and the general social aspects about the place where you may spend many years—perhaps your entire career.

After a long day or two of being "on show," you will be finished. There is nothing more to do but wait. Academic recruiting tends to be notoriously slow. In part this is because a department may be interviewing many candidates staggered over months. In part this may be the delay in getting the departmental faculty together for your visit or subsequent discussions and decision. There may be other factors, such as institutional commitment for space, start up package for your laboratory, and so on. So you will need to be patient.

> **What is the importance of an interview in selecting a faculty candidate?**
>
> "The interview is pretty much the only thing that matters to me. My main interest in interviewing a candidate is to figure out whether that individual will work well with the rest of the team. I would rather have a good hard-working individual who can join the team in an effective way than a genius who cannot get along with people."
>
> —*Paolo Bonato, PhD, Harvard Medical School, USA*
>
> **What questions would you typically ask the candidate?**
>
> "I like to know what candidates are passionate about. I like to make sure that the job opportunity we discuss is an opportunity for them to do what they want. It is not going to work otherwise."
>
> —*Paolo Bonato, PhD, Harvard Medical School, USA*
>
> "From the perspective of a faculty hiring another faculty, I would like to know if someone thinks first about their own career or the career of their students and fellows. In the long term, helping the career of students and fellows is beneficial to everyone. I also want to see that someone takes a realistic perspective of their work within the broader context of science, technology, and life itself."
>
> —*Jennifer Elisseeff, PhD, Johns Hopkins University, USA*

Getting an Offer

Now the good news! You get a call from a department chair or program director, expressing his or her interest in having you join the department or program. The chair (or director) will now redouble efforts to sell the department and the university—remember; now you are being "recruited." The chair will probably describe general terms of the offer, including salary, startup package (an extra one-time shot of funds and other support to help you get started in establishing your laboratory), the space you will have available for your laboratory, and your teaching load. The actual numbers may not be spelled out, and should not be discussed right away. This is where a careful negotiation may be needed. You will be encouraged to visit the campus again, which can be worthwhile, since your first visit and schedule may have been too hectic and stressful to be able to make a fair evaluation of the place. Also, if things have gone well with your job search, you may be evaluating more than one offer, so a return visit is worthwhile for the sake of comparison. On the return visit you can learn more about the university, department, and its people by taking a second, closer look at the campus and department. You can also check out your designated laboratory area, and meet with faculty and students that you may have missed the first time around. You may be interesting in learning about the city or town the university is based in—its amenities, the cost of living, and so forth. This is

> **What is your process of identifying and finally selecting a prospective faculty candidate?**
>
> "The search is advertised and the search committee screens applications. Particular attention is paid to research accomplishments, evidence of leadership, formal training, and letters of recommendation. During this paper review the committee will address the potential fit with the department. Interviews and 'job talk' are next, and several candidates will be asked to visit. Input from all faculty (and students) is solicited for each finalist. The search committee recommends one candidate to the department head, who then seeks advice from the department's personnel committee. Final decision is with the department head who then must seek approval from the Dean and the Provost."
>
> —*Roger Mark, MD, PhD, Harvard Medical School and Massachusetts Institute of Technology, USA*
>
> "Faculty candidates come to our attention through their scientific results. People within our own department or colleagues from outside will point out exciting research being done by the candidate."
>
> —*Elliot McVeigh, PhD, Johns Hopkins University, USA*

now your turn to ask a lot of questions and make your own decision based on both tangibles, such as the terms of the offer, and intangibles, such as the social chemistry within the department and quality of the local school system.

Negotiating the Offer

Most new hires, fresh doctorates and postdoctoral fellows getting their first faculty position are quite inexperienced in the matter of negotiating the offer. Some things are negotiable and others are not. Most may feel very awkward or hold idealistic positions such as that they are scientists and not in it for the money. The awkwardness arises from a lack of experience, while the idealism comes from educational training and upbringing. You need to overcome both of these attitudes and take charge of your negotiations.

You will be offered the going salary. That is, most departments and universities have a reasonably well set salary scale for starting faculty. The starting salary is generally what the market conditions dictates: there is always a healthy competition amongst peer institutions. In addition, there is a competition with industry. There may be an informal internal understanding to maintain parity among the junior faculty and among the faculty in various departments. Hence, generally, you will get a fair and competitive salary offer and you may or may not be in a position to negotiate the salary. You would be in a better position to do so if you have a competitive offer from a peer institution. It is not a healthy idea to play institutions against each other—on the other hand, having competitive offers is a valuable tribute to your credentials. On occasion, universities may sweeten the salary pot through an endowment, consulting, or opportunities during summer; however this is rare for a starting junior faculty.

"We seek candidates who work at the interface of multiple disciplines, especially someone who can use quantitative approaches and engineering perspectives to solve biomedical problems. We also try to assess if the candidate has demonstrated a degree of independence in research thinking during the postdoctoral career."

—Shankar Subramaniam, PhD, University of California – San Diego USA

What are they key attributes you look for in a faculty member you recruit?

"The applicant must have clearly demonstrated bioengineering skills and interests. Possibly the most important factor is a collaborative and encompassing attitude. Our research work is arranged in teams and the group dynamic is of paramount concern. We need faculty members who will be open and sharing with good interpersonal skills."

—Nigel Lovell, PhD, University of New South Wales, Australia

"One of the main expectations from young faculty members is that they develop an independent research program. This requires original thinking to differentiate themselves from their previous advisor and others in the field of their choice. Therefore, one of the key attributes to look for in a faculty candidate is the potential for independent and original thinking."

—Dominique Durand, PhD, Case Western Reserve University, USA

"I look for passion coupled with intelligence and a capacity for hard work. I look for a team player – someone I would want to work with everyday."

—Herbert Voigt, PhD, Boston University, USA

"In addition to accomplishments listed in the CV, people skills are important. Can the person present ideas and data in ways that are understandable and keep the audience in tune? Can the person communicate? Charisma is a big plus."

—Eric Young, PhD, Johns Hopkins University, USA

"We expect the candidate to assume that he or she will compete at the highest level and be able to establish a dynamic laboratory immediately. A junior candidate should be willing to establish a national leadership role within three to four years and influence the growth and thinking in the field."

—Shankar Subramaniam, PhD, University of California – San Diego, USA

You will need to negotiate your start up package. This includes your 12 month salary during the first few years. Most engineering departments provide 9 month "hard" salary (that is the salary is guaranteed by hard money from the university). To get a faculty member started, most universities will give support for summer salary for the first one to three years. In the later years you need to earn this through grants, teaching or consulting. There are many programs, especially those located in medical schools, that are on 12 month salary plan but expect a faculty member to bring in 50% or more of the salary through grants. This type of compensation is known as "soft" salary. Hence, it is helpful to get as much hard money, in terms of number of months supported and for as many years as possible (typically three), as up front as possible.

The next and very negotiable item is your start up research funding package. You will need varying amount of money to set up your laboratory. Theoreticians may need space and computers and support for graduate students or postdoctoral fellows. Experimentalists will, in addition, need funds to buy relevant equipment and pay for supplies. In some fields extremely expensive and unique equipment may be needed. Hence the start up funding profile varies quite widely. You should realistically assess your own needs and professionally convey these, rather than go for a dollar amount to simply get as much money as you could get. If the difference between your need and what the chair is willing to offer is small, it will be bridged through discussions. If the difference is great, the chair may still get what is needed from the dean, departmental reserves or endowment.

Often overlooked is the need for space. All too often the laboratory space is not ideal for a faculty's needs. It may need renovation, or may even be a part of a new building plan. Or perhaps a previous faculty member is still vacating the space. As a consequence of all these factors it is very important to negotiate the actual space: its location, its square footage, renovation needs, expansion possibilities, ease of access (particularly if you will be transporting heavy equipment into the space), and so on. It is best to firm this up before you accept the offer. Otherwise, the space you envisioned may not materialize—there may be delays, perhaps due to a communication gap or some change in institutional environment, or delays in building and renovation. Such delays can be a serious setback for a junior faculty as precious momentum is lost.

The next item is teaching load and responsibilities. It is likely that you have been interested in an academic career because you like teaching and student contact. Hence, you may be eager to teach one or more courses. On the other hand, teaching new courses, even preparing for them, can take a great deal of time and may pose a major distraction when you are attempting to set up your laboratory, recruit students, and get your independent research off the ground. Hence, most programs will offer a gentler start, from none to one course to teach during the first year and increasing to as many as two or three courses each semester in steady state, especially if you are not funded. The start up may also include summer salary support, again giving you time and money to focus on your research. You may ask to teach a graduate course in your area of expertise to share your knowledge and expertise and to generate student contact for recruiting them to your lab. You may show your departmental citizenship and offer to teach undergraduate courses, singly or preferably jointly with other faculty. However, managing teaching responsibilities initially is very critical to having the time to launch a successful research program.

Other negotiable items include moving expenses, support for graduate students, teaching load, travel and other general funds. Access to good graduate students and support for them can be extremely important to launch your research program quickly.

What are the key expectations of the candidates you try to recruit?

"Their ability to discover new knowledge or create new technologies and then to publish that knowledge or bring those technologies to bear on existing problems. All other aspects of a good faculty seem to flow from that."

—*Elliot McVeigh, PhD, Johns Hopkins University, USA*

"I want someone who obviously excels within the rigors of their discipline but in addition is able to easily communicate with experts in other disciplines and can think out of the box – is not afraid of 'crazy' ideas. They also need to be independent."

—*Jennifer Elisseeff, PhD, Johns Hopkins University, USA*

Do you have any advice on how to negotiate the terms of the faculty position?

"I think the most important thing to consider is that the faculty position is a good match for you – schools vary tremendously in the expectations they have of junior faculty, the culture within a department, the quality of graduate students and the level of institutional support. The financial terms of the offer are a very small part of making an important decision. I think it is important to consider carefully what is most important to you as a person and determine which department provides the best opportunity for you to succeed in your own terms."

—*Rebecca Richards-Kortum, PhD, Rice University, USA*

"You need to talk to recent recruits at a number of different places to know about what kind of packages are being offered. I believe in the importance of a collegial and supportive environment – one that will give you the fertile ground to succeed."

—*Jennifer Elisseeff, PhD, Johns Hopkins University, USA*

Getting Started

If all has gone well, you have received a good—perhaps even your dream—offer. Celebrate! Then hurry up and finish your doctoral or postdoctoral work and, in parallel, get ready for some of the most challenging years of your life. Having an offer in hand sets the timetable and provides a great incentive (and pressure) for you and your advisor and committee to wrap up your dissertation and your current research project. You will very quickly realize the necessity to stay focused and not let your research shift too far from what you had originally proposed. You will need to focus on writing and submitting papers. Some papers, if left unfinished, can be completed after you move to your new position.

You will arrive at your new home (i.e., laboratory) for the next few years, or perhaps your life. If all has gone well, you will have a vacant space to move into. You may have used the previous months to complete all the renovations and to order furniture and laboratory equipment. If you did so, you are off to a quick start. If not, you will have plenty to do as you combine setting up your laboratory with finishing off papers, writing funding proposals, initiating new collaborative discussions, preparing teaching material, and recruiting graduate students. You will be busy!

It will seem as if everything happens and needs to happen all at once. You will need to energetically set up your laboratory. Since you may not have recruited students into your lab, or a postdoctoral fellow or a technician you have hired may not have arrived or been trained, you will need to be hands-on. You may have to deal with issues that range from deeply technical and scientific to the purely mundane, for example, ordering supplies and equipment. You will also start meeting with your departmental and institutional colleagues and potential collaborators. In many interdisciplinary and particularly experimental fields, this is critical to building a successful research program. You may want to start giving seminars not only in your department but in other departments and schools in your university so that that your research receives greater visibility and you yourself have greater potential for meeting new collaborators. Making yourself seen and heard can also lead to recruiting of graduate students.

Research

The most important thing hanging over your head will be obtaining funding, preferably peer-reviewed, for your research. You may have a comfortable start up package, but that will last perhaps two to three years and will quickly wind down. Eventually, you will need to support your graduate students, sustain the costs of running your laboratory, and perhaps pay part of your salary on an ongoing basis through research grants. It is therefore *essential* to kick start the grant and funding cycle at the earliest possible moments so that your research can flourish. Fortunately, there are many opportunities for fresh or junior faculty to apply for funding: different private research foundations and government agencies give start up or early investigator grants. Depending on the quality of your ideas, your preliminary data and results, your

How important is grant writing for your academic work?

"Most research requires funding to attract graduate research assistants, so grants are necessary to get the research done."

—John Webster, PhD, University of Wisconsin – Madison, USA

"Grant writing is a very important component for academicians. It represents a challenge to seek competitive funding through grant writing, but it is also a fair mechanism for stimulating creative work and determining the appropriate distribution of limited resources."

—Bin He, PhD, University of Minnesota, USA

"Writing, in my opinion, is the most critical asset to a successful academic career. Because grants provide the financial means to pursue individual research goals, grantsmanship skills are essential. Grant writing can also provide the impetus and the vehicle to carefully hone a research plan. Because my own research is largely experimental, funds are necessary for both support of personnel as well as for acquisition of equipment and supplies. Thus, without a funding stream, much of my academic research would grind to a halt. Writing is also critical in disseminating one's research through such means as journal and conference papers."

—David Mogul, PhD, Illinois Institute of Technology, USA

"Science is funding-driven. The number one priority must be to secure funding. Pick important and fundable research topics. Then devote all the energy to building a program around the research theme, approaching it from an angle that leverages on the background and unique expertise of the faculty."

—Kam Leong, PhD, Duke University, USA

How do you put together a successful grant?

"A successful grant must first have a highly original and creative idea addressing a significant problem. It should then be supported by reasonable preliminary data attesting to the merits of the proposed idea and feasibility of the proposed studies. It is important to include all needed expertise on the team. Finally, doing fine work and publishing in rigorous journals is very important. This is a subtle point and reflected in the evaluation of qualifications of investigators. One starts preparing for a successful grant by doing good work well before starting to write the proposal."

—Bin He, PhD, University of Minnesota, USA

"Go into the lab and get some preliminary data to show that you have actually started on the research. Provide many illustrations to help the busy grant reviewers understand what you want to do."

—John Webster, PhD, University of Wisconsin – Madison, USA

collaborative team and your own confidence and grantsmanship, you may immediately be ready for the big league and may apply for any openly competitive grants.

It is challenging to select good research problems and ideas to pursue. It would seem reasonable to start with the research done as a doctoral student or as a postdoctoral candidate. However, that research is "owned" jointly by you, your advisor and your collaborators. Hence, you should carefully choose a continuation of that work with acknowledgement, or a tacit understanding and acceptance of your advisor and collaborators. However, very likely, you will want to add your own fresh ideas and directions and use your prior work as preliminary data, as well as evidence of your expertise. This may be a safe route. It is also not uncommon that the freshly minted assistant professor wants to pursue quite novel ideas and directions. That may be exciting and may even be very innovative. It is also somewhat risky in that preliminary data may not be available and considerable work during the early faculty tenure would be required to get the work off the ground. Junior faculty members often carry out both safe and risky research, and gradually chart their own path. Charting one's own path is necessary in any case to fully demarcate your own contribution from your mentors' and to fully develop your own research and academic potential.

Indeed, mentoring at this stage can be invaluable. There are many obstacles to navigate and problems to juggle. Which of your many research ideas to pursue immediately? Which one of

"There are many essential qualities of a successful grant but, to me, the most critical is that it needs to be easy for a reviewer to understand what is being proposed and how the research team plans to address the goals of the proposal. One way to promote clarity is to make a proposal modular, with each major section divided into subsections that contain clear heading sentences. Wherever appropriate, the specific question or hypothesis being should be highlighted. Each question or hypothesis needs to clearly state the rationale for addressing it, the methods to be used, and potential problems that may arise with strategies to overcome or circumvent these hurdles."

—*David Mogul, PhD, Illinois Institute of Technology, USA*

What are some attributes of a successful researcher?

"Being meticulous and hard working. Not giving up easily. Also, solving important problems."

—*Ali Khademhosseini, PhD, Massachusetts Institute of Technology, USA*

"Curiosity and life long learning."

—*Reza Shadmehr, PhD, Johns Hopkins University, USA*

"Care about student intellectual growth."

—*Dominique Durand, PhD, Case Western Reserve University, USA*

How do you choose research problems, prioritize them, and gauge their importance?

"I decide based on my expertise as well as my knowledge of the key challenges facing our field."

—*Ali Khademhosseini, PhD, Massachusetts Institute of Technology, USA*

"I try to pay attention to new ideas and data emerging from the literature to fuel my plans for further experiments."

—*Reza Shadmehr, PhD, Johns Hopkins University, USA*

"Ideally, we should choose the most critical problems to be solved in our field of endeavor. The priority should be decided in terms of several factors, including significance, innovation, and feasibility. The importance is gauged from how much difference one's research would make in his or her field."

—*Shu Chien, MD, PhD, University of California – San Diego, USA*

"Rather than thinking about the great tools engineers might apply to biomedical problems, I approach clinicians and ask them, "Do you have some problem that you think engineers could solve?" When the answer is "Yes" then I know I have a client who will work well with me and make use of what I develop."

—*John Webster, PhD, University of Wisconsin – Madison, USA* your ideas will very likely get funded and who funds these types of ideas? What are the prevailing funding opportunities? Having your grant critically read by a senior and experienced faculty is vital. Through interactions with your new mentors (and sometimes friendly junior faculty you can more easily turn to), you can receive positive feedback, as well as criticism, to help you select good problems, write good research proposals, and successfully move your research forward. The senior departmental faculty—even better, a considerate and caring department chair—can be invaluable in helping you juggle your many responsibilities. A good rapport with your chair can result in a better teaching load, exemption from administrative duties, or simply in smoothing out bureaucratic departmental and institutional processes. All too often junior faculty are reluctant, or may be shy or afraid to approach and seek out mentors. But having mentors and friendly advisors is critical to make the right choices at the right time.

"I think the best recipe to be successful in science is to be in predictive mode. Most researchers are attracted by big 'fashionable' topics, which normally spell disaster because you are by definition starting late. Therefore I always try to find topics that I judge important but are unexplored. The way I find these topics is by questioning the fundamental assumptions of methods that are widely applied."

—Jose Principe, PhD, University of Florida, USA

"I tend to identify a need in the medical care domain, usually based on my personal experience. If that problem seems to have a potential solution that is amenable to the approaches within my grasp, then I explore solutions. A good problem to work on would be one that represents a real need, is potentially solvable, stimulates interesting and innovative technology, and is fundable."

—Roger Mark, MD, PhD, Harvard Medical School and Massachusetts Institute of Technology, USA

How do your focus your research and practice it?

"When you address fundamentals, the scientific question is normally pretty clear. If not, I spend a lot of time discussing the scientific question to clarify the core of the problem. I tell my students that a question well formulated is half of the answer. Then you should spend sufficient time thinking about the question and how it addresses fundamental concepts. I think it is a waste of time to pursue 'low hanging fruit', i.e. easy-to-get results that are not of any fundamental nature. Once you have core results, then you should go out and apply those results to important problems and compare your solutions with those obtained using the best techniques available in the field. Most of the time you have to collaborate with other researchers who are experts in the target field, because this saves you a lot of time."

—Jose Principe, PhD, University of Florida, USA

Teaching

What about teaching? It seems everyone talks about research only, at least at the major universities. Of course, interest and love for teaching is why most of us go into academia. Interactions with students and the opportunity to train and mold them with new knowledge and ideas and prepare them for different careers is our primary mission as teachers and professors. At major research universities, faculty members do seem to spend predominant time on research. Still, teaching is a critical mission and requires serious attention, planning and prioritization. Generally, you will have a limited teaching load in the first one to three years. You may initially teach a graduate course in your area of expertise so that you can bring this new educational opportunity to your department and its students. That will afford you the best opportunity to interact and recruit graduate students to your laboratory. You may also be asked to give selected lectures, or to contribute to core undergraduate courses. Eventually, you will teach between one and four courses per semester depending on your department and university, and the nature of your appointment (for example, "soft" money appointments come with very little teaching load while "hard" money appointments come with relatively greater teaching load). It is not uncommon to "buy out" of some teaching; that is, if your research grants can pick up part of your salary, the department may absolve you of some teaching responsibilities.

Of course, it is critical to take teaching very seriously and do an outstanding job. Very good preparation and advanced planning can really help with teaching. At the same time teaching and teaching preparation does compete with your time to do research and other professional activities and development. Hence, you do need to strike a balance and devote time proportionate to your expected priorities and performance.

The Early Years

In the first three years it is expected that the faculty member will have set up his or her laboratory and will have recruited graduate students to the lab. More importantly, the faculty member's own research should be on independent footing, having earned an initial career development grant or, even better, a competitive peer-reviewed research grant or grants. In addition, the young faculty would by this time have wrapped up papers from the doctoral and postdoctoral period and be sending out papers from his or her own lab. These original papers should be published or should be in various stages of submission, review and acceptance. This is like half time in a major sporting event; time to evaluate the overall performance, including successes,

What do you feel are the important attributes of a good teacher?

"The most important element is that one should care about teaching, and care about the students. A good teacher would spend a lot of time preparing for the lecture and think about how the students would perceive it. The ability to speak effectively is a must. New teachers should tape their lectures before giving them and while giving them. Most people do not like to hear themselves, but what they learn by listening is invaluable, and well worth the effort required to break this barrier."

—Shu Chien, MD, PhD, University of California – San Diego, USA

"The ability to empathize patiently with a learning student facing each new concept. A willingness to listen carefully, even when it seems you have heard it all before. A good teacher is always excited about learning new things and is eager to share that joy and new-found knowledge."

—Alan Sahakian, PhD, Northwestern University, USA

"A good teacher imparts the love of learning! Good teachers are able to present material in the context of larger problems and challenges and push the student to think critically about the world around them."

—Tejal Desai, PhD, University of California – San Francisco, USA

How do you interact with students and post-doctoral fellows? How do you guide them in their research? What do you do to motivate and inspire them for academic careers?

"I like them to become independent thinkers. I provide guidance as they require but do not want them to depend on it. I want them to come up with ideas that are better than mine and make those ideas into reality."

—Ali Khademhosseini, PhD, Massachusetts Institute of Technology, USA

"I make sure they know that I will support them even though they may fail in their attempt to solve a problem. I encourage them to go to as many scientific meetings as possible and to interact socially with colleagues, so they begin to feel a part of our scientific community. And I encourage them to view problems from non-traditional points of view, so their work can be set apart."

—Reza Shadmehr, PhD, Johns Hopkins University, USA

"I try to be a role model in the same way my mentors were. I teach by example, describing the thought process I am following in a given situation. Perhaps the best inspiration comes from letting students and fellows experience the excitement and joy of teaching and doing research

(continued)

problems and barriers. With these factors in mind, it is easier to plan, somewhat urgently and aggressively, for the next two to three critical years.

Successful faculty usually select very good research problems, and aggressively seek out research funding, as well as strong collaborators, students, and postdoctoral fellows. They also develop excellent focus and direct their work towards getting peer-reviewed publications out the door. Early success with grants, particularly with the career development grants, really helps. A vibrant lab attracts students and collaborators, which in turn feeds into strong research productivity. This in turn generates ideas and preliminary results that can be used in preparing competitive research grants. Initial success with grants brings new resources and further energy to the laboratory. This is what I call a positive feedback or "virtuous cycle."

themselves in a protected environment, and from giving them enough opportunity and responsibility to carry a project through all its phases."

—*Alan Sahakian, PhD, Northwestern University, USA*

"I strive to promote a sense of working together to achieve the common goal of doing good science instead of a routine employer-employee relationship. Generally I like to offer the big picture of the project, and then provide the best environment and conditions for a lab member to succeed. This includes seeking the best collaboration if the lab does not have the expertise, freeing lab members from funding worries, and sending lab members to conferences. It is important to not stifle creativity. Naturally the best way to mentor is to set an example of working hard and being passionate about what we do."

—*Kam Leong, PhD, Duke University, USA*

"I give my students and post-docs the independence to develop new ideas and to work on something that really interests them, not something that I hand to them. I hope to instill in them the excitement of developing new science and working at the boundaries of disciplines to create something new."

—*Tejal Desai, PhD, University of California – San Francisco, USA*

There can be many missteps and pitfalls along the way. The first misstep is usually in setting up the laboratory. Between getting new space, renovation, and purchasing and installing equipment, precious time can be lost. Experimentalists are particularly slowed when getting techniques and procedures going in their own lab. The approaches and ideas that were worked so easily in your mentor's laboratory are now shown to be related in part to the experienced hands of your mentor—you yourself will now have to develop deep-seated problem-solving skills. Recruiting high quality graduate students can be a challenge. Students may hesitate to join a new lab. You may also have limited resources beyond your initial start up funds to support new students. Postdoctoral fellows also choose to go to well-known institutions or advisors. While this reality favors senior faculty and established labs, the new faculty member's emerging area of research, fresh face and youthful energy may work to his or her advantage. Many junior faculty do not do a good job of balancing teaching, advising, departmental service, and research. It is easy to get caught up

What would be your advice to a junior faculty regarding balancing teaching/research and other demands of the academic life?

"I think this is a very serious problem these days. Many are choosing not to enter academia, as a balanced life is basically impossible with the dramatic reductions in grant success rates. My only advice would be to be sure to have your priorities straight. These may include starting and having a family in what will likely be your pre-tenure years, and expecting that many important but lower priority things to fail."

—*Krishna Shenoy, PhD, Stanford University, USA*

"Both teaching and research are important. Teaching is also a kind of an 'input.' In general, students may not understand a subject during a course of study, but they will clearly understand the subject when they teach it!"

—*Shangkai Gao, PhD, Tsinghua University, China*

with excessive course preparation. It is also easy to get "sucked into" departmental and administrative duties, particularly if you are a female or minority, (although good chairs and mentors may try to protect the junior faculty). Choosing and attacking good problems trips up many junior researchers. Many also drag their feet in getting publications out of the door: some hold out for very high impact results and journals and others lose focus and do not finish their studies and complete the writing. Initial lack of success with grants can also then lead to a reverse of the virtuous cycle, draining more and more time writing grants and seeking funding.

"Junior faculty should have clearly articulated metrics of performance regarding the balance of teaching, research, service, and I will add, industrial activities (industrial sponsored research, consulting, entrepreneurial activities). Untenured junior faculty in science and engineering should, as a general guideline, (a) not aggressively pursue industrially sponsored research (b) should be nurtured to develop their independent scholarly output derived from competitively awarded extramurally-funded research, (c) be engaged in teaching and student advising and do so as well as or better than one's peers (this is an absolutely necessary but insufficient condition for success in academic life)."

—*Anthony Guiseppi-Elie, ScD, Clemson University, USA*

As a professor, how do you divide time between education, research, consulting or entrepreneurial activities?

"Teaching is a very important work for a professor. However, I believe that a good teacher should have abundant research experience. I spend most of my time in research."

—*Shangkai Gao, PhD, Tsinghua University, China*

"As a professor, my principal focus is on education, which is the training a new generation of creative thinkers who have depth of appreciation for the fundamentals that guide and govern their field of study. Research is but a vehicle to accomplish that educational goal. University-based research involving high school students, undergraduates, graduate students and post-doctoral fellows affords one the opportunity to convey the use of the scientific method in creative problem solving on many levels. The forgoing accounts for approximately 85–90% of my working time. Consulting and entrepreneurial activities occupy approximately 10–15% of my working time and tend to occur in short periods of intense activities."

—*Anthony Guiseppi-Elie, ScD, Clemson University, USA*

Tenure

Tenure is a major career step-or hurdle-depending on how you look at it. The tenure decision is generally made five to six years after starting as an assistant professor. Generally there is a yearly review; at about the three year time frame, there is an internal or intermediate review of the faculty member's credentials and accomplishments. The intermediate review may be done by the department chair or by senior faculty members. They will evaluate your research productivity and secondarily your teaching and departmental service and citizenship.

The tenure decision steps are fairly similar, although not identical, at various universities. The decision may begin with the chair, who, upon reviewing the candidate's credentials, arrives at an informal judgment that the candidate is promotable or likely to receive tenure. Often the chair may consult other senior faculty. In some very large departments this task may be distributed. At some point, the chair needs to prepare, with the help of the candidate, a strong dossier. This would include the CV and the supporting document or letter from the chair outlining the candidate's credentials and highlighting the major accomplishments. The chair would review research, teaching, advising and administration, general citizenship and national or international leadership. The specific scientific accomplishments may include highlights supplied by the candidates, including a selected few major results and publications. The letter would spell out research grants, especially the ones that are highly selective and peer reviewed. Further, the chair and the review committee members evaluate the teaching performance and evaluations by the students. The chair outlines the work on behalf of the department, participation in committees, and other efforts the candidate has made on behalf of the university. The candidate should provide information on leadership activities, such as chairing or organizing programs at conferences, giving invited talks, and serving on major federal panels. The chair may also request three to five external references from the leaders in the field that the candidate works in.

The most important next step is the assessment by an independent *ad hoc* committee or an academic council (these groups may have different structures and names in different settings). The group will seek out recommendations and references from a dozen to two dozen scholars in the field of the candidate. These letters are crucial: they should provide an objective evaluation of the candidate's accomplishments. The letters should look beyond simply counting papers and grant dollars to instead assess the impact of the candidate's work and leadership. It does *not* help if the candidate is not known to the referee—it is most helpful if the evaluator is aware of the candidate's work through conference presentations or journal submissions. A non-committal letter from the referee, or one with outright negative comments, can be quite damaging. However, usually the candidate receives glowing reviews for specific research accomplishments and leadership. It is common for evaluators to be

What are the successful strategies, or "do's and don'ts" in preparing for tenure/promotion?

"Promotion with tenure is best achieved if you establish a research reputation for yourself, and have a reasonable to excellent teaching record. Research reputation or standing is best measured by (a) external funding; (b) publications; and (c) an intangible which is akin to branding i.e., external reviewers clearly being able to associate your name with a field or a contribution or an approach. For (c) it might be useful to consider using consistent language in referring to your work so that people can think, for instance, "Cancer Nanotherapeutics – Prof. John Smith at ABC University". Also, starting from the second to third year, keep a mental list of possible external referees for tenure promotion. Remember—these will be people you have NOT collaborated with."

—*Ravi Bellamkonda, PhD, Georgia Institute of Technology, USA*

"There are three essential components: Strong journal articles, lead investigator status on peer-reviewed grants, and strong references. The last part can be a problem if the field is large and the research is not earth-shaking. Go to international meetings, give strong, well-rehearsed talks, and ask perceptive questions. You want people asking 'Who is that person?'"

—*Gerald Loeb, MD, University of Southern California, USA*

How do you prepare for academic promotion? What do you need to present to the promotion committee?

"Pursuit of excellence in research and teaching. If the faculty member can publish significant papers in first-rate journals, receive extramural support (especially from the NIH), and do an excellent job in teaching and some service functions, the promotion will take care of itself. Of course, the faculty member needs to organize documentation for presentation to the promotion committee, but that is easy when the substance is there. I have never done anything special to prepare for my promotions and never thought of a promotion as a goal. My goal was pursuit of excellence."

—*Shu Chien, MD, PhD, University of California – San Diego, USA*

"Promotion from assistant to associate professor is made on the basis of promise, whereas promotion from associate to full professor is based on fulfillment of that promise. Sometimes this latter promotion can be made on the basis of educational or service leadership, but in my experience this takes longer and one must also demonstrate leadership in education through such avenues as publication in the field of education. Awards and election to Fellow status in professional societies can also assist with promotion to full professor. Essentially, you must demonstrate leadership in your discipline – in research and/or education or service."

—*Ravi Bellamkonda, PhD, Georgia Institute of Technology, USA* asked to compare the candidate with his or her peers and leaders in the field at that stage of the career.

Tenure recommendation is made by the council or the committee appointed by the dean or the provost, although the final decision may rest with the provost, president or the trustees of the university. While all aspects of academics should matter, it is well-known and accepted that the overall assessment is tied to the faculty member's research and scholarship. The external letters and opinions are used to affirm the research scholarship and accomplishments. Internal reviews are used to determine teaching performance and contributions to the institution. Poor teaching performance or citizenship in the department and the university can have a subtle but damaging effect. Hence, successful candidates should present a well rounded portfolio that includes outstanding original research and evaluations, sound teaching, and adequate contributions to the institution. The department chair presents all these credentials to the academic council, which then evaluates the candidate's CV, chair's recommendation, and external references and makes its final decision and recommendation to the dean or the provost.

Succeeding in Academia

Success in academia takes many shapes and forms. The tremendous intellectual freedom and diversity of education and research pursuits take so many divergent paths for all faculty members that there are no simple metrics of success and no single path to travel. The joy and beauty of an academic life is to find one's own path, develop one's own metric and research ideas, and to explore the boundless horizons of scholarship. That being said, successful academicians do many things right and do them well. First and foremost is scholarship. Scholarship is predominantly judged by the research that has been accomplished, which in turn is most easily assessed by the publications and the respect of the peers in the field, and less easily evaluated by such factors as degree of innovation and impact of the scientific work, and the candidate's perceived degree of leadership in the field. The number of publications expected per year, and the overall total of papers "required" to receive tenure varies widely. It is a better practice to assess the impact of the work, whether by looking at the selectivity of the journals in which the work is published or the citations the article has received in other publications. Since the ability to obtain funding in a competitive environment is so difficult, particularly peer reviewed research grants from major federal grants, funding can take the form of a yardstick by which to measure a candidate's success. However, all of this may be superseded by the recognition and impact the scholarly work receives. This is not easily judged by counting papers or citations or grant dollars. It may be judged, more qualitatively than objectively, by peer recognition. It may be judged by the accolades received, such as "Best Paper Award", "Outstanding Junior Faculty Award", "Young Innovator Award," "Early Career Development Award," or "Best Teacher Award" given by journals, society or the institution itself. In the long run scholarship, discoveries, and innovations of the faculty member may well be judged by their lasting impact on the scientific discipline.

Successful academicians often teach well by taking their expertise and scholarship into the classroom. They develop innovative curricula, novel approaches to teaching, and their teaching and books become widely accepted. Student evaluations are gathered at almost all universities and reviewed within the department and presented to the tenure decision committee.

An ultimate way to be recognized and deemed successful is through the success of the students you have trained. Nothing gives an academic more pleasure and pride than the success of his or her own protégé in whatever field he or she chooses to commit to and excel in.

"Be true to yourself. Make sure you understand each element of your work completely, exactly, deeply. Struggling to do this will not prevent all mistakes, but your mistakes will be fewer, and the advances you make will be greater."

—*James Bassingthwaighte, PhD, University of Washington, USA*

"My only advice to students and young professionals is "never, ever give up." The opportunities are abundant in bioengineering; sometimes it is necessary to look for them, even passing through a long, tortuous path. Eventually, you may be lucky enough to put an important idea into practice. I feel happy and gratified when I say: "I work in bioengineering." There is no other field that could give me this level of motivation and satisfaction."

—*Sergio Cerutti, PhD, Polytechnic University, Italy*

Concluding Remarks

Finally, everything devolves to academic excellence and leadership: in research, teaching, curriculum development, disseminating science and contributing to organized scientific enterprise, training and mentoring. Successful academics excel at one or more, if not all, areas. Commitment to excellence is the hallmark of leading academic programs and their faculty. Stellar faculty strive to do their best in teaching and mentoring, because this helps produce successful future scientists and academics. Most academics will agree that the passion for teaching and research, doing something original, and making a contribution to science and the society is what drives them. It is easy to get caught up in the daily grind of teaching and running the lab, or the challenges of juggling research, teaching and administrative duties, "publish or perish" mentality, or the stresses of research grants. However, it is important to keep in mind the reasons you joined the academia: to learn, be creative and innovative, interact with and teach young students and develop the careers of your protégés, carry out exciting research ideas and generate new knowledge, and leave a lasting impact to the society and the world in general. Pursue your dreams with a high degree of passion, commitment to excellence, and with highest standards of integrity.

Representative Reading Resources

1. J.A. Goldsmith, J. Komlos, and P.S. Gold., *The Chicago Guide to Your Academic Career: A Portable Mentor for Scholars from Graduate School through Tenure*, University of Chicago Press; 2001.
2. M.M. Heiberger, J.M. Vick., *The Academic Job Search Handbook* (3rd Edition), University of Pennsylvania Press; 2001.

3. E.R.B. McCabe, L.L. McCabe., *How to Succeed in Academics*, Academic Press; 2000.
4. S.L. Barnes, *On the Market: Strategies for a Successful Academic Job Search*, Lynne Rienner Publishers; 2007.
5. P.J. Feibelman, *A Ph.D. Is Not Enough: A Guide to Survival in Science*, Perseus Books Group; 1994.
6. R. Boice, *Advice for New Faculty Members*, Allyn & Bacon; 2000.

6 Teaching Colleges and Universities

Maria E. Squire, PhD

Department of Biology, The University of Scranton, Scranton, Pennsylvania, USA

It was not that long ago that I was in the position that you are currently in, trying to decide what career path I wanted to pursue. Upon graduating with my Ph.D. in biomedical engineering, did I want to pursue a career in industry? Was a career working at an intellectual property law firm for me? Or did I want to pursue a career in academia? The choices seemed endless and at times overwhelming, yet my decision to become a faculty member at a small teaching-focused university became the clear choice for me.

You might be wondering how, with all of the choices available, I came to a decision to pursue an academic career at a teaching university. My decision was based on my undergraduate experience at a small teaching university, and on my experience as a graduate student at a large research university. I received my B.S. in biology and Spanish at a Catholic and Jesuit University, where class sizes were small and there were significant opportunities for professors to interact with undergraduate students. Asking professors questions outside of class was not a problem, and many professors even gave out their home phone numbers. As undergraduates, we worked one-on-one with faculty in their laboratories, and I developed close mentor-mentee relationships with a few professors whom I am still close to today. When I was in graduate school, I observed the faculty in my department focusing their efforts on directing research laboratories, writing grants and manuscripts, and attending meetings. With all of the pressure to maintain active research programs, they seemed to have little time to devote toward interacting with and teaching undergraduates. In fact, faculty generally taught no more than one undergraduate course per semester, and graduate students and postdoctoral scientists trained many of the undergraduates working in the laboratories. My goals and aspirations were more in line with the career path at a university similar to the former rather than the latter.

I have a vivid recollection of the endless questions that I had once I had made my decision to pursue an academic career. If you are even thinking of considering an academic career at a teaching university, I am sure that you have similar

G. Madhavan et al. (eds.), *Career Development in Bioengineering and Biotechnology*,
DOI: 10.1007/978-0-387-76495-5_6, © Springer Science+Business Media, LLC 2008

questions to those that I asked myself. What is a career at a teaching university like? Am I really prepared for teaching? What are the expectations for research? Am I going to have the resources that I need for my research? It is my aim to pose answers to some of these questions, and, in addition, I will share the knowledge and experiences that I have gained as an assistant professor in the Biology Department at the University of Scranton.

What Do I Need to Know if I am Considering a Career in Academia?

When deciding on a career in academia, there are a variety of different positions that are available. There are non-tenure track positions that may be either part-time or full-time, and may be either a short-term appointment to fill a particular departmental need for a finite period of time (i.e., a semester), or a long-term appointment. Then there are the tenure-track faculty positions, which are full-time appointments. New tenure track faculty members are hired at the rank of assistant professor, with the opportunity for being granted tenure and promotion in rank to associate professor after a probationary period. There is the eventual opportunity for promotion to full professor.

What is tenure, you ask? Tenure is a status granted to a faculty member after a probationary period, typically six years, which ensures lifetime job security for the faculty member (barring major misconduct on the part of the faculty member or financial exigencies at the university). This assures not only economic stability but also academic freedom in research and teaching [1]. A faculty member is awarded tenure based on satisfactory accomplishments in three areas: scholarship, teaching, and service. The primary difference between requirements for tenure at a teaching university versus a research university is the *emphasis* that is placed on each of the three criteria mentioned above. While research universities place more weight on the scholarship category, teaching-focused universities typically place more emphasis on accomplishments in teaching. As I will discuss shortly, faculty scholarship is also important at almost all teaching universities, but the benchmarks are generally not as rigorous as those at research universities.

What are the Expectations/Demands at a Teaching College?

Teaching

As a faculty member at a "teaching university," the majority of working hours during the academic semester are consumed by classroom teaching and teaching-related responsibilities. A faculty member's teaching load is usually determined in terms of credit hours, which, for lecture courses, is equal to the actual number of hours spent teaching in the classroom (contact hours). This is not the case for laboratory courses, where partial credit is awarded for each contact hour spent teaching in a laboratory [2]. For example, Dr. Jones teaches a lecture that meets three hours per week and receives three credit hours for that work, whereas Dr. Miller teaches a laboratory that meets for three hours per week and receives only one or one and a half credit hours for that work. If you are wondering what this means in terms of time spent on teaching, consider the following. In general, of faculty at four-year colleges who participated in the Higher Education Research Institute's (HERI) 2004–2005 national survey of college and university faculty members, 47% report that they spend nine to twelve hours per week in the classroom teaching (contact hours), and 26% report spending more than twelve hours per week teaching. My teaching load consists of one nine credit semester and one twelve credit semester per year, but I am fortunate to receive three credits of release time for research so that I teach nine credits each semester. These nine credits equate to nine contact hours and consist of two different lecture courses, in one of which I teach two sections of the same course. Over the past two and a half years, I have taught five different courses, three of which have been introductory courses for non-majors, and two upper level courses for majors.

Outside the classroom, teaching-related responsibilities include designing course syllabi, writing and revising lectures/laboratories, preparing PowerPoint slides and handouts, creating and grading all assignments, quizzes, and exams, holding office hours, and answering student questions via e-mail. In many cases, you will NOT have a teaching assistant to help you with these tasks. Seventy percent of faculty members (four year teaching colleges) who responded to the 2004–2005 HERI study reported spending *at least* nine to twelve hours of time per week on activities related to "preparing for teaching" [3]. This figure includes faculty members at all stages of their careers, and, from personal experience as a newer faculty member, I can attest to spending approximately five hours per day preparing for lectures, creating and grading papers, and interacting with my students. In total, I spend approximately 40 hours per week just on teaching and related responsibilities. Although this number will decrease as I establish my courses and gain more experience, teaching and related responsibilities are the primary focus of faculty members at teaching universities during the academic semesters. Therefore, heed this warning: *if you are not passionate about teaching, this career path is NOT for you*!

Research

Do not let the caption "teaching university" fool you. One should NOT enter this academic career expecting it to be a "refuge from research" [4]. Research, in some capacity, *is* expected and conducted at almost all teaching colleges and universities. Many share the belief that "we teach better when we are up to date in our discipline" [5]. Furthermore, if you do not have an active research agenda, you will likely be unsuccessful in obtaining tenure. Most faculty members at teaching universities, from those early on in their careers to the more veteran faculty, will agree that it is extremely difficult to find time during the academic semester to do research. Of those surveyed in the 2004–2005 HERI report, 40% responded that they spent only one to four hours per week on research, 20% spent between five and eight hours, and only 16% responded that they spend more than eight hours per week on research [3]. With so little time during the semester, faculty members make most of their progress on research during the summer and winter breaks. Yes, I did say during the summer and winter "breaks." Do not decide on a career in academia thinking that you are going to have vacation for three months out of the year. This philosophy will certainly derail your path to tenure.

As a faculty member at a teaching university, I do almost all of the research-related tasks that faculty at research universities do, but on a smaller scale. In my laboratory, I am the research director and I am also doing the research in the lab, with the assistance of undergraduate students, but without graduate students (our department does not have a graduate program), a post-doc, or a technician. Grant writing is part of my research responsibilities, as I need to apply for funding for my research projects. I have the opportunity to apply for some internal funding opportunities to get my research projects started, and I did receive some start up money to purchase general laboratory supplies and some smaller equipment. However, for the more expensive equipment and my long-term research projects, I also seek out external funding opportunities. Fortunately, the National Science Foundation and the National Institutes of Health, as well as some foundations, have grant programs that are exclusively designed for small, non-Ph.D. granting colleges and universities [6]. Finally, I present the results from my research at annual scientific meetings, and write up my results and submit them to peer-reviewed journals. From personal experience, I can attest to the fact that the research demands seem overwhelming, and it is extremely difficult when you are trying to set up your lab space and get started with research while preparing and teaching courses for the first time. If you are trying to imagine what it is like, think of what it is like for you as a graduate student when you are juggling your academic classes and a 20-hour per week teaching assistantship, while trying to find time for research. Being a faculty member at a teaching university and trying to find time for research is not that much different.

Service

The third component of being a tenure-track/tenured faculty member is the requirement for service to the students, the department, the university, and the community. Service to the students may include providing academic and career advising, writing letters of recommendation, or serving as a faculty advisor for a student club (i.e., the biology club) or honor society (i.e., Beta Beta Beta, the National Biology Honors Society). At the level of the department, service activities may include working on curriculum development or assessment, coordinating and ordering new library books and journals, or interacting with prospective students and their families at open houses. Service to the university community is also important, and, in fact, faculty members play an important role in university governance. Although the precise structure of committees varies from university to university, some examples may include committees that work on university planning, university mission and identity, the general education curriculum, the faculty handbook, and faculty development (including research and travel). Additionally, a faculty member can sit on the faculty senate or the board of rank and tenure [7]. Finally, it is important to participate in your local community by doing things such as volunteering and/or working with a local organization related to your professional discipline.

How much time is dedicated to service, you ask? On average, the majority (60%) of faculty members surveyed by the HERI study reported spending between one and four hours per week on student advising, 70% reported spending one to four hours on committee work and attending meetings, and 46% reported spending between one and four hours on community service activities [3]. New faculty members are generally not expected to start service work right away. In my case, my service activities in my first year entailed interacting with prospective students at the University's open house, becoming the faculty advisor for an honor society, and participating as a judge at a local high school science competition. I was not given any students to advise, nor was I asked to be on any university committees. In my second year, I got more involved, but I heeded the advice of fellow faculty members who told me not to seek out committee work because it would find me, and it did. I was asked to replace a colleague as a departmental representative to a committee during her sabbatical, asked to join two university committees, and was recently elected to another committee for this academic year. I have heard, though I have not personally had this experience, that women and minorities are prone to being called on more often to do service than others. There is a danger to becoming involved in too many committees and service activities, and if you are not careful, it will take valuable time away from your research, so it becomes important to know when and how to say no. However, becoming involved in committee work is important not only to fulfill your service requirement, but also to get to know faculty members from other departments, as well as the university staff and administrators.

Working Towards Tenure

As just described, faculty members must strive to achieve excellence in teaching, scholarship, and service in order to be awarded tenure. How is this measured? Well, the truth is that the emphasis on each area and the criteria used for evaluation vary somewhat from university to university. Excellence in teaching is evaluated based on student evaluations and peer evaluations from fellow faculty members. Scholarship is generally measured by a number of factors, the most significant being evidence of progress and development in research, and the accrual of presentations and peer-reviewed publications. Additionally, evidence of scholarship includes involving undergraduate students in your research projects, writing for external grants, giving invited talks, writing book chapters, etc. Finally, service to your department, the university, and the community is also examined during your probationary and tenure reviews. The best advice I can give you is to try and get a feel of what the expectations for tenure are at a particular university during the interview process. Once you are hired and on campus, speak to a number of colleagues both inside and outside your department to get a better idea of what the expectations are at the university. Make sure that you ask multiple people, as opinions may differ. Finally, if your progress is not regularly reviewed (I undergo an annual review process that monitors my progress toward tenure), seek out feedback on your progress so that you can work on areas that need improvement before it is too late to do so.

What Should I Do to Best Prepare Myself for this Career Path?

So, you like what you have read so far and think that an academic career at a teaching university is for you, but you are not sure how best to prepare yourself. First and foremost, you are going to need to pursue and complete a Ph.D. in your discipline. From scanning job postings on various websites and from what I have read [8], there are very few opportunities for applicants with less than a Ph.D. (or the equivalent) to obtain a faculty position at a teaching college or university. Whether or not you need a post doc really depends. It seems that, particularly if you are applying for positions in engineering departments, the more common path is to go directly into a faculty position upon completing a Ph.D. [8]. Additionally, in the biology department of which I am a member, myself and one other faculty member were both hired in the last five years without having done a post doc. My advice would be to research job openings, preferably in the fall when they are more numerous, and start to get acquainted with the minimum requirements for the positions.

During your graduate training, you should do everything you can to gain teaching experience [8]. It is one thing to observe teachers from the perspective of a student and to admire excellence in teaching, but it is a completely different situation to teach a classroom full of students and do it well [8]. There are many benefits to gaining teaching experience during your graduate studies. First, it will allow you to see if teaching is really for you, and if you can see yourself spending your career (or a good part of it) in the classroom. Additionally, it gives you some experience, which will be helpful, if not required, when applying for positions. Finally, the experience will help you in writing your philosophy of teaching statement, which is often required with your application for faculty positions. So how can you gain teaching experience? First, become a teaching assistant (TA). This will give you an introduction to some of the duties of a college professor, including creating and grading assignments, quizzes, and exams, and holding office hours. Alternatively, or additionally, you can arrange to give a few guest lectures for a course, or even teach an entire module for a course. I did this, and, in addition to the experiences from being a TA, I gained the experience of choosing the topics and objectives, writing my lectures, preparing the slideshows and handouts, and delivering the lectures. Engage in teaching opportunities outside of the classroom as well by mentoring students (high school, undergraduate, or junior graduate students) in the laboratory. This too will help you learn how to work with students, give you a perspective into how people think and learn, and also help you to gain the skills that you will need to train students who will be working with you in your own research laboratory. Finally, if you have the opportunity to do so, give talks or lectures for student or professional societies that you belong to.

You are undoubtedly going to gain significant research experience in graduate school, but there are certain activities that you should be sure to gain experience in while in graduate school. First, publish! Learn how to write scientific papers and respond to reviewers' comments and criticisms as a graduate student. Second, be involved in your mentor's grant writing, or even write your own grant. Learn the ins and outs of developing hypotheses and specific aims, designing the experiment, and writing the budget. Finally, attend scientific meetings and present your data so that you can start getting to know colleagues in your field. These may be people who you can seek advice from or even collaborate with in the future. My research efforts have been significantly enhanced through continued collaborations with colleagues who I initially worked with when I was a graduate student. You will likely be more successful if you have experience to fall back on and individuals from whom you can seek advice, and with whom you can collaborate.

Finally, in preparing for a career in academia, as with any career preparation, seek out advice. You can read books such as this one. Other sources that I have found helpful include the career advice column found on the *Chronicles of Higher Education* website (www.chronicle.com/jobs/news/), and a book entitled *Advice for New Faculty Members*, by Robert Boice. Importantly, speak to faculty members who have positions at teaching colleges or universities similar to the one that you hope to obtain. Perhaps your dissertation mentor has a colleague that you can speak to, or maybe a family member or a friend of a family member knows someone at a teaching university. You might even have a former professor from your undergraduate *alma mater* who you can contact. These individuals know the day-to-day routine at a teaching university, and will be able to share their experiences with you and provide advice on how they have been successful in balancing the teaching, research, and service requirements of faculty members at teaching universities.

References

1. DeFleur M. Raising the Question #5 What is Tenure and How Do I Get it? *Communication Education* 2007 Jan; 56(1):106–12.
2. Jacobson J. "Do Science Professors Get Enough Credit?" *The Chronicle of Higher Education Chronicle Careers* 2002 [cited 2007 June 1] Available from: www.chronicle.com/jobs/news/2002/12/2002120401c/careers.html
3. Lindholm J, Szelenyi K, Hurtado S, Korn W. "The American College Teacher: National Norms for the 2004–2005 HERI Faculty Survey," Los Angeles: Higher Education Research Institute, UCLA; 2005.
4. Hall D. "Interviewing at a Teaching-Focused University," *The Chronicle of Higher Education Chronicle Careers* 2003 May 28 [cited 2007 June 1] Available from: www.chronicle.com/jobs/news/2003/05/2003052801c/careers.html
5. Krebs P. "Colleges Focused on Teaching Too Often Neglect Research," *The Chronicle of Higher Education* 2005 Sep 23; B14.
6. Woolston C. "Small-Scale Science," *The Chronicle of Higher Education Chronicle Careers* 2002 October 22 [cited 2007 June 1] Available from: www.chronicle.com/jobs/news/2002/10/2002102201c/careers.html
7. Gibson G. *Good Start: A Guidebook for New Faculty in Liberal Arts Colleges.* Boston: Anker Publishing; 1992.
8. Reis R. *Tomorrow's Professor: Preparing for Academic Careers in Science and Engineering.* New York: IEEE Press; 1997.

7 Industry Research and Management

Mark W. Kroll, PhD

Taser International, Inc, Scottsdale, Arizona; Mark Kroll & Associates, Crystal Bay, Minnesota; Department of Biomedical Engineering, California Polytechnic State University, San Luis Obispo, California; Department of Biomedical Engineering, University of Minnesota, Minneapolis, Minnesota; and Anderson School of Management, University of California, Los Angeles, California, USA

Medical devices are a critical part of modern diagnostics and therapeutics. A career as a bioengineer in the medical device industry can be highly fulfilling when it involves developing devices that dramatically improve the quality and longevity of life—usually without the need for a patient to remember to do anything on a daily basis but live.

Background

I was in the ninth grade when the first human heart transplant was performed in South Africa. That event made international news and I became inspired. I was energized to the point that I decided that I wanted to make a career in medical device engineering and management. With typical adolescent idealism, I assumed that with medical devices we could make people live forever.

I ended up working in the electromedical device industry—specifically, in the realm of cardiac rhythm management. This domain includes pacemakers, implantable defibrillators, and ablation devices. The market is about US$10 billion in annual sales worldwide. In spite of its size and respectability today, the electromedical device industry had very scandalous and questionable origins. Pacemaker research came from attempts to reanimate guillotined criminals; defibrillation research began with fibrillating murderers for statutory electrocution in New York State.

Other product ideas can be found in the electromedical quackery that claimed to cure many people via electricity 100 years ago. Quacks claimed that they could

G. Madhavan et al. (eds.), *Career Development in Bioengineering and Biotechnology*, DOI: 10.1007/978-0-387-76495-5_7, © Springer Science+Business Media, LLC 2008

use electrical stimulation to cure everything from cancer to depression, sexual dysfunction, and incontinence. During the 20th century, such charts were typically good for a robust laugh, but now very serious researchers are working to use electricity in every one of those areas. Electrical current is being evaluated for cancer therapy, and electrical stimulation is being evaluated for everything from depression to seizure disorder and obesity.

There is no degree better matched to the medical device industry than that of bioengineering.

> "Bioengineering training is very helpful in the medical device industry. Bioengineers tend to have a better understanding of human biology and physiology, as well as disease, which allows them to apply their engineering knowledge in this context. Additionally, bioengineers tend to understand the demands of developing technology in a regulated environment, that is, in using development, pre-clinical, and clinical testing processes that will be acceptable to the Food and Drug Administration. Bioengineers have a leg up on other engineers in medical device development."
>
> —*Ross Jaffe, MD, Versant Ventures, California, USA*

Many graduating bioengineering students go directly into the medical device industry. However, there are many bioengineers who began in academic research, or aerospace, or other technical industries such as communications. The great pleasure of being in the medical device industry is that one can truly know that he or she is making products that make people either live longer or feel better. While I am clearly not objective, I believe it is accurate to say that there is very little over-marketing of medical devices such as there is with the pharmaceutical industry. Medical device companies do aggressively promote their products to physicians and reimbursement agencies. However, they do not engage in the practice of drug companies that get drugs, such as statins, approved with studies of very high-risk patients and then market them to be over-prescribed to the general population. The frustration is that this is a highly regulated industry.

> "The industry today is far different from when I began my career. Before the Food and Drug Administration regulation era, one could use good judgment about what was a design concern and what was not. The government decided they could pass a law to keep people from making mistakes, but all it did was raise the cost enormously and we still have mistakes because we are all humans. Now a career in a medical device company means you follow a cookbook procedure that makes the process years long and dragged out and boring to anyone that likes to see things happen on a timely basis."
>
> —*Daniel Adams, SciMed Life Systems (now Boston Scientific Corporation), California, USA*

How to Succeed?

There are three types of bioengineers who succeed in this industry. The first is the producer who simply does his or her job, shows up to work on time, and is dependable while not being outrageously courageous or provocative. This is the type who gets the work done and probably comprises 80–90% of the engineers. We also have the two troublemaking types who are eager for greater fame, glory, and power. These are those that want to have the high level management positions, and, in the second case, these are the engineers who have the highest positions of technical leadership.

> "Proceeding up the ladder of success requires one to be selective about the projects that they do at work. If they accept non-challenging projects then progression will be slow or non-existent. Yes, you can choose what projects you will do for a company, but you have to be assertive about it. Sure, not all projects will be challenging but having challenging projects is an overall goal. Do not worry so much about remuneration; work hard, be dedicated, be challenged and the former will result."
>
> —*John Poore, Cardiac Rhythm Management Division, St. Jude Medical, Inc., California, USA*

It is not easy to decide if you would be happier getting your power, glory, and recognition through the management route or through the technical route. Medical device companies recognize the critical importance of the highly creative and insightful technical talent, and are very happy to reward this talent handsomely. Their pay grades are not formulaic, as they might be in a commodity manufacturing company or a post office where grades are based on management responsibility and seniority.

> "Consider specialization in trendy technologies. Invest in refining communication skills and collaboration skills. Refine your problem solving skills and ability to apply your available resources to solve challenges. Embrace teaming and collaboration."
>
> —*James Causey, formerly with Minimed, Inc (Medtronic Corporation), and Lifescan, Inc (Johnson and Johnson Corporation), California, USA*

I strongly counsel the extremely bright and creative engineers to not necessarily assume that they have to be a big boss to make a lucrative salary and to have a substantive influence. In fact, if someone is extremely talented technically, it will be far easier for them to have a major influence and make good money by staying on the technical track.

> "To be successful as an engineer in a technical ladder program, the engineer needs to be recognized as an expert in their field within their company and, for some, within their field of business. This high level engineer can be critically important to the technical success of a company."
>
> —*Randall Nelson, Evergreen Medical Technologies, California, USA*

If you are thinking of the management track, you should ask yourself some difficult questions. Am I a natural leader? This is more than being able to organize a softball team. Do people naturally come to you for advice and decisions? Are you inherently organized (is your car trunk clean, your desk clean)? Do you have the organizational skills to not only manage yourself, but to manage people underneath you? Do you budget household expenses? Do you know how much money is in your checking account? If you can answer *yes* to all those questions, you might do very well in management.

So many excellent books have been written about management that this chapter will not go into further details for this track. Presentation skills are critical, and will be touched on in the next section. Sophisticated complaining skills can also be helpful for this track. I am not talking about daily whining that will destroy any career. However, a talented performer—desiring the management track—should let his/her supervisor and those one level above them know on an annual basis that he or she is planning a management career.

If, on the other hand, you look around in meetings and realize that you have the best grasp of the technical issues of the problem under discussion, and this happens on a regular basis, then you might very well be more successful pursuing a technical track.

Minding Your P's and P's

If you decide to pursue a technical track, you should focus on four things besides just doing your job. Obviously you have to do your job and you have to get your *projects* done on time. This is one of the five P's that is so obvious that we will not spend time on it. However, the four P's that will have the greatest influence on your career—outside of your first job—are *papers, patents, physician relationships*, and *presentations*. Let us begin with the physicians. Into whatever medical practice specialty your device falls, your physician contacts will be more important to your career than your engineering contacts. Forming relationships with the top physicians in your specialty area will allow you to get fast answers to questions about physiology, anatomy, and clinical applications—answers that will be critical to the successful development of a medical device. These relationships will be critical to your ability to have studies performed quickly. Finally, should you decide to go off on your own and do your own start-up someday, these relationships will be essential in providing medical and scientific advice and for obtaining financing.

> "Develop your physician relationships as those will be more important to your career than your engineering relationships."
>
> —*Ted Adams, Vice President of Research & Development, St. Jude Medical, Inc., California, USA*

The papers you publish and the patents you receive live on forever. They are a service to society, as they share your original ideas with your peers of today and tomorrow. They will also give you the satisfaction of a certain type of immortality.

"When I have fears that I may cease to be
Before my pen has gleaned my teeming brain,
Before high piled books, in character,
Hold like rich garners the full-ripened grain;"
—*John Keats*

These tangible peer-reviewed items are called "verifiers" that establish that you actually did something new and useful. However, most engineers fail to take the time to write up patent disclosures. Here is a common situation: I see an engineer who solved a tricky problem in his development project, and I recommend a patent disclosure be forthcoming. A few months later, the patent disclosure would still not be done. I would be told that the deadline was just too tight on the project, and the boss just would not give him any time off to write up the patent disclosure. In most companies, this would be true. The engineer's supervisor does not benefit from the junior engineer filing patents. The main evaluation criterion for the supervisor is that the development project was completed on schedule. Another patent disclosure from a junior engineer on the project is generally not counted in the evaluation for that project. Thus, the supervisor will not encourage the engineer to produce a patent application or a paper on some developmental issue.

However, there are deeper issues in the supervisor/engineer relationship. The typical engineer assumes that the supervisor is the same as the professor. When someone does a good job for the professor, the professor gives them an "A" and they go on to the next class. However, in industry the supervisor is somewhat disincentivized to promote the engineer. If the supervisor promotes the engineer, several negative things happen. One, he or she loses a good engineer. Two, the engineer is now his or her competition for the next promotion. Thus, the things that would tend to give the junior engineer high visibility—the papers, patents, or physician relationships—are not seen as favorably by the supervisor as they are by the engineer.

For this reason, the engineer needs to take the time to write up the patent disclosures and write up papers on their work. These do not have to be medical papers; they could be papers in basic science, engineering, or trade journals, for example. They could be very specifically restricted to their own area of technical expertise. The important thing is that the engineer is differentiating himself or herself from the masses who merely do their job in their cubicle without any public recognition. The following table will illustrate the primary effect reach and the time constant of effect in years for each of the P's.

Measurable	Primary Effect Reach	Time Constant of Effect (years)
Projects	Internal	1–2
Physician Relationships	Internal & External	10
Patents	Internal & External	20
Papers	External	15
Presentations	Depends on Audience	1

Local and International Relations

Another critical factor for success is understanding the market. If you are writing software for a medical device or working on certain material for an implantable device, do not assume this all you need to do.

> "Immerse yourself in the market segment your company competes in. Understand your company's competitors and their strengths and weaknesses. Go to school on your market. Focus more on the market, finding and understanding the mechanics of making development happen. Study your market and understand the genesis of failure in companies that have existed in your market and failed. Learn what mistakes they made and do not repeat them."
>
> —*James Causey, formerly with Minimed, Inc. (Medtronic Corporation), and Lifescan, Inc. (Johnson and Johnson Corporation), California, USA*

Internally, you will find that the personnel in the clinical studies group and the marketing group will be very grateful for your input and feedback. Nothing frustrates them more than to have a technical question that they need answered, and find that they have to send two to three email requests to finally get an answer two weeks later. Why? Because the engineer did not think it was important enough. After it wasn't as if it was the engineer's own boss asking. However, if you respond quickly and accurately to the requests for help from other departments, you will soon be recognized as the "go-to person," and your value in the company will rise significantly—independently of your relationship with your own supervisor.

Product development is now often a global effort. Since English is well established as the international language, engineers from the Western nations often assume they can communicate with their global peers just as they would with those in the next office. But just because English may be your first language does not mean it is the first language of your peers. Be sensitive. Learn a few basic greetings in their language. Do not use slang or uncommon words in your emails. In presentations and phone conferences, speak slowly, with careful enunciation and some intentional redundancy. You want the listener to understand the sentence even if they only understand 70% of the words.

On Education and Training

Post-graduate education offers difficult choices. With the increasing credentialism in our society, it is a significant handicap not to have some graduate degree. For the biomedical engineer, the common choices are typically the PhD and the MBA. If you feel that you would want to go into marketing or general management, then an MBA is probably the wise decision. A warning: many engineers are disappointed when they immerse themselves in an MBA program and compare that to the intellectual rigor of genuine engineering classes. Do not let this bother you. You are getting an MBA for the credential and to learn the terminology. However, you should not feel that the MBA is the *sine qua non* for management.

> "I strongly believe that in the medical field, it is essential that personnel in management possess a technical degree related in some way to the company's business. From my observation, most non-technical people do not have the logical, fact-based, methodical problem-solving skills needed to solve problems regardless of the business problem faced."
>
> —*John Berdusco, former Vice-President of Administration, St. Jude Medical, Inc., California, USA*

For a PhD or MS in Engineering, one would need to decide between bioengineering and a more traditional core curriculum. The advantage of staying with the bioengineering curriculum is that there are usually fewer prerequisites, and there is presumably a faster path to an advanced degree. However, bioengineers are always faced with the criticism that their knowledge is wide but not deep (note that this directly contradicts Ross Jaffe's advice, as mentioned earlier, and thus is not a universally held opinion). Of course, the same could be said for electrical engineering today, as no electrical engineering program could deeply cover integrated circuit design, digital communications, *and* power distribution. Thus, it is really unfair to suggest bioengineering is lacking in sufficient depth compared to one of the classic engineering fields. However, an MS or PhD in a classical core field such as mechanical, chemical, electrical engineering, or even the more modern computer science gives the credibility of an advanced degree, but also eliminates the issue of the occasional prejudice against the depth in bioengineering.

> "I think the best advice I could give would be to get a solid background and understanding of human physiology and anatomy and if possible a basic understanding of the major disease states and how they are managed. I think our most successful and valuable engineers are those who really understand the application and how the devices or features will be used in patients. That level of practical knowledge will serve them well."
>
> —*Eric Fain, MD, Cardiac Rhythm Management Division, St. Jude Medical, Inc., California, USA*

An underutilized path to the medical device industry is that of medical school. One can finish four years of medical school in the United States (which is slightly more in other countries), but may not go on to pursue fellowships, medical residency, or obtain licensure. The advantage of this approach for a career in the medical device industry is that it provides solid training in anatomy, physiology, and pathology. It is also faster than a typical doctoral training program.

Where Not to Start

One of the biggest mistakes people can make in this business is to try to do a start-up right out of school. This is not a problem with the typical student, but is a problem with brighter and more confident students. I continually try to counsel students away from this, but the common refrain is "Well, I do not have family responsibilities now and I can afford to be poor." In fact, the student is not just lacking family responsibilities—the student is lacking the credibility to gain any kind of financing, and really does not know what he does not know. My advice to such confident and bright people is to get a job at the biggest company that they can and try to continue to live on a student budget. That is, they should not run out and buy a new car and a fancy loft, but rather save 30% of their income. After doing this for several years, they will not only have the credibility to do a start-up, but they will have the financial resources to survive for several years with the challenges of a start-up. They will also know something about the business. It is assumed—usually correctly—that the hiring standards and training of a large company are superior to those of a smaller company.

However, the big negative with a large company is the common enforcement of indentured servitude. Here is how this works. You take your first job, and you have a mountain of paperwork to sign. One form is medical insurance, one form is dental insurance, and one form is a patent assignment. Another form might be for the bicycle club. However, one form in most states is the "non-compete" agreement. This seems innocuous enough. It is explained as merely preventing you from going to a competitor within one or two years of leaving your present job. Nothing could be further from your mind. You are going to work at this wonderful company. Your pay has just gone up by a binary order of magnitude from your minimum wage college job. You could not imagine ever being so disloyal as to leave to work for a competitor. Then you put in a good ten years in the big company. You do everything right. You get your products developed on time, you develop a good patent portfolio, you develop a good portfolio of papers, and, most importantly, you develop a good portfolio of physician friends in your area of specialty. Now you get recruited to join a startup. You would like to leave, and someone reminds you of the non-compete agreement. Well, that is no problem—it is only for two years. What are you going to do for two years? Are you going to flip burgers? The sad reality is you can only support your standard of living by working in the industry that you have grown up in.

Now you protest that the startup actually is not in the area that you have worked in, it is a medical device, but it is not similar to the ones you have been working on. Unfortunately, every year you have gone to the internal science fair in which you saw the research projects from all of the groups within this large company. This was not done only to inspire you and to build loyalty and cross-fertilization of ideas. It has been admitted in court that these science fairs are also done so that you will be exposed to research in all the areas of the company so that you specifically cannot work with any competitor or start up that has anything to do with any part of your present employer.

How do you avoid this problem? One suggested expensive solution (in terms of real estate and taxes) is to have your first appointment in California. California does not allow indentured servitude. In fact, this is probably one of the reasons why that state—in spite of its notorious anti-business taxes and regulations—continues to have good job growth. America succeeded largely because of the allowance of the free flow of capital. California is one state that allows free flow of human capital and talent.

"Look for a small company to get broader experience and a public company so you can get some stock options and the company has to meet shareholder expectations. Make sure the company is working in a new growth area rather than in a mature industry. The company should be one of the first two or three in a new field and have a good patent position. Most important is to understand the culture of the company and how they treat employees. If the company has a bad reputation or high turnover, avoid it. If they do not want to hear your opinion, avoid the company."

—Daniel Adams, SciMed Life Systems (now Boston Scientific Corporation), California, USA

Another thing you can do to protect yourself—after your first decade with the big company—is to take a vacation day and intentionally miss the science fair. It should be documented that you were out that day so that you were not exposed to projects from other departments or even other divisions in the company. This will make it easier for you to go to a startup with a medical device that is not directly related to your present job. Now you are ready to go out and enjoy the excitement of a smaller company and participate in the stock option lottery.

Concluding Remarks

A career in the medical device industry can be psychologically and financially rewarding. Most bioengineers do not optimize their career path, and thus miss out on many of these rewards. Take the extra time and effort to optimize your career, and the payoff will be large. It is worth the investment to do it right.

8

Independent Research Laboratories

David J. Schlyer, PhD

*Brookhaven National Laboratory,
US Department of Energy, Upton, New York, USA*

As I came to the end of my time in graduate school, like all graduate students, I was totally consumed with two things. The first was finishing my thesis, and the second was finding a place to work when I got out. In those days, there were no word processors or graphics programs, and computers occupied whole rooms. As a result, each sentence of my thesis had to be hand-written and then typed carefully onto the cotton rag paper required for a thesis. No mistakes or corrections were allowed on the final version. All of the figures for my thesis (of which there were many) had to be hand drawn and then inked onto vellum paper. My wife was doing the typing and I was doing the drawing, which left little time for looking for a job. I had decided (after extensive consultation with my thesis advisor) that I wanted to get a post-doctoral position that would expand my horizons beyond my thesis area. I applied to several prestigious universities, but there were no openings available at that time (or at least that is what it said in the rejection letters). A professor from another university was visiting my mentor and mentioned that he thought there was an opening at Brookhaven National Laboratory (BNL) in New York. Since I was living in San Diego and had never been to the eastern part of the United States, this seemed like a good place to visit for a year or two. I applied for the position, finished my thesis, and got the job without an interview or even a visit. The lesson I learned was that it is very often the personal connection that gets the job. Everyone hopes their resume will outshine all of the others, but it is often more important to have someone your prospective employer knows to personally vouch for you.

As my wife and I drove across the country from San Diego to New York, I was in high anticipation of what was in store for me in this new environment. Since there had been no interview at Brookhaven, I had no idea what the place was going to be like. From the mental picture I had of New York, I was expecting all of Long Island to be one vast city. As we crossed the George Washington Bridge, I felt

G. Madhavan et al. (eds.), *Career Development in Bioengineering and Biotechnology*,
DOI: 10.1007/978-0-387-76495-5_8, © Springer Science+Business Media, LLC 2008

confident that my picture was correct. However, the map indicated that we needed to go farther. As we kept driving, the buildings got shorter and the bright lights were replaced by houses surrounded by trees, until all we could make out was a towering wall of green leaves with no houses or people to be seen anywhere. When we finally arrived, I was sure I had gotten lost because the place was in the deep woods and very far from the city lights.

The next day, I met my new boss for the first time and he explained the project that he wanted me to work on. It was poles apart from what I had been doing in graduate school and required learning a whole new field. I could not wait to get started. As time went on and the year I was planning on spending at Brookhaven became thirty years, my interests have evolved and my research has covered many areas, but I have been very fortunate in always being able to have interesting things to investigate, and to have the support I needed to do them.

My broad field of study is Positron Emission Tomography (PET), although under that umbrella, there are many particular areas in which I have been interested. The idea behind PET is that we can use radioactive isotopes attached to a specific molecule to visualize a process occurring inside the body. It is a complicated process and requires expertise in a wide variety of skills. The field of PET imaging research is relatively young. I was very privileged to be in one of the places where it all started. During the 1950s and 1960s, there had been considerable work done on medical imaging, and some on positron emission imaging, but the path leading to routine clinical applications began about the time I arrived at BNL in 1976 (this was completely coincidental). During those early years, there was so much to do to bring the then immature science into the hospital where it could be used to help people. In the early days, we were trying to find radiotracers that could be used to identify or quantify some physiological process. It was the work of my boss at BNL, Al Wolf, in collaboration with Louis Sokoloff, Abass Alavi, Martin Reivich, and other colleagues from the University of Pennsylvania, to try to assess the amount of energy the body was using with an analog of sugar labeled with the radioactive isotope fluorine-18 called fluorodeoxyglucose (FDG). This has turned out to be one of the defining moments in the clinical application of PET, and today this molecule is used all over the world to detect cancer.

There are many tasks involved in a single PET study, and to do all these tasks requires people with a very wide variety of disciplines. The radioisotope has to be made using a particle accelerator, which requires physicists and engineers. Once the radioisotope is made, it must be very quickly incorporated into the tracer molecule, which requires organic chemists and radiochemists. Since the half-lives of the isotopes ranges from 2 minutes to 2 hours, this synthesis must be fast and efficient. The radiotracer preparation has to be tested to be sure it is pure and safe for injection, which requires analytical chemists and pharmacists. Finally, it is injected into a patient and then a PET scan taken, which requires nurses and physicians. It may also be necessary to draw blood during this process to help in the analysis, which requires nurses or technicians. Finally, the images which are produced have to be analyzed, requiring mathematicians, nuclear medicine physicians, and radiologists.

Although I am using PET as an example, this kind of organizational structure is common in other areas of research as well, with each task carried out by a specialist who works as part of the whole team. Interdisciplinary science is the methodology of the future, and working as a team is the only way to be on the cutting edge of science. It has been said that as late as 1890, if one studied hard for their whole lives, it was possible to know everything then known about science. Those days are gone, and now it is impossible for any one individual to be an expert in all fields.

The clear conclusion is that a wide variety of people are needed to make interdisciplinary research move forward. One of the joys, and challenges, of this field is trying to communicate with all of the other people involved. Since our backgrounds are so different, we often have to learn to speak to one another in a common language. A classic example came when we started talking about NET. The chemist got up and said "We need to look at the NET." Everyone agreed. Of course, the computer programmers thought we were talking about the Internet, while the physicist thought we were talking about a noise equivalent test (NET), the neuroanatomist thought we were discussing a norepinephrine transporter (NET) receptor site in the brain, and the oncologist was focused on a neuroendocrine tumor (NET). As the discussion went on, no one could figure out why the other people were using the word NET. It took considerable discussion to straighten this all out.

The benefit of this diversity is that there are many opportunities in this field, regardless of one's background. There are always ways to contribute to the overall effort of carrying out research in this area. There are very few people in this field who started out studying this in graduate school. Graduate school should teach you how to think about a new problem, do the background research, and to make a contribution to the understanding of the problem. It should not limit you to one topic, and it should make you feel confident enough to explore new areas that interest you. This should be true regardless of your particular area of specialization.

My particular working environment is the National Laboratory within the United States. There are six laboratories which make up the core of the National Laboratory system. They are Argonne, Brookhaven, Lawrence Berkeley, Oak Ridge, and Pacific Northwest National Laboratories, and Idaho National Engineering and Environmental Laboratory. They are managed and operated for the US Department of Energy (DOE) by contractors that perform basic scientific research and environmental remediation. At present, all the national laboratories combined employ about 21,000 staff, of which approximately 66 percent are managers and professionals.

In 1946, representatives from nine major eastern universities – Columbia, Cornell, Harvard, Johns Hopkins, Massachusetts Institute of Technology, Princeton, University of Pennsylvania, University of Rochester, and Yale – formed a nonprofit corporation to establish a new nuclear-science facility. They chose a surplus army base "way out on Long Island" as the site. Brookhaven National Laboratory was established in 1947 in Upton (which was an old Army camp),

New York. Brookhaven is a multi-program national laboratory operated by Brookhaven Science Associates for the DOE. Six Nobel Prizes have been awarded for discoveries made at the laboratory. Brookhaven has a staff of approximately 3,000 scientists, engineers, technicians, and support staff, and over 4,000 guest researchers annually. Brookhaven National Laboratory's role for the DOE is to produce excellent science and advanced technology with the cooperation, support, and appropriate involvement of our scientific and local communities. According to the BNL homepage, the fundamental goals of the laboratory are as follows:

- To conceive, design, construct, and operate complex, leading edge, user-oriented facilities in response to the needs of the DOE and the international community of users.
- To carry out basic and applied research in long-term, high-risk programs at the frontier of science.
- To develop advanced technologies that address national needs and to transfer them to other organizations and to the commercial sector.
- To disseminate technical knowledge, to educate new generations of scientists and engineers, to maintain technical capabilities in the nation's workforce, and to encourage scientific awareness in the general public.

These overall goals are shared by most independent research laboratories. BNL is spread out over 5,000 acres, and includes divisions in Basic Energy Sciences; Life Sciences; Energy, Environment and National Security; National Synchnotron Light Source; and Nuclear and Particle Physics. The national laboratory environment is an especially good place to explore your interests with people of similar natural curiosity.

There are some distinct advantages and some disadvantages to working in a place such as this. Some of the advantages include:

- You do not usually have teaching duties so you have more time to do science.
- There are many departments at the institution with well-known people working in them who are almost always willing to talk about their work and give you insights on ways to approach a new problem.
- There are large facilities available which offer unique analytical or investigatory techniques which are not available at universities.
- You have the freedom to follow your ideas about research, as long as you can convince someone to supply funding so that you can pursue your ideas.
- The environment is very pleasant, since these institutions are usually somewhat removed from populated areas.
- There are usually close ties with nearby universities where one can find collaborators or students to work on new projects.
- Employees are allowed, and even encouraged, to be adjunct professors at nearby universities, and sometimes give lectures in classes or at seminars.
- Students usually enjoy working at the laboratories, since they get their names on papers and usually are able to be first author on at least two or three papers by the time they graduate.

- Patenting inventions is encouraged (each inventor gets $250 for filing a patent application). Intellectual property rights are owned jointly by the inventor and the national lab. This means that the inventor gets some percentage of the profits after the inventions are licensed and earn enough to pay the lab back for the patent processing and application fees.
- It is a lot easier to work in inter-disciplinary fashion, and collaboration between departments is usually encouraged.

There are also some disadvantages to this environment. Some examples include:

- This is a federal laboratory, which means we are subject to governmental regulations from which the universities are often exempt. This tends to require one to spend a lot of time writing procedures and documenting that the work is well planned and safe.
- Scientists working at the lab have to provide their own funding. For instance, our support comes from two major sources. One is government funding in the form of grants from the National Institutes of Health (NIH) and Department of Energy. The other is private companies which support our work through cooperative research programs. In order to get funding, we have to compete with all of the other scientists who are also trying to get money. The competition is stiff, so we need to make sure we are at the forefront of science or we will never get any funding for our research.
- Because we have to supply our own funding, we must spend a significant fraction of our time writing research proposals to NIH, DOE, and other funding agencies, and getting the applications filed properly.
- Graduate students are paid out of the grants that we receive. The cost to the grant of a student is about twice what they are paid, due to laboratory and university overheads.
- There are no guarantees of job security, and if you cannot stay competitive, you may be let go. Usually, another job can be found within the lab, but there have been several cases where people have left the lab due to a lack of funding. Usually people see the writing on the wall and leave voluntarily.

There are several career pathways into the laboratory environment. Some of our people came as students to do a research project here while enrolled at another institution. Although this is a rare situation, it is a way in for some. A more common pathway is to come here as a post-doctoral fellow. If these people turn out to do a very good job, they are sometimes hired as staff. One can think of the one- to two-year post-doctoral position as a trial period, after which progression to staff is a possibility. Of course, this also depends on the funding available at the time of transition. Sometimes it is just not possible, for financial reasons, to bring someone up to the staff from a post doctoral position. The third common route is to be hired in mid-career. This is common if there is a specific program that needs a specific skill.

Once in the laboratory, the career advancement path is similar to academia. One usually enters as an *assistant scientist* and holds this rank for two to three years.

At the end of that time, one is usually promoted to *associate scientist*, which goes for another two to three years. Then the person goes on to *scientist*, with a term appointment of another two to three years. At Brookhaven (and some other national laboratories), a decision must be made at that point, with three possible outcomes. The person's department assigns a committee to evaluate the candidate. Based on the evaluation and review by a laboratory-wide committee and the laboratory director, the person may be let go, put onto continuing status, or granted tenure. At Brookhaven, the difference between continuing and tenure is rather small, but most of the group leaders have tenure. It is possible to move from continuing to tenure later in one's career, but this is less common.

The other part of this equation is the particular research project one is working on. It is often the case that this changes over time as interests and program needs change. In my particular case, I came to BNL as a post-doctoral fellow to do research on the kinetics of atomic chlorine species created by a microwave discharge in a gaseous plasma. It was new for me, and very interesting, and I was having a great time learning about gas phase chemical kinetics and vacuum ultraviolet spectroscopy. During this time, the method for the synthesis of FDG had just been worked out in Al Wolf's group and was ready to be used in a human volunteer. There was no PET tomograph at Brookhaven, so the FDG had to be transported down to Philadelphia to be used. It was an exciting time. Although I was working pretty much on my own in my research, I had lots of interaction with the PET group helping with small problems, and I felt like we were all working together to push back the frontiers of science. After some time working on a series of different projects, but in close association with the PET group, I gradually merged with the larger group and changed my focus to doing PET research on new radioisotopes, new radiotracers, new PET instrumentation, and doing biochemical kinetic analysis. All of this has kept me active for the last 30 years.

To this day, I look forward to my day at work. I usually have some problem that I am trying to solve at the moment, and spend the thirty minute drive thinking about how it might be approached or what the results of our latest experiments might mean. I enjoy coming to work early so that there is some quiet time before the usual chaos of the day begins. It is always necessary to keep up on the literature, so I usually spend this early portion of each day doing searches for the latest research articles in my area. This has become a lot easier with the advent of electronic journals and libraries where I can peruse the latest journals with a cup of coffee firmly in hand. One of my other jobs when I was a new staff member was to go to the library each week, read the table of contents of all the relevant journals, and mark the articles which I thought might be of interest to all the people in the group. This alone was an education, since I was exposed to a lot of articles on subjects I knew nothing about. These often piqued my interest, and I made an effort to find out what they were talking about.

After I do a little literature reading, I get started with the day. There are always meetings to attend, which is usually a necessary evil since many times these meetings are concerned with administrative details, budgets, safety planning, records review, or some other somewhat mundane task. These are necessary, but not necessarily enjoyable. I do look forward to research meetings where we can discuss the latest results and plan for the next experiments. These are often very lively meetings, since we are all dedicated to what we do and want to express our own views on the order in which things should be done. In our group, we have full group meetings once every two weeks, and smaller meetings of two to four people about every other day. These small meetings are to discuss particular experiments and specific results. They are often between a graduate student and his or her mentor. On the alternate weeks when we are not having a group meeting, we have a journal club where someone brings an article from the recent literature and we all discuss the methods used, the results, and how we would have done the experiment differently. These can also be lively as each of us brings our views to the table. There are no bounds on the topic of the paper, it is usually very educational for all of us, particularly the students.

The rest of the time is taken up with fixing things that are broken, and administrative duties such as time cards, work planning, writing safety procedures, and ordering equipment.

The particular area of research on which our group is focused is probing the chemistry of the brain using nuclear medicine techniques and highly specific radiotracers. One of the common misconceptions that people have about working in a field like this is that it is very difficult to learn. There are so many kinds of people needed to do this research that nearly anyone can make a contribution. The most common pitfall for people coming in is that they want to learn everything about all aspects of PET and medical imaging, and that has become very difficult as the field has expanded. This is true regardless of the particular area in which one is working. In any interdisciplinary field, it is sufficient to learn one area well, and then learn how to talk to people from other disciplines who can complement your knowledge and form a complete team. As time goes by, you will learn more and more aspects of the research and come to appreciate the knowledge other people have. The opportunities are unbounded, and if you spend the time learning a little about all aspects of this field of science, it will be possible to speak with more authority on the direction the research should take. The leaders in this field come from all disciplines, and there is not a single area of study which will prepare you for all aspects of this field. This is real on-the-job training. You would have to spend many years in school to learn as much as you can from the other people who are working with you on a daily basis.

To get into this medical imaging field, you need to have an in-depth knowledge of one aspect of the research. This could be in the medical applications, physics of radioisotope production, radiochemical synthesis, pharmacology, or any one of ten other areas. All that is required is that you be able to contribute something to the overall research effort. The same is true of all other research in bioengineering, biotechnology, or any other field in the life sciences.

It is my opinion that this field of interdisciplinary science has a very bright future. If I had to choose a single word to describe the next frontier in science, it would be "interactions." We are just beginning to scratch the surface of being able to understand how the body and brain work, how they interact with one another, and how they are affected by our genetics and the environment. There is so much left to learn that we can all stay busy for at least thirty more years. The national laboratories have the resources to carry out this research. There is also a wealth of clinical applications which have yet to be developed. Once we understand the basic chemistry and chemical dynamics of these interactions, we can go about the task of using that knowledge to develop new clinical procedures or diagnostic tests to identify, and perhaps cure diseases.

I believe that the national laboratories, and other independent research environments where people from vastly different backgrounds can be brought together to work on important problems, offer a truly unique and fertile ground for new ideas and research directions. The luxury of being able to devote a large fraction of your time only to research is a real advantage to this type of environment. Often the big machines and user facilities available at places like this are the key to making new discoveries, and applying new techniques to old problems. I also believe that the ability to have close interactions with a university is crucial to this process, since students often have not developed the prejudice that sometimes comes with experience, and are willing to entertain novel and apparently silly ideas that lead to new perspectives. These are very exciting times indeed.

9

Public Sector Research, Development, and Regulation

Jove Graham, PhD

Food and Drug Administration, US Department of Health and Human Services, Center for Devices and Radiological Health, Office of Device Evaluation, Division of General, Neurological and Restorative Devices, Orthopedic Spinal Devices Branch, Rockville, Maryland; and Geisinger Center for Health Research, Danville, Pennsylvania, USA

"Industry, or academia?"
"Industry, or academia?"
"Industry, or academia?"

This was the question I heard over and over again as I was preparing to graduate with my PhD in bioengineering in the spring of 2002. Everyone wanted to know what career path I was going to choose after graduation, and the question was always phrased as if there were only two possible choices for a bioengineer: industry or academia. Was I going to look for an engineering job at a medical device company? Or would I apply for a post-doc, or perhaps an assistant professorship at a university?

It is easy to feel as if those are the only two options when you are immersed in the everyday experience of being a student. On one hand, you may be a poor, starving student at the moment, but, all in all, academia may seem like a pretty good lifestyle. You might look at your professors and think, "I could be like them." On the other hand, perhaps you are tired of the academic scene. Perhaps you have friends who have graduated and are out in the real world of industry, making big bucks. You may feel frustrated by the pace of academic research, and eager to have a daily impact on a product, medical technology, or company. Chances are, most bioengineers you know are in academia or industry, and it may be difficult to imagine yourself in any other environment.

As I discovered, however, there are many other choices out there. One of those choices is public service. After graduation, I did not want to stay in academia, so I thought that industry was the only other option available to me. I sent out emails

G. Madhavan et al. (eds.), *Career Development in Bioengineering and Biotechnology*, DOI: 10.1007/978-0-387-76495-5_9, © Springer Science+Business Media, LLC 2008

and resumes to medical device companies, and asked my friends to help me look for job prospects. I didn't really know, however, what type of company I wanted to work for—big or small? Well-established or startup? Frankly, I was also having some trouble finding jobs whose descriptions seemed to fit me. I had a PhD in bioengineering, with lots of experience in research, laboratory testing, publication, and presentation. I did not know how to design something in CAD, however. I would not know how to implement or supervise a manufacturing process. I felt like none of the engineering job descriptions in the advertisements I saw really described *me*.

Nevertheless, I had received some positive responses to my resume, and I was in the process of interviewing at some of these companies when I was alerted to a job posting on a bulletin board at the Society for Biomaterials annual conference. The employer was looking for a MS or PhD-level bioengineer. It said they were looking for someone with a background in mechanics, especially fracture and fatigue, and that laboratory testing experience was "a plus." The job description said it would be approximately "50% laboratory research and 50% review of *new medical devices*." Wow, I thought! Now that sounds like an interesting job for *me*. The employer was the FDA.

The FDA

What is the FDA? The FDA, or Food and Drug Administration, is a science-based law enforcement agency within the executive branch of the US federal government. There are similar agencies in countries around the world, such as the MHRA (Medicines and Healthcare Products Regulatory Agency) in the United Kingdom, the TGA (Therapeutic Goods Administration) in Australia, or the MHW (Ministry of Health and Welfare) in Japan. The FDA is an unusual place, because how often do scientists and engineers like you or me also get to be law enforcement officers? The FDA does not write or create laws as Congress does, but its job is to enforce laws pertaining to food, drugs, medical devices, and other related products. Part of enforcing those laws is guaranteeing to the American public that companies are manufacturing and selling medical products that are safe and effective, and that they are representing those products to the public in an honest, accurate way.

Since its very beginnings, the FDA has had a long, proud history of scientists with a desire to use their skills and expertise to protect the health and welfare of the public.† The origins of the FDA, even before it existed as an agency, can be traced to a group of chemists in the US Department of Agriculture in 1883. Led by Dr. Harvey Wiley, a physician and professor of chemistry from Indiana, these scientists were interested in the problem of detecting fraud in the food industry. Transportation and storage of food was still a challenge in the days when ice was

† For a more thorough review of FDA history, I recommend Phillip J. Hilts' *Protecting America's Health: The FDA, Business, and One Hundred Years of Regulation* (New York: Knopf, 2003).

still the principal means of refrigeration, and milk was still unpasteurized. Chemical additives and preservatives to prevent food spoilage (or, more nefariously, to mask it) were becoming more common, and the effects of these chemicals on the human body were largely unknown and untested. Fortunately, as the science of chemical additives advanced, so did the ability of science to detect the presence of these chemicals and to study their effects on the body. Wiley's team, in fact, sponsored a series of unconventional experiments where volunteers would eat increasing amounts of chemicals being used as additives by the food industry (including borax, formalin, and sulfuric acid) to look for medical symptoms. These "Poison Squad" experiments helped to focus the public's attention on the issue of food safety, and eventually contributed to the passage of the first federal law regulating food and drugs in 1906. (It is important to note that subjects *were* informed of what they were doing, and no one ever died or was seriously injured.) Dr. Wiley himself resigned six years later to take over the laboratories at *Good Housekeeping* magazine, where he continued working to inform consumers about the quality of their purchases by creating the famous "Good Housekeeping Seal of Approval."

To this day, FDA scientists and engineers take their responsibility very seriously: to use the best scientific methods available to ensure the safety and effectiveness of medical products. Bioengineers are particularly involved in the regulation of medical devices. The FDA is divided into different centers which oversee each type of product. Most medical devices are regulated by the Center for Devices and Radiological Health (CDRH), unless they also contain biological material or drugs, in which case they might be regulated jointly by CDRH and another center. CDRH's databases (available via www.fda.gov/cdrh) currently list over 4,800 categories of medical devices, and approximately 100,000 individual device brands or models. These devices are marketed in the US by over 20,000 companies, the majority of which are small businesses with fewer than 500 employees (or, in many cases, fewer than 50). It is the job of the approximately 1,100 employees of CDRH to regulate this entire industry, overseeing how these products are manufactured and marketed.

What Engineers do at the FDA

But what do engineers at CDRH do, exactly? How does an engineer "regulate" the medical device industry? Engineers work in the different offices of CDRH that handle different aspects of the regulatory process. The Office of Device Evaluation (ODE) reviews and approves new medical devices before they are marketed. The Office of Surveillance and Biometrics (OSB) monitors devices that are already on the market for unanticipated problems. The Office of Compliance (OC) performs site inspections, ensures that manufacturers are complying with the law, and takes action against firms that violate it. The Office of Science and Engineering Laboratories (OSEL) performs laboratory research, helps to develop testing standards for medical devices and materials, and supports the other offices' activities by providing technical expertise. The job I applied for, after seeing that flyer on the bulletin board, was working as a mechanical engineer in this fourth office (OSEL). I was hired and worked there full-time for three years, and I now spend 20% of my time working as a consultant for ODE in addition to a separate job at the Geisinger Center for Health Research, part of a nonprofit healthcare system. I can tell you the most about OSEL and ODE, where the greatest number of bioengineers have an impact.

Device Review (ODE and OSEL)

Although the Office of Device Evaluation (ODE) is primarily responsible for the *premarket review* of new medical devices, engineers in both ODE and OSEL are actively involved in this process. A medical device is defined by law as just about anything (except a drug) that is used to diagnose, treat, mitigate, cure, or prevent a disease. When you go to a doctor's office, the exam table that you sit on and the stethoscope around your doctor's neck are medical devices. Medical devices can be much more complex, however: MRI machines, hip replacements, pacemakers, and artificial heart valves are medical devices, too. When Congress passed the first law in 1976 that established our current system of medical device regulation, it recognized that it is much simpler to evaluate the safety of a stethoscope or a tongue depressor than it is to evaluate the safety of a pacemaker. For that reason, all medical devices are grouped into three classes, according to their intended use and risk profile, so that the FDA can apply an appropriate amount of scrutiny to each device type.

Depending on the classification of the device, manufacturers are required to submit different types of information to the FDA for review before they can sell (or *market*) a new device in the US. Many of the simplest device types are largely exempt from this premarket review process. If not, manufacturers must submit information that shows their device is safe and effective. For some of these devices, it is sufficient to show that the device is *substantially equivalent* to a device that was already on the market prior to 1976 (or, by logical association, a device that has been linked through equivalence to a pre-1976 device). It is assumed that if a new device is equivalent to an existing device, it must be equally safe and effective. These devices are commonly described as "510(k)" devices, a term which refers to the section of the law that applies to them, and the type of application they require. For other device types, it is not enough to show that the device is equivalent to something else. Instead, the manufacturer must submit scientific evidence that the device is safe and effective in its own right. These devices are commonly referred to as "PMA" devices, because they require a "premarket approval" application. Reviewing the evidence in both of these types of application is something that engineers do at the FDA. Note that FDA engineers are not actually testing the devices themselves—we are not *Consumer Reports* magazine! It is the manufacturers' responsibility to test their own devices and submit information to the FDA, and the FDA evaluates that information.

When designing any new device—not just a medical device—a design engineer generally has two sets of constraints in mind. First, there are performance specifications, or things that he or she *wants* the device to do. When designing a cardiovascular stent, for example, he or she wants the stent to deploy correctly and to hold a blood vessel open. *Effectiveness* relates to how well the device meets these specifications. A manufacturer must tell the FDA the intended use of the device. What is it for? What is it expected to accomplish? If the device is not effective in achieving its intended purpose, the patient may be putting themselves at risk (for example, by having an unnecessary surgery or by receiving treatment with an ineffective device when they could have been treated with an effective one). Therefore, FDA engineers want to know whether the device will meet its performance specifications and be effective.

Second, the design engineer must consider the potential failure modes, or the things that he or she does not want the device to do. If it is an implantable device, for example, he or she probably does not want it to break, wear out, harm the tissues around it, or cause an adverse biological reaction. *Safety* relates to how well the device prevents these things from happening. So, in addition to evaluating effectiveness, FDA engineers also want to know that the device is unlikely to fail, and is therefore safe. In fact, FDA engineers probably spend more of their time evaluating safety rather than effectiveness. The effectiveness of many devices depends greatly upon interactions with the body's environment, which can only be observed during a human clinical trial. Before a manufacturer can conduct a clinical trial on an unapproved device, however, they must first demonstrate to the FDA that the device should be safe for human use, based on laboratory, bench, or animal testing results. These are the kinds of tests that bioengineers are usually best qualified to interpret and evaluate.

So, if you are an engineer like me at the FDA, you may receive an application (or a section of an application) from a company that wishes to market a new medical device. It is your job to determine whether that application shows that the device is safe. Where do you start? Normally, you begin by thinking of many of the same questions as the device designer. What are the possible failure modes of this device? What could go wrong? How might the device harm the patient? Or, conversely, how might the patient's body damage the device? These are bleak, pessimistic questions, but they are the ones that engineers must start with if they are thinking about safety. Of course, you generally try to limit your list of failure modes to things that could go wrong *if the device is used correctly*. Even a stethoscope could pose a risk to patients if a doctor decided to wrap it around someone's neck and strangle them! But that would certainly be an unusual situation, and probably not the responsibility of the engineer to guard against. Still, you should ask, *will a well-intentioned person be able to use the device correctly and without difficulty*? Are the instructions confusing or unclear? Are there ambiguous aspects of the design that might lead a doctor to accidentally use it backwards, upside-down, or otherwise incorrectly? If so, then such potential errors should be considered.

Otherwise, your list of failure modes consists of things that could go wrong even when the device is used properly. As already mentioned, an implantable device could potentially break, wear out, harm the tissues around it, or cause an adverse biological reaction. Each of these categories can be divided into even more specific failure modes. A device could break in many different ways and at different locations, due to compressive forces, tensile forces, shear forces, or torque. A device with multiple components could wear out at different interfaces. The device could harm the body's tissues in a number of different ways, either physically (by cutting or abrading tissue), or chemically, depending on the biocompatibility of the materials. Next, consider the potential consequences of each failure mode by asking, "What would happen to the patient if ____ occurred?" Some types of failure may be more serious or harmful to the patient than others. Any failure mode that would seriously injure or worsen the health of the patient, especially if it would require a surgical procedure to correct the problem, should probably be considered a critical safety issue.

Identifying how or under what circumstances each failure mode could occur helps determine what kind of tests or analyses are needed to guard against those failures. If one potential failure mode is that the device could fail under compressive force, you know that some kind of mechanical bench testing is probably needed to measure the compressive strength of the device. The more details a design engineer can specify when thinking about *how* exactly a device would fail, the better they can envision the type of test that addresses that failure mode (including boundary conditions, loading, environment, and other parameters). For any failure mode that would be considered a critical safety issue, a device manufacturer should perform some kind of testing or analysis to show the failure mode is unlikely to occur.

If the manufacturer has performed tests to address all the important failure modes, then you (as a reviewer) finally get to dive into the nitty-gritty technical details of the testing reports. The FDA does not require that any test or analysis be performed according to one prescribed method. The methods and assumptions that a company uses to test their device should be reasonable, however. Your engineering and laboratory experience are particularly needed in this area to interpret test reports and decide whether they seem reasonable to you. If you are reading a mechanical testing report, for example, you want to know how the loads were applied. How were the devices fixtured in the testing apparatus? What environment were they tested in? Were the devices used for testing the same as the finished devices that will be sold? Were enough samples used to ensure repeatability of the results? If a device will be sold in a range of sizes, were the smallest, largest, or worst case sizes tested? Before you even look at the results section of a test report, all of these questions about the methods are important. The method determines whether the results are actually going to be meaningful and relevant to the failure mode in question.

Assuming the methods are reasonable, the last step in evaluating a test report is to look at the results and determine how strongly they demonstrate that the device will not fail. This involves not only the results themselves, but also the *acceptance criterion* that the manufacturer has defined for the test. For example, a company may report that a mechanical test of their orthopedic implant demonstrated it can withstand 800 Newtons (N) of compressive force without breaking. Well, that is great! But what does that number mean? By itself, it means very little. To be meaningful, we need at least one other number—an objective acceptance criterion—that we can compare it to. It is the manufacturer's job to identify such a number and explain why it shows that their device is safe. Perhaps a textbook or journal article has reported that the maximum force on that region of the body is only 500 N. Or perhaps another implant of the same type, previously approved by the FDA, has been shown to have a maximum strength of only 700 N. Better yet, perhaps the manufacturer tested a similar, previously approved implant themselves using an identical method, and showed its strength was only 700 N. In each case, these numbers could be used as acceptance criteria to justify that the test result (strength of new device $\geq$ 800 N) demonstrates that the device is safe. It is the FDA reviewer's job to assess where such numbers came from, and how relevant they are to the current test results. How were the numbers derived? Do they apply to this device, given its intended use and function? Do they apply to the type of patients who will be using or receiving the device? If there is a strong connection to the actual expected conditions under which the device will be used, you can be confident that the device really did pass the test.

Being a bioengineer at the FDA, and reviewing new applications for medical devices, is a bit like reading a mystery novel in reverse to see if all the clues add up to the final solution. The end of the "novel" is always the same: the author (that is, the manufacturer) always concludes that their new medical device is safe and effective for human use. The interesting and important part, however, is in seeing how they arrived at that conclusion. Did they start by considering all reasonable failure modes? Did they perform tests to address those failure modes? Did those tests use reasonable methods and assumptions? Did the results compare favorably to objective acceptance criteria? If so, then you can conclude that the device should be safe. Notice that I use the word "reasonable" a lot when I am talking about reviewing new devices. Even in the cut-and-dried world of engineering, there is rarely one right answer for how a company can evaluate the safety and effectiveness of a medical device. Marketplace competition and intellectual property issues demand that every new device is different than the ones that came before it. Being an FDA engineer means more than just shuffling papers, checking boxes, or applying predefined formulae or equations. Every device and application is slightly different. You must use your skills, experience, and judgment to read and interpret what other engineers have done, and to decide whether they have used reasonable scientific methods in reaching their conclusions about device safety and effectiveness.

Laboratory Research, Standards Development and Postmarket Testing

As I mentioned earlier, not all FDA engineers spend all their time reviewing devices. In the Office of Science and Engineering Laboratories (OSEL), where I worked for three years, time was also spent in the laboratory conducting research or developing standard test methods. The majority of laboratory research projects are what we refer to as "regulatory research." That is, the FDA constantly wants to know if it is asking manufacturers the right scientific questions when it comes to safety and effectiveness of their devices. If we neglect to ask important questions, we may be putting patients at risk. On the other hand, if we are asking questions that are frivolous or not really important, then we are wasting time and resources (both the FDA's and the manufacturers'), and possibly hurting patients by delaying their access to new, beneficial devices. Most of the time, you can judge whether a scientific question is important or not based on past experience or literature. However, sometimes a question is too specific or a device is too novel, and the best way to decide whether something is important is to test it in the lab yourself.

One example of an OSEL project involved testing the resorption rate of bioresorbable polymers under varying loads. The question that the FDA was asking was, "Should the manufacturers of bioresorbable medical devices be applying a load to the device when they measure and report its rate of resorption?" This was an important question, because the devices are often loaded *in vivo*, and if a device resorbs too quickly or slowly, it could affect the device's effectiveness. An objective answer to this question wasn't really available, however. Many people had their own theories about how load affected resorption rate, but there wasn't much published information to definitively back up anyone's theory. So, engineers at the FDA built a machine to apply static or dynamic load to resorbable devices and tested several different resorbable materials. They concluded that yes, the FDA ought to be requiring manufacturers to use an applied static load when they report resorption rate because load has a measurable effect on that rate. This finding was incorporated into future reviews of those products, and this experiment is a good example of what's considered a successful lab project at the FDA.

Research is also funded primarily from the FDA's annual operating budget, so most projects are more modest than a multi-year, million dollar project that might typically be funded by a National Institutes of Health (NIH) R01 grant. (FDA employees are not actually eligible for NIH funding.) However, the labs are very well-equipped, and most employees are on salary so that they do not have to depend on soft money research grants to support themselves. Engineers in OSEL have good access to equipment, space, and other people. There are also opportunities for hiring student interns, especially during the summer. Ultimately, the limiting factor on getting laboratory work done at the FDA is really *time*. You must manage your own time well, and manage other people's requests for your time. Scientists in ODE, in particular, often request assistance from OSEL in reviewing specific sections of new device applications. The engineers in OSEL are considered consultants who have some freedom to accept or refuse a request to help review something, but device review is part of the job. It can also be hard to refuse someone when they ask for your help.

In addition to original research, the laboratories at OSEL are also used for two other major activities: standards development and post-market failure analysis. The FDA does not establish its own standards for devices or materials, but participates as a member of several national and international voluntary consensus standard organizations, such as the American Society for Testing of Materials (ASTM) and the International Standards Organization (ISO). These organizations bring together people from industry, government, academia, and consumer groups who write standards and standard test methods for devices and materials. They are *voluntary* organizations because the members volunteer their time and efforts, and because the standards they write are not mandated by law (unless a government decides to officially adopt them later). They are *consensus* organizations because their rules for writing standards encourage agreement among all parties, and prevent one person or organization from having too much control over how the standard is written. Development of standards is beneficial for the FDA's mission, because standards simplify the process of evaluating test methods and results from different manufacturers. If a manufacturer is able to use a standard test method instead of developing their own test protocol from scratch, it can potentially save them time and resources. It also saves the FDA's time, because if a reviewer is already familiar with that standard, it is quick and easy for them to understand and interpret how the test was performed. Many of the basic method-related questions that the engineer would normally ask (as we discussed previously), will already have been answered.

FDA scientists and engineers take part in the writing and revision of these standards, but also use their laboratories to contribute to the process. Sometimes, as a standard is being written, preliminary lab work is needed to perform pilot studies, to establish test parameters, or to see if the test method is even feasible. After a standard test method has been written, it may be necessary to "test the test method." If the standard is well-written, then different people and labs should get the same results if they perform that test on the same material or device. To verify such robustness, FDA engineers will often participate in what is known as *round-robin testing*, where several labs receive samples of the same test material, perform the test according to the standard method, and then compare their results. If different labs get wildly different results using the same method, then perhaps it is not such a good method. In that case, it may be necessary to rewrite the language of the standard to make the instructions clearer or revise technical aspects of the standard to improve the method itself. ASTM Standard F2129, a test method for measuring the corrosion resistance of small metal implants, is one example of a standard that was largely developed by OSEL engineers who also took part in a round-robin test of the standard after it was written.

Finally, engineers in OSEL sometimes use their laboratories to examine devices that have been turned over to the FDA because of some problem or complaint. A doctor may suspect there is something wrong with a device, or perhaps a device fails outright. In these situations, the FDA wants to determine the cause of the problem and what action should be taken to prevent it from happening again. Perhaps there was a manufacturing defect, in which case the FDA should work with the manufacturer to fix the problem. Perhaps the device was used incorrectly, either due to an isolated incident (for example, someone who just didn't follow directions) or because the manufacturer needs to provide better instructions or training to the users. In the latter case, another office within CDRH (the Office of Communication, Education, and Radiation Programs, or OCER) may also want to take a more active role in educating patients and doctors about the proper use of that general type of device. Perhaps the problem was caused when the device was reprocessed or resterilized by a third party, in which case the FDA should work with that firm to fix the problem. In rare cases, the FDA may suspect that a manufacturer is intentionally breaking the law by incorrectly manufacturing, processing, or labeling their devices, in which case the FDA could request or confiscate samples of the device to confirm these suspicions. While these types of testing are not considered routine work for engineers in ODE or OSEL, they are examples of how we can use our laboratory capabilities to assist in the regulation of medical devices.

What My Day-to-Day Work was Like

As I mentioned at the beginning, my job at OSEL was divided between laboratory work and device review. Was my time equally divided, "50/50," between these two activities, as the job originally advertised? Probably not, but people's jobs rarely compartmentalize themselves into such neat little segments. Some weeks, I spent almost all my time reviewing devices. Other weeks would go by when I was in the laboratory almost all the time. Finding and maintaining balance between the different tasks you are supposed to be doing is a skill you will need no matter where you work. All in all, I probably spent more time working on device reviews, because it interested me, I was often being asked for assistance, and I felt like I was making the most valuable contributions there.

When I was involved in the review of a new device, my day might begin with the delivery of a big, thick document to my desk. Applications to the FDA can sometimes be hundreds or thousands of pages long. Was it my responsibility to read every single page? No-of course not. Part of what is nice about working at the FDA is that you are a member of a larger team that may include other engineers, physicians, chemists, biologists, and statisticians. A lead reviewer may be ultimately responsible for each submission, but he or she typically asks for assistance from such a team. Everyone works together to evaluate the device, and everyone is responsible for a different aspect of the review. Animal data might be reviewed by a biologist. Clinical trial data might be reviewed by a physician, with the quantitative analysis of the data reviewed by a statistician. Bench data are usually handled by engineers. Everyone's input is important, and the decision to approve a device is reached based on everyone's reviews.

I would read my section and consider all the questions about failure modes and testing that we discussed earlier in this chapter. Then, most importantly, I would summarize my conclusions in a written review. I had to be able to describe any safety or effectiveness issues clearly and succinctly. If my summary was too long or confusing, the other team members might have to ignore my work completely and go read the original submission themselves, which would be a waste of everyone's time. It was also not sufficient for me to just critique a device. I had to think proactively and describe what actions needed to be taken by the manufacturer to resolve each issue. Otherwise, I would put everyone (the FDA and the company) in an awkward state of limbo with no idea of what to do next. My recommendations would be included in a letter to the company, and I might be asked to help again when the company responded to those recommendations. Sometimes, a teleconference or a face-to-face meeting with a company would be held during the course of a review. On one occasion, I was even called upon to present details of my review at a public hearing. In any case, the ability to clearly communicate my ideas—to other FDA scientists and to the device manufacturers themselves—was very important to successfully performing my job.

When I was involved in a laboratory project, my days would be much more varied, depending on the project and the phase it was in. As with any laboratory job, I was either in the laboratory, setting up and running experiments, or at my desk, planning an experiment, budgeting a project, or analyzing data. Most importantly, I was fortunate to work in a collaborative environment where I could interact with other lab members, summer students, physicians, epidemiologists, and statisticians during each project. Remaining active in research with a laboratory was important to me because it keeps your hands-on skills fresh, helps your ability to understand others' lab reports, and helps keep you up-to-date with current knowledge of your field. It is also fun to roll up your sleeves and get your hands dirty in the lab.

Job Skills and Qualifications

It is not a requirement to have a PhD to work as a bioengineer at the FDA; while the FDA employs a large number of PhD's and MD's, the majority of reviewers hold bachelor's or master's level degrees. What is more important than your degree are the skills you have. The skills needed to work as a bioengineer at the FDA are not that different from the skills you will need anywhere. First and foremost, even more important than your technical skills, is that you need solid communication skills. You can be a brilliant engineer or scientist, but if you cannot communicate your ideas—in writing, in person, and over the phone—to other people who need to hear those ideas to do their jobs, you are not really doing *your* job well. Communication skills require practice, so you should take every opportunity you can find to write, speak, and present, particularly if you are still a student. Remember that whatever your career, you will need to discuss technical subjects not only with other engineers, but with people from non-engineering backgrounds. At the FDA, we work with doctors and scientists in different disciplines, and communicate constantly with company representatives who have many different backgrounds. Your work and your ideas are only as good as your ability to share them with others.

Next, it is important to feel confident in your knowledge of a particular technical subject so that you can apply that knowledge to whatever medical devices you are asked to regulate. Notice I did not say you need to be an expert on any particular type of *medical device* before you can start working at the FDA. While it certainly helps to have a prior interest and some knowledge of devices, the device industry is constantly expanding and innovating. It is more important that you have a solid foundation in some branch of bioengineering, whether it is related to mechanical, electrical, chemical, or another branch, so that you are prepared to understand and evaluate any type of new device that comes your way. While it does not hurt to have a broad understanding of other disciplines as well (for example, biology or chemistry), you want to be sure that there is some area of engineering in which your knowledge runs deep, and where you can feel confident that you will be making a unique and significant contribution to a team, especially if that team already includes doctors, biologists, and chemists.

Finally, your critical thinking and reasoning skills can be just as important as your technical expertise. Remember that assessing the safety or effectiveness of a new medical device is not like solving a math problem or applying a formula. You must be able to interpret and follow a manufacturer's logical arguments about their devices from start to finish, and determine for yourself whether those arguments make sense and are scientifically supportable.

Working in the Public Sector

Working for the federal government is a comfortable lifestyle for many people. In my experience as a PhD-level engineer in OSEL, I was rarely expected to work more than 40 hours per week, and I could save up extra hours worked one week and apply them to a future week. I received 13 days of sick leave per year, and 13 vacation days per year. I could choose from several health insurance plans, and although I had to pay for a portion of the health insurance premium myself (which is not always the case at a private company), that money was deducted from my paycheck before being subject to income tax. Similarly, I could voluntarily make pre-tax contributions to a 401(k)-type retirement program called the Thrift Savings Plan, and the federal government matched those contributions up to 5% of my salary.

"But what is the salary like?" you might ask. "Won't I make less money working for the government than I would in private industry?" I believe the answer is, "Perhaps, but it depends." Searching for *private vs. public salaries* on the Web shows you a wide range of varying opinions on this topic. Groups that advocate a limited government, such as the Cato Institute, correctly note that, according to data from US Bureau of Economic Analysis, the average federal worker earns more than the average private-sector worker, especially when benefits are considered as well as wages [1]. It is probably more sensible, however, to compare salaries based on individual jobs than averages across the entire workforce, since the federal government has a larger percentage of white-collar workers than the US workforce as a whole (roughly 87% vs. 60%)[2], and a larger percentage of engineers and scientists (roughly 11% vs. 4%) [3]. The US Bureau of Labor Statistics estimated that in May 2005, the median annual salary for bioengineers across the US was $71,840, and the median annual salary for bioengineers working for the federal government was $79,310 [3]. In the same report, the federal government was listed as the fourth top-paying "industry" sector for bioengineers, ranked behind semiconductor/electronics manufacturing, research and development, and industry management.

These comparisons, of course, do not take into account several confounding factors. Many people trained as bioengineers may be working under job titles other than "Bioengineer": for example, my original job title at the FDA was "Mechanical Engineer." Also, even among bioengineering jobs, job responsibilities, training, and experience, which affect salary, vary widely. The Congressional Budget Office (CBO) has suggested that overall, the federal government may hire people for jobs where they have greater responsibilities than they would normally have in private industry, given their background and experience [4]. The government may offer lower pay than private industry would for these senior-level jobs, however. The net effect, according to CBO's analysis, suggests that government pays people with similar qualifications and background about the same as private industry.

One sign of satisfaction among federal employees is the fact that workers quit or resign from federal service at a very low rate of only about 1–3% per year [4]. Salary, benefits, and the federal retirement system probably contribute to this phenomenon, but there are other factors as well. Some people may be attracted to the challenge of more responsibility, job security, and opportunity for variety and advancement through internal transfers than a similar job in the private sector would give them. And the strong desire to serve the public should not be underestimated. I have been continuously impressed by the sincerity and seriousness with which FDA employees take their responsibility to protect and promote the health of patients across the United States. As with anyone involved in healthcare, we take it very personally when we see patients suffer, and we take great pride when we see our efforts having a positive impact on people's health.

Conclusions

Despite the complexity of the job, working as an engineer for the FDA can be very rewarding. You get to see all the newest and latest technology that companies have developed; you get to interact with device companies; and you get to see the nuts and bolts of how medical devices are designed, manufactured, sterilized, tested, approved by the FDA, and marketed. If you think you would like to work in the medical device industry some day, it is an incredible opportunity to gain all kinds of insight and understanding into how the regulatory process works, and how the industry operates. (Note that there are, however, very strict rules governing conflicts of interest, financial ties to the industry, and leaking of confidential information, which may restrict you from certain activities during and after employment at the FDA.)† If you wonder whether working for the government could offer intellectual stimulation and exposure to new technology, I can tell you emphatically that the answer is, "yes!"

† During employment, the primary restrictions relate to financial interests (e.g., investments) in any company regulated by the agency, or activities that would appear to cause a conflict of interest. Post-employment, there are temporary restrictions on your ability to appear before the government on a company's behalf, and permanent restrictions on your involvement with any matter you influenced as an agency employee. For more details, consult the Code of Federal Regulations 5 CFR XLV or the Federal Register, Volume 61, Number 147, pp. 39755–39767, available online at www.gpoaccess.gov/fr/index.html.

In over three years at the FDA, I have learned a lot. I really enjoy the people I work with, and I like the fact that every day is different, and that I never know what question or challenge is going to land on my desk on any given day. I've met a lot of people in other research labs and in the device industry through my work with standards development organizations. I realize that the skills I developed at the FDA are very desirable from the industry's point-of-view. So whether you are interested in a lifetime career in public service, or a different long-term career related to medical devices, it's hard for me to imagine a better or more rewarding career move than working for the FDA.

References

1. Edwards, C., "Federal pay outpaces private-sector pay," *Cato Institute Tax & Budget Bulletin*, No. 35, May 2006.
2. Congressional Budget Office, "Changes in federal civilian employment: an update," May 2001.
3. Bureau of Labor Statistics, "May 2005 National Occupational Employment and Wage Estimates," available at www.bls.gov.
4. Congressional Budget Office, "Comparing federal salaries with those in the private sector," CBO Memorandum, July 1997.

10

Clinical Medicine and Healthcare

Leann M. Lesperance, MD, PhD

Department of Pediatrics, State University of New York – Upstate Medical University, Syracuse, New York; Department of Bioengineering, Thomas J. Watson School of Engineering and Applied Science, State University of New York, Binghamton, New York; and United Health Services Hospitals, Johnson City, New York, USA

It is not surprising to find clinical medicine and healthcare within the domain of traditional career pathways in bioengineering. Since the first bioengineering programs started almost forty years ago, many of their graduates have gone on to medical school and other health professions. Some students planned to enter medical school all along, while others (myself included) discovered their interest in clinical medicine during their bioengineering studies.

The fact that you are reading this book suggests that you are undecided or may be open minded about your future plans. Now that you are deep into it, however, you have learned that most of the contributing authors also experienced major periods of indecision at some point in their careers. Career paths are rarely smooth and often have confusing forks and even hairpin turns!

Mine certainly has been circuitous, but also fortuitous. As a high school senior, I applied to the College of Engineering at Marquette University in Wisconsin, USA because I enjoyed and was good at math and science, and because my dad was an engineer from Marquette. I further chose to major in bioengineering because I had always been interested in health, and because I wanted to help people. Although my reasons were not the most compelling (and even erroneous – I now understand that *all* engineers help people), I did make an excellent choice. I loved every minute of that bioengineering program, from calculus, chemistry, and biology, to circuits, thermodynamics, and telemetry. I stayed on at Marquette to get my master's degree and then went to Massachusetts Institute of Technology in Boston to get my doctorate in medical engineering from the Division of Health Sciences and Technology (HST). In an effort to train engineers who can communicate better with doctors, students in the unique HST program were required to take basic science courses (for example, anatomy, pathology, and pharmacology) *with* medical students. After a year of coursework, the engineering students then

G. Madhavan et al. (eds.), *Career Development in Bioengineering and Biotechnology*,
DOI: 10.1007/978-0-387-76495-5_10, © Springer Science+Business Media, LLC 2008

spent six weeks seeing patients as if they *were* medical students. I loved this clinical experience and decided to go to medical school[1].

Perhaps *you* are interested in becoming a physician. This chapter is a place to start your research. It will begin with a brief overview of various healthcare careers, and then will focus on that of physicians – reasons to consider this career, reasons to be cautious, and the steps needed to get there.

Types of Healthcare Careers

More than six million people work in the healthcare field in the United States today, and the number is constantly increasing due to our aging population. As discussed in the rest of this chapter, bioengineering students generally are well-prepared for a career as a physician. However, the majority of healthcare workers are *not* physicians; they work in all healthcare fields, with and without direct patient contact, and in a variety of settings, including medical offices and hospitals. Bioengineering graduates would be overqualified for some of these non-physician positions and not well-qualified for others.

A few positions, such as home health aides and nursing aides, require only a high school degree with on-the-job training. Other positions, such as opticians, medical office assistants, radiology technicians, dental laboratory technicians, and nursing assistants, require a high school degree, plus completion of a formal training program that may range from a few months to a few years in length.

Many healthcare positions require undergraduate study in a specific field, such as dietetics, medical technology, speech pathology, social work, pharmacy, or physical therapy. (Few of the required courses would overlap with the bioengineering curriculum.) Further graduate work and completion of a certification process often are necessary.

Nurse practitioners and physician's assistants provide diagnostic, therapeutic, and preventive care under a physician's supervision. Nurse practitioners typically are registered nurses (with bachelor's degrees) who have completed several years of advanced training. (Some schools offer accelerated bachelor's programs for students who have a degree in another field.) Physician's assistants programs generally require at least two years of college along with significant healthcare experience prior to admission.

Reasons to Consider a Career as a Physician

Students who have chosen to major in bioengineering may want to consider a career as a physician because the two fields share many attributes. Bioengineering and medicine are both intellectually challenging, science-based, technology driven, service-oriented, financially secure, and well-respected fields.

---

[1] The story does not end there nor does the career path straighten. I became a pediatrician, practiced for several years, cut back after having children, and now have returned to academic engineering as faculty.

Intellectually challenging. From start to finish, engineering and medicine are intellectually challenging careers. Engineering and medical students are among the most hard-working on campus, spending long hours in classrooms, laboratories, and libraries. To be successful, both types of students must develop excellent study habits, and also must learn to think critically and solve problems. Engineers and physicians continue to work hard throughout their professional careers. Both use critical thinking and problem solving skills every day, and must keep up-to-date on professional literature as well as contemporary issues.

Science-based. Medicine and bioengineering both draw heavily from the natural sciences, especially chemistry, biology, and physics. Some physicians and bioengineers use the scientific method to do research in their fields. However, medicine and bioengineering are considered applied sciences, and most people in these fields apply their knowledge to practical problems.

Technology-driven. Many engineers are technophiles who love to try the latest gadgets, tinker with toasters, or rebuild engines and hard drives. Most physicians today rely heavily on technology for administrative, diagnostic, and therapeutic tasks. For example, computers are used for scheduling, billing, and medical records, personal digital assistants (PDA) help with clinical decision-making, and MP3 players and the internet provide continuing education; patient vital signs are checked using electronic scales, digital thermometers, automatic blood pressure cuffs, and infrared pulse oximeters. Some specialties, such as radiology and nuclear medicine, are almost entirely technology dependent.

Service-oriented. Some students choose bioengineering over other engineering disciplines because they want to help people. Of course, they could have chosen *any* field of engineering because humanity benefits greatly by advances in all the engineering disciplines. For example, our health has been dramatically improved by the invention of power transmission lines (electrical), water treatment plants (environmental), and automobiles (mechanical). Likewise, many students want to be doctors so that they can help people. The field of medicine offers unlimited opportunities to improve the health and well-being of people of all ages, from all walks of life.

Financially secure. Starting salaries for engineers are typically among the highest on campus. However, the number of engineering positions open each year varies, and there are a limited number of jobs available in the United States that specifically require a bioengineering degree. In contrast, physicians, especially those trained in the United States, generally do not have difficulty finding a job. Furthermore, starting salaries for physicians are well above those for engineers – for example, starting salaries for the lowest paid physicians (pediatricians or family practitioners) are twice that of entry-level engineers. Salaries for the highest paid medical specialists are several times higher. Of course, one must take into account the substantial debt physicians accumulate during medical school.

Well-respected. Your parents may groan if you tell them you are joining a rock band after college; they will probably cheer if you tell them that you landed an engineering position or a spot in medical school. Throughout history, engineers and doctors have been well-respected members of society.

Path to Medical Career

There are several steps involved in becoming a physician – undergraduate school, medical school, and training.

Undergraduate school. Medical schools do not require a particular undergraduate degree. However, there are specific course requirements that must be satisfied (see Table 10.1). In general, these must be completed by the time a student applies for medical school, which is typically in the summer between junior and senior year.

Will majoring in bioengineering *help* you get into medical school? The short answer is maybe yes, maybe no. Applicants to medical school outnumber offers of admission by nearly two to one, so many students look for things that set them apart from the other applicants. Although it has become more common in recent years, having a degree in bioengineering is still far less common than the more traditional pre-med majors. According to statistics from the Association of American Medical Colleges (AAMC) Data Warehouse for the 2000–2001 entering class, almost 36% of total applicants majored in biology, 9% majored in other biological sciences, and 11% majored in chemistry or biochemistry. Only 1% of the applicants majored in bioengineering. However, applicants from bioengineering were more likely to be accepted. The same AAMC data set showed that while only 47% of applicants were accepted overall, more than 65% of bioengineering applicants were accepted. In fact, bioengineering had the highest percentage of accepted applicants for that year.

On the other hand, engineering coursework can be more challenging for some students. A bioengineering major with a 3.5 grade point average (GPA) will not necessarily look better than, or even as good as, a biology major with a 3.9 GPA. Bioengineering students may have to take a few additional courses to complete all the requirements for medical school. However, they may have to use all of their free ("easy") electives to do so, thus making a challenging curriculum even more challenging. Students interested in medical school should register as soon as possible with their college pre-med or pre-health advisor to stay informed regarding these issues.

Medical school. There are two types of medical schools, allopathic and osteopathic. Allopathic schools grant an MD (medical doctor) degree, while osteopathic schools grant a DO (doctor of osteopathy) degree. Allopathic and osteopathic doctors study similar subjects, undergo similar training, and are both "real" doctors, but osteopathic medical students receive additional training in osteopathic manipulative medicine, a hands-on technique that can help with pain relief and healing. Approximately five percent of practicing physicians are osteopathic doctors.

Table 10.1 Courses required by most medical schools

One year biology
One year physics
One year general (inorganic) chemistry
One year organic chemistry
One year English

Table 10.2 Common medical specialties. Numbers in parentheses are years of training required. Those in italics are primary care fields in which physicians provide preventive care, diagnose and treat common medical problems, and refer to specialists when necessary

Aerospace medicine (3–4)	Occupational medicine (3–4)
Allergy and immunology (5–6)	Ophthalmology (4)
Anesthesiology (4–5)	Otolaryngology (5–6)
Cardiology (6–7)	Pathology (4–5)
Critical care (4–6)	*Pediatrics, general* (3)
Dermatology (4–5)	Pediatrics, subspecialties (4–7)
Emergency medicine (3–4)	Physical medicine & rehabilitation (4)
Endocrinology (5–6)	Psychiatry (4)
Family medicine (3)	Pulmonary diseases (5–6)
Gastroenterology (5–6)	Radiology (5–6)
Hematology – Oncology (4–6)	Rheumatology (5–6)
Infectious disease (5–6)	Surgery, colon and rectal (6)
Internal medicine (3)	Surgery, general (5)
Internal medicine-Pediatrics (4)	Surgery, neurological (6–8)
Medical genetics (4–5)	Surgery, orthopedic (5–6)
Neonatal – Perinatal medicine (5)	Surgery, pediatric (7)
Nephrology (5–6)	Surgery, plastic (5)
Neurology (3–4)	Surgery, thoracic (7–8)
Nuclear medicine (3–4)	Surgery, vascular (8)
Obstetrics and gynecology (4–5)	Urology (5–6)

In the United States, medical students usually spend four years getting their degree (MD or DO). The first two years are spent in the classroom studying basic science, while the last two years are spent in clinics and hospitals caring for patients.

Training. After completing medical school, physicians must spend three to five (or more) years training in their area of interest (specialty). The years immediately after medical school are called a residency – probably because the trainees practically live (reside) at the hospital where they learn to care for patients. Some medical subspecialties require a fellowship, involving one to three additional years of training, after the residency. Table 10.2 shows various specialties and the number of years of training required for each.

Types of Physicians

Deciding which type of physician to become may be even tougher than deciding whether or not to go to medical school. Physicians can be broadly classified into three categories based on the type of therapy they provide – medical, surgical, and mental health. There may already be one category in which you are more interested. Within each of these categories, however, there are a myriad of specialties and subspecialties. The American Board of Medical Specialties currently lists 24 different specialties, and, within these, 100 subspecialties. Forty are listed in Table 10.2. These specialties differ widely – for example, in their years of training, type of patient problems encountered, skills required, contact with patients, on call responsibilities, hours worked, and salaries earned. It will be important that you choose a specialty that fits your needs, wants, and interests.

Reasons to be Cautious

Although there are compelling reasons to consider becoming a physician, there also are reasons to be cautious. You may not want to choose a career in medicine because it is challenging, service-oriented, highly regulated, and costly.

Challenging. Physicians spend much of their time thinking about problems, even when they are not at work. Evenings and weekends "on call" add to the hours spent deep in thought. Moreover, the decisions one has to make as a physician are frequently life and death decisions. These intellectual challenges make the job more interesting, but can also lead to exhaustion and burn-out.

Service-oriented. Physicians spend their days helping people, which can be incredibly rewarding, but exhausting, too. It takes a great deal of energy to listen to people's problems and find solutions all day. Furthermore, physicians have to put their best selves forward during each patient encounter, even on days when they do not feel their best.

Highly regulated. The field of medicine is highly regulated, regardless of specialty. Physicians must cope with mounds of paperwork, for example, from government agencies, patient insurance companies, and physician malpractice carriers. Rules designed to help patients, such as the one that protects patient privacy (Health Insurance Portability and Accountability Act or HIPAA) and the one that ensures patients with emergency medical conditions are treated regardless of ability to pay (Emergency Medical Treatment and Active Labor Act or EMTALA), have created some additional challenges for physicians.

Costly. Although the dividends are high, becoming a physician requires an enormous investment of money and time. According to a 2006 report published by the AAMC, the median debt for a public medical school graduate is now almost $120,000, and that for private school graduates is over $150,000. Furthermore, most physicians spend at least seven years in medical school and training.

Is the Road I Traveled for You?

Being a physician is an excellent career choice, and medical school is an option that bioengineering students may want to consider. While some of the issues involved have been raised here, students definitely will need to do more research before embarking on this career path. Students may find the reading list at the end of the chapter useful in this regard.

The decision to become a physician certainly was not one that I made easily, and the years spent in medical school and doing a residency were definitely tough.

However, I truly enjoy the time I spend with my young patients (and their parents), and I feel privileged to be part of their lives. If I had it to do over again, I would; and if either of my daughters (Teresa or Catherine) wants to be a doctor someday, I will support their efforts wholeheartedly.

Suggested Reading

Informational

- *A Career in Medicine: Do you have what it takes?* 2nd edition, by Rameen Shakur, Royal Society of Medicine Press, 2006.
- *Doctors and Discoveries: Lives that Created Today's Medicine,* by John Galbraith Simmons, Houghton Mifflin, 2002.
- *Medical School Admission Requirements 2008–2009,* by Julie Chanatry, Association of American Medical Colleges, 2007.
- *Planning a Life in Medicine: Discover if a Medical Career is Right for You and Learn How to Make it Happen* (Career Guides), by John Smart, Stephen Nelson, and Julie Doherty, Princeton Review, 2005.
- *The Yale Guide to Careers in Medicine and the Health Professions: Pathways to Medicine in the 21st Century* (The Institution for Social and Policy St), edited by Robert Donaldson, Kathleen Lundgren, and Howard Spiro, Yale University Press, 2003.
- *Top 100 Health-Care Careers: Your Complete Guidebook to Training and Jobs in Allied Health, Nursing, Medicine, and More,* by Saul Wischnitzer and Edith Wischnitzer, Jist Publishing, 2005.

Introspective

- *A Life in Medicine: A Literary Anthology,* edited by Robert Coles and Randy Testa, The New Press, 2002.
- *A Measure of my Days: The Journal of a Country Doctor,* by David Loxterkamp, University Press of New England, 1997.
- *Complications: A Surgeon's Notes on an Imperfect Science,* by Atul Gawande, Picador, 2003.
- *How Doctors Think,* by Jerome Groopman, Houghton Mifflin, 2007.
- *Kitchen Table Wisdom: Stories That Heal,* by Rachel Naomi Remen, Riverhead Books, 1996.
- *Other Women's' Children,* by Perri Klass, Random House, 1990.

Humorous

- *House of God,* by Samuel Shem, Dell, 1979.
- *So You Want to be a Doctor,* by Stuart Zeman, Ten Speed Press, 1992.

11 Intellectual Property Law

Kenneth H. Sonnenfeld, PhD, JD

King and Spalding, LLP, New York, New York, USA

With the cost of bringing a new chemical entity to market as a drug reaching about one billion dollars, and annual sales of many drugs reaching over one billion dollars, strong intellectual property rights are essential. The acquisition, enforcement, and challenge to such intellectual property rights provide varied and challenging opportunities for persons interested in pursuing a career in intellectual property law.

A career in intellectual property law provides a stimulating and rewarding opportunity to participate in ensuring that the fruits of creative thinking are appropriately protected. Persons working in intellectual property law in biotechnology or bioengineering become involved in both cutting edge science and the relatively recent history of science. The protection of biotechnology and bioengineering inventions often requires one to master complex scientific facts and principles along with equally complex and evolving doctrines of intellectual property law, particularly patent law. Accordingly, a person involved in practicing intellectual property law is constantly learning about technology as well as the developing law.

In practicing intellectual property law, one may work to define and protect the intellectual property created by another, or seek to limit the scope of protection claimed by another in order to maintain the unfettered use of subject matter that is already in the public domain or not adequately described in a patent. Thus, one involved in a career involving intellectual property law participates in promoting the progress of science and useful arts by securing for limited times, to authors and inventors, the exclusive right to their respective writings and discoveries. (*Article I, Section 8, Clause 8, U.S. Constitution.*)

This chapter provides an overview of career opportunities in intellectual property law. The various forms of intellectual property are first briefly described, followed by a description of career opportunities for one interested in this area of law.

G. Madhavan et al. (eds.), *Career Development in Bioengineering and Biotechnology*, DOI: 10.1007/978-0-387-76495-5_11, © Springer Science+Business Media, LLC 2008

Intellectual Property – What is it?

Intellectual property is created through the mental activity of individuals, and results in intangible ideas that may be embodied in one or more forms. These forms are protected through enforceable legal rights that prevent others from using the intellectual property. There are generally four types of intellectual property: patents, trademarks, copyrights, and trade secrets.

Patents: A patent is a grant of rights by a government, that allows the patentee, for a limited period of time, to exclude others from making, using, selling, or offering for sale the subject matter within the bounds of the claims of the patent (the invention). Patents are limited to inventions that are novel and not obvious over the prior art. Since the early 1980s, patentable subject matter includes genes that are isolated and purified so as to distinguish them from genes existing in nature. The explosion in biotechnology has encompassed isolated genes, transformed cells, modified genes and proteins, and various forms of nucleic acids that may modulate protein expression, among many other areas.

Trademarks: "A trademark is a word, phrase, symbol or design, or a combination of words, phrases, symbols or designs, that identifies and distinguishes the source of the goods of one party from those of others." (*www.uspto.gov.*) Trademarks may relate to the name, color, and distinctive features of a product or service used in commerce. In the biotech/pharmaceutical industry, trademarks are particularly important because they can help distinguish drugs from one another. Infringement of a trademark occurs when a competitor uses a mark that is sufficiently similar to a preexisting mark, thereby leading to confusion between the *bona fide* product and the infringing product.

Copyrights: Copyrights relate to the expression of an idea, and not the idea itself. Provided that an expressive work is "new" and "fixed in a tangible medium of expression" (e.g., on paper or saved on a computer), the creator of the expressive work has a copyright in that work. Copyrightable subject matter includes the language used to express an idea, illustrations, works of art, music, and even choreography. The text and artwork of an advertisement are examples of material that is subject to copyright protection.

Trade Secrets: A trade secret is confidential information that provides a company with a business advantage. Trade secrets may include formulas, compilations, patterns, methods, techniques, and other similarly valuable business information. A company that does not demonstrate sufficient effort to maintain the confidentialty of its trade secrets risks losing them, and the ability to enforce them.

Pharmaceutical products may therefore that subject to each form of protection discussed above. A new chemical entity is discovered useful in treating cancer, for example, may be protected by patents relating to the chemical entity itself, its method of manufacture, as well as its method of use. Trade secrets may relate to aspects involved in scale-up production of the drug, marketing strategies, and other valuable business information. Trademarks and copyrights may then be obtained for the look and name of the drug, as well as written text describing the drug both in product literature (copyright), or advertisements (copyright and trademark). Persons pursuing careers in intellectual property law are concerned with securing, enforcing, or transferring these various intellectual property rights.

Intellectual Property Involves Evolving Doctrines

Intellectual property law is constantly evolving to accommodate advances in science and technology. In the biotechnology and bioengineering arts, this evolution has been particularly significant regarding the scope of patentable subject matter and issues relating to patent validity. As a professional working in intellectual property, one not only must continually learn and apply these changes, but may also be actively engaged in bringing about such changes. It has been only twenty-eight years since the Supreme Court in *Diamond v. Chakrabarty, (447 U.S. 303 (1980)*, held that recombinantly modified cells are patentable. Since then, patents have been issued to isolated and purified genes, gene fragments, modified cells, transgenic animals, and other products of the biotechnology revolution. Biotechnology has had profound effects on medicine, agriculture, and several other industries, and will continue to do so. Yet debate on patentable subject matter continues, and persons involved in intellectual property law are at the heart of this debate.

Intellectual property law has been concerned with whether one has a property right in their surgical waste or discarded cells (*Moore v. Regents Cal., 51 Cal.3d 120, 271 Cal. Rptr. 146, 793 P.2d 479 (1990)*; whether expressed sequence tags (ESTs) without a specific use unique to the individual sequence satisfy the statutory utility requirement (*In re Fisher, 421 F.3d 1365 (Fed.Cir. 2005)*; what constitutes conception of a specific gene (*Amgen v. Chugai, 921 F.2d 1200 (Fed.Cir. 1991)*; and many other important issues. Whether one is drafting a patent application or litigating a patent, one must also consider issues of enablement, and whether the text of the patent enables one skilled in art to make and use the claimed invention. Similarly, identifying when an invention was conceived and by whom, and what actually constitutes conception, are issues that must be examined and resolved by the patent professional. Since inventions must not be obvious over the prior art, patent professionals are often focused on what might be small but significant differences between a claimed invention and the prior art. For example, is a single point mutation in a known gene obvious over the known gene itself, or does a gene from one species render obvious the corresponding gene of a different species, that is, what degree of sequence similarity renders one gene obvious over another?

Some of the current intellectual property issues relate to fundamental questions of ownership of inventions based on genetic material, and whether the patenting of such material may limit access to diagnostics or therapeutics. Because the ability to obtain meaningful patent protection is critical to making significant investments in the research and development necessary to develop new diagnostics and therapeutics, these debates continue. Other current issues include finding ways of providing economically depressed countries access to new medicines while providing incentives for research and development. The patentability of isolated or recombinant genes has been criticized for various reasons, including the increased cost it may incur for various diagnostics. Gene chips that include means for detecting patented nucleic acid sequences for screening in the presence of mutations within an array of different genes may require multiple patent licenses to obtain rights to use methods and probes to detect the specific mutations. These and other issues are dealt with by people working in intellectual property law as they confront these situations and seek to resolve them.

Types of Legal Careers

Careers in intellectual property law are varied, and may involve different training and specialization to become an attorney, patent agent, patent examiner, licensing specialist, scientific advisor, or paralegal. Persons with legal training in intellectual property law may also become involved in management, in executive positions that do not necessarily involve the practice of law.

Intellectual Property Attorney: An attorney at law is someone licensed to practice law in a particular jurisdiction. Only an attorney may provide legal advice. Questions of patent infringement or patent validity are examples of questions about which only a licensed attorney can provide an opinion. A patent attorney is an attorney licensed to practice law, as well as being registered by the United States Patent and Trademark Office (USPTO) to represent patent applicants before the USPTO. Registration to practice before the USPTO requires meeting certain requirements discussed below, which include passing an exam to test proficiency in patent law and rules of practice.

In the United States, an attorney may provide counsel, draft legal documents, provide legal opinions, and represent a party in court. Thus, there is no formal separation between attorneys who litigate and those who only represent clients before the USPTO. This contrasts with certain foreign countries where patent attorneys are not licensed to represent clients in court, but represent clients before the relevant patent office. A further separation may also occur between solicitors who may prepare a case for trial, and barristers who present the case in court. In the United Kingdom, for example, a European patent attorney would represent a client to secure a patent through prosecution of the application in the European patent Office. Enforcement of the patent against an infringer would involve a team of individuals, including European patent attorneys who have expertise in analyzing the patent and the patent law of the country in which the patent is being asserted, as well as solicitors and barristers to prepare and put the case on trial, but who also are experienced in intellectual property law.

The career of an intellectual property attorney in the United States may take many different paths. People often focus their practice predominantly on trademark or patent law, especially if the attorney has an advanced technical degree. Within each of these categories, focus may be placed on litigation, prosecution (securing the patent or trademark through correspondence with the USPTO), or licensing.

Litigation involves an interparty dispute under the jurisdiction of a court or administrative agency. Both litigation and prosecution intellectual property attorneys may be involved in planning the initial strategy of the case, whether for the plaintiff or the defendant. After filing the complaint in court, the parties embark on the process of discovery, which in the United States involves the review and exchange of sometimes massive numbers of documents, followed by review of the opposing side's documents. Discovery also includes preparation, taking, and defending depositions of witnesses to seek relevant information regarding the issues in the case. Another often fascinating aspect of litigation is working with scientific experts. These experts may work with attorneys to assist in the development of legal positions, as well as to testify at trial to explain the technical aspects of an invention, for example, or the state of the art at the time the patent application was filed.

As stated above, only a registered patent attorney or patent agent may represent an inventor in matters before the USPTO. This process includes the preparation of the application and responding to the examiner at the USPTO, who usually rejects the initial application on one or more statutory grounds. Responding to the examiner may include providing legal arguments and scientific evidence to the examiner to refute the grounds of rejection. Arguments often assert that the examiner failed to appreciate the significance of some data or scientific publication, misapplied the law, or used an improper legal standard.

Prosecution of a patent application provides one with an opportunity to work closely with the applicant/inventor, in order to understand the invention in sufficient detail to draft the application with claims of sufficient scope to prevent others from obtaining the benefit of the invention without infringing the claims. In addition to working with the inventor to understand and define the invention, the attorney also typically searches and analyzes the prior art so as to be able to distinguish the invention from what may have existed in the past. Patent prosecution can involve complex issues, which require mastery of both the technical information and the relevant legal doctrines to overcome grounds of rejection made by the examiner. Overcoming such grounds of rejection to obtain the issuance of a patent covering a product with annual sales of over a billion dollars provides the prosecuting attorney with an opportunity for making a significant contribution in securing the intellectual property rights surrounding such a valuable invention.

Licensing of intellectual property rights is another significant activity of an intellectual property attorney. An attorney involved in intellectual property transactions often must first undertake a review of the underlying intellectual property to assess its validity and scope. This process of due diligence, if involving a patent, would typically include an analysis of the patent and its prosecution history, confirmation of ownership, a search and analysis of prior art, and consideration of third party patents to assess whether there is freedom to operate the invention claimed in the patents that are the subject of the transaction. Negotiating terms of the license may be done by both the principals of the parties involved in the transaction, as well as the attorneys representing the parties. Attorneys may draft the transaction documents, or review and revise as necessary the documents drafted by licensing specialists. Whether done by attorney or specialist, the process that typically involves substantial negotiation, and a certain amount of creativity, to provide for an appropriate set of enforceable obligations of the parties to accomplish their desired goals. For example, "field of use" restrictions, that limit the fields in which the licensee may practice the invention, are important to the pharmaceutical and biotechnology industry, since a licensor may be able to license specific fields of use of an invention to parties best suited to commercialize those fields.

Counseling clients regarding the validity and scope of intellectual property rights of a third party is another important role of the intellectual property attorney. A trademark attorney may provide counsel on whether or not another company's mark is sufficiently different to avoid confusion. The attorney may also advise whether the client's mark is too similar to the registered mark of another. The biotech/pharma patent attorney would also provide counsel on infringement and validity issues with respect to patents owned by another party.

A considerable part of a patent attorney's practice may be focused on providing legal opinions regarding the scope and validity of third party patents to determine whether there is freedom to operate for a particular product, or whether the product would infringe another party's patent. In the biotechnology industry, this may include an analysis of patents relating to the synthesis of a product, the product itself, and formulations of the product, as well as its method of use. This task also involves an analysis of prosecution histories of other patents, relevant prior art, and consideration of whether the patent of interest satisfies the statutory requirements for patentability.

Invalidity due to lack of enablement or lack of written description have been extremely important defenses to charges of infringement of patents relating to biotechnology inventions. In *Genentech v. Novo Nordisk, 108 F.3d 1361 (Fed. Cir. 1997)*, for example, Genentech's patent on expression of human growth hormone (hGH) in using enzymatic cleavage was invalidated because the court held that Genentech's patent failed to enable one skilled in the art to obtain hGH without undue experimentation. Attorneys for both sides worked closely with their experts to understand the state of the art during the early days of recombinant protein expression, and to assess the adequacy of the patent disclosure in providing instruction for how to make hGH.

For an intellectual property lawyer in biotechnology, the importance of having a strong technical foundation and the ability to learn and apply that foundation to new technology cannot be overemphasized. After all, the nature of biotech intellectual property law is to constantly be exposed to new scientific concepts and ideas, which become the intellectual property being protected.

Patent Agent/Scientific Advisor: A patent agent is a person who has passed the registration exam administered by the USPTO, and is therefore qualified to prosecute patent applications in the USPTO on behalf of clients. As stated above, only an attorney can give an opinion as to the validity, enforceability, or infringement of a patent. Like the attorney, however, the agent works closely with inventors to draft and prosecute the application in the USPTO. Analysis of prior art and application of patent law are used to prepare and to prosecute patent applications.

To take the examination to be a patent agent, one must have obtained a degree in one of several accepted technical fields of study, or have obtained sufficient academic credits in such fields. Acceptable fields of study include biology, biochemistry, chemistry, physics, pharmacology, botany, marine technology, molecular biology, microbiology, biomedical engineering, and chemical engineering. The agent's exam tests for proficiency in substantive patent law, as well as the rules of practice before the USPTO. Upon passing the exam, paying the appropriate fees, and demonstrating good moral character and reputation, one is given a registration number by the USPTO, allowing him or her to represent clients in patent prosecution matters before the USPTO. Like an attorney, a patent agent works with inventors to define their inventions and prepare the patent application to claim the invention. After filing the application, a patent agent corresponds with the examiner to obtain allowance of the claims and issuance of the patent. At times, amendment of the claims may be necessary to overcome the grounds of rejection.

Patent agents may be employed by law firms or corporations, or may work independently. In law firms, patent agents may work with attorneys on matters besides prosecution. Due to their expertise in understanding relevant technology, patents, and prosecution histories, patent agents may assist attorneys in litigation, in connection with providing legal opinions, and other similar matters. In litigation matters, patent agents may conduct searches and analyses of prior art, and identify and work with scientific experts and other technical witnesses. They may also analyze technical information from scientific notebooks or other relevant sources. Thus, they may play a crucial role in the enforcement or defense of a patent claim.

Typically, scientific advisors are persons with advanced degrees, hired by a law firm, who have not yet become patent agents. They may work with attorneys on litigation or prosecution matters to provide their scientific expertise, and function like a patent agent. However, they are unable to represent clients before the USPTO until they pass the agent's exam.

Licensing specialist: A person interested in transactions and putting together deals may choose a career as a licensing specialist. As a licensing specialist, one may be involved in identifying licensing opportunities for either in-licensing or out-licensing technology. Having identified the opportunity, the licensing specialist may interact with management to decide the terms of the transaction. A licensing specialist may, or may not, have a law degree. As stated above, an attorney will often review the terms of the agreement after it has been negotiated but prior to signing.

Management: Often the senior intellectual property counsel of a company may be well suited to assume the responsibilities of management. Management of the intellectual property estate is a significant, and sometimes even dominant, function of management. One may, therefore, have an active career involving intellectual property law through his or her activities in managing company where intellectual property represents a significant company asset.

Employers

Opportunities for employment, or a career involving biotechnology and bioengineering intellectual property law, may be found in government or the private sector, and may provide employment in any of the professions described above.

Government

Government provides many different opportunities for employment in intellectual property law. These opportunities may be with federal agencies such as the USPTO, government technology transfer offices such as at The National Institutes of Health (NIH), as well as Congress and the executive branch.

Federal Agencies

USPTO: The USPTO provides an excellent opportunity to obtain training and experience in patent law through by an examiner. Through the process of becoming an examiner, the USPTO provides in-depth training in both patent law and rules for practicing before the USPTO. Historically, training at the USPTO was largely through apprenticeship. Junior examiners would be assigned to more senior examiners, who reviewed their work. The USPTO has recently stated that it plans to rely less on the apprenticeship form of training, and to expand the duration and depth of its formal training.

The USPTO provides employment opportunities for both lawyers and non-lawyers. Besides the examining corps, the USPTO also employs persons in the Board of Appeals and Interferences, the solicitor's office, the petitions branch, and other sections. Although a law degree is not required to be an examiner or supervisory examiner, other positions, such as being an administrative patent judge presiding over interferences, or representing the USPTO as a solicitor, for example, do require a degree in law. Many of the examiners in the art units that examine patent applications directed to biotechnology inventions have an advanced degree such as a doctorate.

An examiner reviews filed patent applications to determine whether they meet the statutory requirements for patentability, that is, that the invention is novel and not obvious over the prior art, and the description of the invention in the patent application provides an enabling disclosure, a written description of the invention, and the best mode of carrying out the invention as of the date the application is filed. As part of the examination, the examiner conducts a prior art search, and may rely on identified prior art to support a ground of rejection of the patent application. Through the process of issuing examination reports, Office Actions, and the applicant's responses to them, an official record is created of the patent prosecution in the USPTO. This record, or prosecution history, is highly relevant to understanding the metes and bounds of the invention, and forms a legal document which may be relied on by a court to interpret the scope of claims, and whether the patentee may have dedicated certain subject matter to the public without it being claimed. In addition to the written record, the examiner may conduct interviews with the attorney or agent representing the applicant, and sometimes the applicant, to discuss issues which may have arisen during prosecution. Due to the complexity of certain biotechnology applications, such interviews may be particularly important for the examiner to understand the invention and to advance prosecution.

A person may decide to pursue a life long career in the USPTO, or decide to continue their career outside the USPTO in a law firm, or in the private sector as a patent agent or as an attorney, if they also go to law school. Persons with experience in the USPTO often provide invaluable insight into dealing with problems that may arise during prosecution of an application. Such individuals may be very desirable to a law firm or company prosecuting patent applications in the USPTO. Because of the high competition to obtain employment at a law firm, spending a few years as an examiner may greatly facilitate one being able to obtain employment at a law firm.

Government Technology Transfer Office: Many agencies of the federal government that are involved in research and development have technology transfer offices to identify and manage their intellectual property. For example, the National Institutes of Health generate large numbers of inventions for which applications are filed. Outside law firms are hired by the NIH to prepare and prosecute these applications. To oversee this process, and to commercialize the inventions, the NIH has an active technology transfer office, in which individuals pursue careers in intellectual property law by being a liaison between the NIH and the law firm, and by functioning as licensing specialists. Many of the responsibilities of intellectual property professionals managing the intellectual property in a government technology transfer office would be similar to those of a person working in a university technology transfer office, as described later.

Congress and Executive branches: Besides the USPTO and the government technology transfer offices, the government provides many other opportunities for someone interested in intellectual property law. Committees of Congress address legislation relating to intellectual property issues, and may seek staff members who are sufficiently knowledgeable about intellectual property issues to advise them on certain legislation that may relate to intellectual property issues. Various departments of government are also involved with intellectual property issues, such as the Department of Commerce, in connection with various trade negotiations and treaties, and the Department of State, which also addresses intellectual property issues with foreign countries.

The International Trade Commission (ITC) has jurisdiction to halt imports of goods into the United States that infringe on a United States patent. Staff members with expertise in intellectual property law participate in proceedings before the ITC with the patentee and importer, and assist the commission in gathering and analyzing appropriate information to determine whether the import infringes on the US patent, and if the ITC should issue an order to the Customs Department to keep infringing imports out of the United States.

Private Sector

Law firm: Law firms provide various ways to pursue a career in intellectual property law, as attorneys, patent agents, scientific advisors, and paralegals.

Attorney: As discussed above, an attorney in a law firm practicing intellectual property law may be involved in litigation, patent prosecution, licensing, due diligence, and counseling. Each of these areas may be rewarding, and an attorney's career may focus on one or more of these particular areas. Working at a law firm provides opportunities to work with different types of clients who have matters involving intellectual property. Law firm clients may include individual inventors, start-up biotech companies, or mature multi-national companies. Clients may also include parties who invest in intellectual property and require counsel on the strength of the intellectual property in which they are investing.

An attorney's career at a law firm typically begins as an associate. Associates are exposed to the different forms of intellectual property law, and may gravitate to practicing predominantly litigation or prosecution. In most firms, offers for partnership are made between seven to ten years of being made an associate. As a partner, one is expected to be able to manage and oversee matters. The ability to generate business, by either bringing in new clients, or expanding the work for existing clients, is an important ability to succeed as a partner. Intellectual property attorneys may focus their careers on litigation or non-litigation matters, such as prosecution of patent applications before the USPTO, providing legal opinions on infringement and invalidity to clients, or on transactional matters. Because of the time demands associated with litigation, and the deadlines associated with prosecution for filing documents in a patent office, most persons who develop a strong litigation practice tend to have few, if any, prosecution matters. Early in one's career, however, experience in both prosecution and litigation is invaluable. For a litigator, it is extremely helpful to have a thorough knowledge of prosecution before the USPTO, as it leads to having an understanding of potential areas of vulnerability that may be useful when attempting to invalidate a patent. Also, for a person prosecuting a patent, it is extremely important to understand how a patent and a prosecution history will be scrutinized and attacked during litigation.

To have a strong biotechnology and bioengineering practice, law firms must demonstrate that the professionals practicing in that area are capable of understanding the technology. Accordingly, such practices usually have several attorneys, agents, and scientific advisors with advanced scientific degrees. It is also common for clients to require that persons working on these matters have such a degree. Although many litigators trying biotechnology cases may not have advanced degrees, attorneys with an advanced degree are often key members of the litigation team. Attorneys with advanced degrees are often particularly well suited for identifying and working with technical experts.

Transition from scientist to attorney, or to patent agent, requires that one recognizes and adopts his or her role as advocate for the client. Although this may seem self evident, this transition is not always easy or successful. For example, what may seem obvious to a scientist may not meet the legal test for obviousness used by a court, and it is sometimes difficult for a former scientist making a transition to law to distinguish between what may seem obvious to them from what would be obvious under the law. Also, one must be able to avoid a form of scientific arrogance where one believes that as a scientist, one can make certain assertions as fact without relying on documentary or expert evidence. An attorney is not a witness, and must use his or her knowledge and ability to corral evidence to prove certain facts, and to advocate the relevance of those facts to the outcome they are trying to obtain for their client.

Industry: Opportunities to practice intellectual property law in industry are also diverse, providing a wide range of skills and level of company involvement. Trademarks, trade secrets, and patents may all be critical to the business of a particular company, and intellectual property legal professionals are responsible for obtaining and enforcing the various rights associated with these forms of intellectual property. Besides protecting these intellectual property rights, handling licensing transactions and providing counsel to avoid infringing the intellectual property rights of others are often significant responsibilities of the corporate intellectual property professional.

In-house counsel: As in-house counsel, the corporate attorney has the company as a client and provides counsel to management. Although many of the responsibilities and tasks of the in-house counsel are similar to that of an attorney in private practice, the in-house counsel may be close to strategic planning and other significant corporate decisions. In-house counsel is often sought by management to provide freedom to operate analyses for products under development. Other important responsibilities may include managing litigation, negotiating licenses, and prosecution of patent and/or trademark applications. Attorneys working on prosecution matters would also work with the corporate scientists to identify inventions for possible patenting or for keeping as a trade secret. Depending on the size of the company, the in-house attorneys may either perform some of these tasks themselves, or manage the activities of outside counsel. When managing outside counsel, the in-house attorney plays a critical function as the liaison between management and the outside counsel executing the goals of management. For relatively small biotech companies, it is not unusual for the intellectual property attorney to function more as a general counsel, advising the company on employment, real estate, or other matters.

The role of in-house counsel managing litigation may involve conducting the initial factual review of the matter to develop strategy and assess likelihood of success. A critical function is to communicate with management the strengths and weaknesses of the matter so that management can make the ultimate decisions as to whether to pursue litigation or settle the matter. During litigation, the in-house attorney may coordinate production and review of evidence such as document production. Attending depositions is often an important way for the in-house attorney to assess how well the outside counsel is conducting the litigation, and to observe first hand how witnesses may perform at trial. In addition, facts learned about the case during the deposition may be important to consider in assessing litigation strategy and the overall merits of the client company's positions.

As an in-house attorney at a biotech/pharmaceutical company, one may be very involved in developing and executing corporate transactions with other companies. By working closely with management to develop terms and conditions, the in-house attorney may play a dominant role in achieving corporate partnerships that may be critical to the company's growth. Besides conducting the negotiations and drafting documents, the in-house attorney may also coordinate the due diligence process to confirm that the potential partner can meet the obligations contemplated by the agreement. During the due diligence process, the in-house attorney would investigate the strength of the intellectual property, including patents, trademarks, and trade secrets, to confirm that the company owned or had rights to the intellectual property, and that it could be used and transferred as contemplated by the parties to the transaction.

Although in-house counsel may not need to log billable hours, one should not assume that their hours will necessarily be any better than that of the outside counsel. Demands of managing litigation or closing a time-sensitive transaction may be extremely demanding on the time of the in-house attorney. However, the strong ties to management and participating in developing and executing corporate strategy can make an in-house position very rewarding.

Management: Persons with expertise in intellectual property may hold high positions in management that allow them to contribute to the development and use of these assets. Being knowledgeable of the science as well as the law protecting some of the most valuable company assets, an intellectual property attorney may be well suited to hold a non-legal position in management, including even the position of CEO. In some companies, the director of intellectual property is a corporate director partner, which gives an attorney opportunities to be part of the company management team.

Patent agent: In industry, a patent agent typically is involved in US and foreign patent prosecution. As discussed above, this would include working with the scientists, and sometimes others, to define the invention and then draft and prosecute the application. Patent agents may also be asked to assist in freedom to operate searches and their analysis. This could include obtaining in-depth knowledge of product specifications and other technical information which is then used to ascertain the relevance of one more patents to the product, and developing strategies to avoid infringement. Except for certain aspects of litigation matters and the inability to provide legal opinions, a patent agent may function in industry in ways that are very similar to patent agents in law firm private practice.

University Technology Transfer Office: Another source of employment for the intellectual property professional is the university technology transfer office. Since the passage of the Bayh Dole Act in 1980, which provides for universities to own inventions developed through federal government funding, many universities have built sophisticated offices to identify and protect intellectual property developed by

their faculty. The technology transfer offices manage the intellectual property by identifying which inventions should be patented, managing outside counsel hired to prosecute the patent applications on behalf of the university, and licensing the technology. Universities have also asserted their patents through litigation. Persons working in the technology transfer office therefore manage the various inventions assigned to them through working with the inventors, outside counsel, and licensees. The requirements for working in technology transfer offices to manage intellectual property assets vary among universities and positions. Thus, opportunities may be available to become a technology transfer officer with an undergraduate degree, an advanced scientific degree, or a law degree.

Getting There

There are several ways to pursue a career in intellectual property law. With respect to practicing intellectual property law in the area of biotechnology, pharmaceuticals, and bioengineering, it is important to have a strong technical background to be able to fully understand the client's technology, as well as its value and utility. To practice trademark law, one does not need a special background or to register with the UPSTO. However, as stated above, one must take and pass the USPTO registration exam to prosecute patent applications before the USPTO, whether as an agent or attorney.

Undergraduate Education

The importance of solid technical knowledge cannot be overstated. The practice of biotech patent law requires one to understand both what happened in the past, and new developments as they are made. It is therefore crucial that one have a solid grasp of the underlying scientific principles upon which the inventions are based. Undergraduate degrees in chemistry, biology, chemical engineering, physics, biomedical engineering, cell biology, or other life or chemical sciences are among those which provide strong foundations for a career in biotech/bioengineering intellectual property law. As discussed above, one must have a sufficient number of credits in specific areas to take the patent agent exam.

Advanced Degree

Also as discussed above, many persons practicing biotechnology patent law, either as agents or attorneys, have advanced degrees. Although not a requirement, many clients require that the persons handling their matters have such an advanced degree. Due to the high cost of preparing and prosecuting a biotechnology patent application, clients strongly desire persons working on their matters to have a strong background in the relevant technology. Such individuals do not first have to learn scientific background fundamentals, and can make more meaningful contributions to describing and claiming the invention if they understand the underlying scientific principles involved.

Degrees in chemistry, biology, pharmacology, molecular biology, neuroscience, cell biology, immunology, physics, physiology, and microbiology are among those that are particularly relevant to practicing biotechnology and bioengineering intellectual property law. In addition, due to the often varied nature of practicing intellectual property law in either a law firm or company, one must be able to apply his or her knowledge to subjects and inventions not necessarily directly related to previous formal training. Where a scientific researcher may devote their career to a relatively narrow field of science, the practice of intellectual property law typically involves a vast array of different technologies over the course of one's career.

Scientific Advisor Programs and Law School

Many law firms that have a strong biotechnology/pharmaceutical intellectual property law practice have also developed scientific advisor programs. These programs hire persons with advanced degrees, typically doctorates, and then train them to become patent agents and attorneys. Some law firms may pay for the person to attend law school while they work full time at the law firm as a patent agent or scientific advisor. A scientific advisor who has not yet passed the USPTO registration exam may have essentially the same responsibilities as a patent agent, except that they cannot represent clients before the USPTO. The scientific advisor program is an excellent way to begin a career in biotechnology intellectual property law, as it exposes one to many of the issues and tasks one may use whether he or she pursues a career in a law firm or a corporation.

The administration of the scientific advisor programs differs between firms, and it is important to understand the obligations and commitments before accepting a position. Some firms allow persons to remain as patent agents, while others require the advisors to attend and graduate law school. Also, requirements for positions as an associate and the class assigned for salary and consideration for partnership can vary among firms.

Although the scientific advisor program provides access to a career in intellectual property law and can pay for law school, for some individuals, it may be desirable to go directly to law school and then seek employment at a firm or company with the law degree in hand. Personal economics, age, and the ability to work and go to school are important factors to consider when deciding which direction to follow. Working full time while going to law school can put significant stress on an individual and their family relations. However, with appropriate support and understanding, one can eventually graduate law school and then concentrate on a career as an attorney.

Becoming an examiner at the USPTO and then going to law school, or seeking a position as a scientific advisor, is also an effective avenue to enter the profession.

Rewards

Having chosen a career in biotechnology/pharmaceutical intellectual property law, one can expect to be challenged and stimulated. Whether dealing with cutting edge technology or unraveling the past history of an invention involved in litigation, one

is constantly learning science and law, and participating in their development together. In addition, if one is interested in science, one has the opportunity to work with leading scientists in their fields, whether as inventors or experts to provide evidence in litigation or in matters before the USPTO.

Through a career in biotechnology intellectual property law, one is also contributing to the development and commercialization of products that provide significant benefit to mankind. Whichever side of the political spectrum one may be on, significant investment into new drugs and medical devices does not occur unless there is a path to protecting the investment through securing enforceable intellectual property rights.

The importance of intellectual property law and the specialized knowledge that one acquires provides significant job opportunities in private practice or industry.

Caveats

As with most highly compensated positions, a career in intellectual property law can be extremely demanding and require significant time commitments both in training and practice. The competitive nature of securing intellectual property rights and the legal requirements for securing those rights require high levels of diligence and timeliness. Missing a statutory filing date can result in the loss of intellectual property rights worth hundreds of millions of dollars. In one recent case, a law firm was subject to a 30 million dollar claim because it missed a date to file international patent applications for an invention. It is therefore necessary that one be extremely responsible and able to handle the stress of meeting deadlines and the demands of clients.

Conclusion

Intellectual property law as it relates to the biotechnology and pharmaceutical industries provides varied and rewarding opportunities for careers. Career opportunities may be found in government, industry, or private practice, and range from paralegal to senior management to law firm partner. With proper dedication and commitment, one can make a significant contribution to the transition of science to practical application through a rewarding career.

Clinical Engineering

Jennifer McGill, MEng

ECRI Institute (Health Devices) and American College of Clinical Engineering, Plymouth Meeting, Pennsylvania, USA

Remember those career aptitude tests in high school? Well, they may not be as useless as they seem, as that was how I first thought of exploring the field of engineering when I was a junior in high school, and subsequently attended an information session on clinical engineering. I met a Clinical Engineer (CE) who worked at the Toronto General Hospital, one of the largest hospitals in Canada, and I became fascinated with the concept of helping hospitals manage their medical technology. I feel that I was lucky to have come across this somewhat obscure field within bioengineering before applying to university. It helped me to decide between becoming a doctor and becoming an engineer, and ultimately has led to a very fulfilling career.

What is Clinical Engineering?

The short definition from the American College of Clinical Engineering (ACCE) is as follows: "A Clinical Engineer is a professional who supports and advances patient care by applying engineering and managerial skills to healthcare technology." This encompasses a large and expanding set of responsibilities for the profession, including (from the Clinical Engineering Certification Body of Knowledge 2006):

Technology Management Responsibilities

- Project Management
- Product Selection/Vendor Selection
- Technology Assessment
- Interpretation of Codes and Standards
- Device Integration Planning
- Device/System Upgrade Planning
- Healthcare Technology Strategic Planning
- Usability/Compatibility Assessment
- Clinical Device Use and/or Application
- Capital Planning
- Electromagnetic Interference (EMI)/Radio Frequency Interference (RFI) Management
- Pre-clinical Procedure Set-up/Testing

G. Madhavan et al. (eds.), *Career Development in Bioengineering and Biotechnology*,
DOI: 10.1007/978-0-387-76495-5_12, © Springer Science+Business Media, LLC 2008

- Clinical Systems Networking
- Life Cycle Analysis
- Coordinating Device Interoperability/Interfacing
- Return on Investment (ROI) Analysis
- Clinical Trials Management (Non-investigational)
- Water Quality Management
- Participation in Clinical Procedures (e.g., surgery)

Service Delivery Management

- Technician/Service Supervision
- Equipment Repair and Maintenance
- Service Contract Management
- Equipment Acceptance
- Equipment Performance Testing
- Computerized Maintenance Management Software (CMMS) Administration
- Parts/Supplies Purchase and/or Inventory Management
- Develop Test/Calibration/ Maintenance Procedures
- Technical Library/Service Manuals Management

Product Development

- Regulatory Compliance Activities
- Documentation Development/ Management
- New Product Testing & Evaluation
- Human Factors Engineering
- Medical Device Concept Development/Invention
- Product/Systems Quality Management
- Device Modifications
- Medical Device Design
- Product Research and Development
- Product Sales/Sales Support

Information Technology/Telecommunications

- Integration of Medical Device Data
- Information Technology (IT) Management
- Server and/or Network Management
- Help Desk/Dispatching/Call Tracking
- Telecommunications Management

Education

- Technician Education
- Device User/Nurse Training
- Develop/Manage Staff Training Plan
- Engineering Education
- International Healthcare Technology Management

Facilities Management

- Facility Emergency Preparedness Activities
- Building Plan Review
- Emergency Electrical Power
- Building Design
- Medical Gas System Testing
- Facility/Utility Remediation Planning
- Supervise/Manage/Direct Facilities Management

Risk Management/Safety

- Patient Safety
- Product Safety/Hazard Alerts/ Recalls
- Risk Management
- Infection Control
- Root Cause Analysis
- Medical Device Incident Reporting under the Safe Medical Devices Act (SMDA)
- Failure Mode and Effect Analysis
- Workplace Safety Practices (OSHA)
- Hazardous Materials
- Incident/Untoward Event Investigation
- Engineering Assessment of Medical Device Failures
- Fire Protection/Safety (Life Safety Code)
- Radiation Safety
- Expert Witness
- Industrial Hygiene
- Investigational Research (Human Use)
- Forensic Investigations

General Management

- Personnel Management/Supervision
- Budget Development/Execution
- Staffing
- Performance and Quality Improvement
- Policy/Procedure Management/ Development
- Staff Skills/Competency Assessment
- Committee Management
- Business/Operation Plan Development/Management
- Revenue Producing Activities

A CE in a small hospital may be responsible for most of the activities outlined above, and may have a few Biomedical Engineering Technicians (BMETs) to directly repair and maintain medical equipment. However, in larger hospitals, some of the responsibilities may be distributed to a staff of engineers due to the scale of the institution or group of facilities. An experienced BMET may eventually become a Clinical Engineering department manager or director as well, and usually their technical education is supplemented with management training. This creates some competition between experienced BMETs and CEs for the management positions in hospitals, so even engineers with a master's degree may need to supplement their university education with some project management courses and/or technical seminars. Look for teleconferences and online courses that can be done on your own time.

Another good source to help understand Clinical Engineering is the Clinical Engineering Standards of Practice developed by the Canadian Medical and Biological Engineering Society (CMBES). CMBES started a volunteer peer review process where CEs would visit a colleague's facility and perform an evaluation of their service. It is a useful learning process for both the reviewed and reviewer because it provides a forum for an open exchange of ideas, as well as providing a tool for the reviewed hospital to measure improvements over time and lobby for resources to bring their service up to the standard.

How do I Become a CE?

If you are lucky enough, like me, to have discovered this interesting field in high school, then the first step to becoming a CE is to get a bachelor's degree in engineering, preferably in the biomedical specialization. Electrical, mechanical, systems, and computer engineers can also enter the clinical engineering field quite easily; however, since the competition for jobs is fierce, a master's degree specifically in clinical engineering is a huge asset. Look for a program with clinical engineering internships in hospitals, as this will give you invaluable work experience and may lead to an entry-level position in the same hospital.

Hospitals are often stretched for funding, so they are usually less concerned with research projects than practical experience. Volunteer work is a great way to supplement your internships and even get some experience at the undergraduate level. It is sometimes difficult to find a volunteer position in a hospital that is connected to the clinical engineering department, so you may need to build a network of contacts first. I have often asked for "informational interviews" early in my career after meeting CEs at conferences. You can learn from more experienced colleagues and hear about their issues or challenges within their own clinical engineering departments. If a director or manager is willing to spend even twenty minutes with a student, that is a good sign. You may learn enough to create a proposal for a specific project that you would be willing to do for them as a volunteer or on contract to get more experience.

Professional conferences are essential for learning skills and getting information on topics in clinical engineering that are not taught at universities, and for developing job connections and contacts for volunteer work. In Canada, the Canadian Medical and Biological Engineering Society (CMBES) includes the field of clinical engineering, and holds an annual conference. In the United States, you should find out where the American College of Clinical Engineering (ACCE) is having their annual meeting. If you become a member of these societies, you can also get access to the membership list to see if there are contacts in your area, and you can get discounted rates on the conferences, reference materials, teleconferences, and courses.

Clinical engineers typically manage large installation projects, investigate incidents, and work with manufacturers to solve problems with equipment or systems, as well as provide expertise to various hospital committees. Project management skills are essential, and if you want to manage a department, then some management and strategic planning training is a good idea too. Computer networking, interfacing, and server management skills are also becoming more valuable in the medical equipment realm. There will always be a need to manage the maintenance of medical devices, but we are starting to emerge from the basements into high-level technology management. Clinical engineers are finding new niches in their organizations, including Chief Technology Officer. Many experienced CE's also enter the consulting arena; however, it is difficult to find enough business to maintain a full income, so often consultants are retired CEs (or very brave and resourceful younger CEs).

Once a CE has obtained three to five years of experience in the field (number of years can vary by state, province, and country), they would be eligible for either their professional engineering license or certification in clinical engineering. In the United States, if the CE's original engineering degree was in a traditional discipline (mechanical, electrical, chemical, civil, etc.), then he or she may be able to obtain a professional engineering license. A license aptitude exam is not typically available for biomedical or clinical engineering, so many CEs have opted to become a Certified Clinical Engineer (CCE). The CCE process and examination was completely revamped in 2004 to reflect the current body of knowledge for the profession, and approximately 30 new CEs have been certified as of 2006. In Canada, by law, you must have a professional engineering license if your job title is "engineer". This has forced the larger professional engineering organizations to recognize clinical and biomedical engineering, and to offer licenses in these specialties. Since the Canadian system does not require an aptitude test if you graduated from an accredited engineering program, these specialties were more easily introduced than in the US, where aptitude tests are required for the engineering licenses. There used to be a certification program in Canada, but it is currently under review and has been suspended until it is determined if it will be resurrected, and in what form. Since, I have spent most of my career in the US, I became a CCE here; however, if I return to Canada, I would need to apply for my P.Eng. if I decided to work in an engineering position.

What is it Like to be a CE?

One thing you are assured of as a CE – there is no "typical" day. Your day may consist of a medical equipment management committee meeting, talking to a manufacturer about a persistent equipment problem, following up on some purchase orders for an installation project, and meeting with clinicians on another equipment replacement or expansion project. All planned tasks can change in an instant if an incident or major problem happens involving medical equipment in the hospital. As more medical devices interface with hospital information systems, expect to spend hours (and sometimes even days or weeks) troubleshooting intermittent issues with interfaces and database servers. Hospital information systems can be so complex that it can be very challenging to identify precisely what is causing a problem.

I have fond (tongue-in-cheek) memories of spending six weeks and several late night calls to the hospital trying to find out what was causing one of my databases to intermittently lose network connectivity. We even had the server rebuilt (a two-day process) to no avail. I had asked the IT network engineers if anything changed on the date that the problems started to occur and was told that nothing changed with the network, only to discover that the server management team had implemented a new network backup system that they broadcast across all ports, knocking my sensitive system out of whack. A simple reprogramming of the backup system finally solved the problem.

CEs can also work for medical equipment manufacturers and for research organizations like ECRI Institute. ECRI Institute is the only place in the world that publishes comparative evaluations of medical devices. As an engineer in the Health

Devices Group at ECRI, I am also involved in investigating reported hazards and problems, writing guidance articles, performing accident and forensic investigations, providing expertise on consulting projects, health technology assessment projects, and reviewing quotation analysis reports, market intelligence reports, product comparison charts, and many more. Once again, there is no day exactly like another, and you are constantly learning to keep up with changes in technology.

Human factors laboratories are also being started by clinical engineers; for example, at the Beaumont Hospital's Technology Usability Center in Michigan, USA, and the Healthcare Human Factors group of the Centre for Global eHealth Innovation at the University Health Network in Toronto, Canada.

One area of clinical engineering I am very interested in is the managed implementation of new health technologies into health systems. This combines rigorous research from traditional Health Technology Assessment (HTA) systematic reviews with practical analysis of the local or regional needs, finances, and impact. Large hospital groups and regional health systems are starting up internal services, sometimes called *health innovation groups*, that perform analysis, conduct post market trials, make recommendations for funding of new technologies, and plan implementation projects for their organizations.

Where is the Profession Heading?

Traditionally, clinical (or biomedical) engineering departments in hospitals were associated with facility maintenance departments, and there are many CE's who are still relegated to the basement underworlds of hospitals, amongst the boilers and the generators. The typical clinical engineering department for a medium to large hospital has lab space for BMETs and/or CEs to assemble, repair, and maintain medical equipment, and some area for storage of spare equipment, spare parts, tools, and supplies. You will definitely see fewer soldering irons and more computers than 20 years ago, since many repairs are at the board level instead of the component level, and medical devices and systems have become increasingly computer and software driven. Due to this convergence of information technology and medical technology, hospitals and health organizations are recognizing the need for expertise in managing the medical technology at the system and communications level, as well as recognizing the need for evidence-based approaches to implementing new technologies into their health systems. CEs have always been involved in managing medical technology, but the technology is no longer comprised of thousands of individual devices – it is now an intricate web of interconnected devices and data.

How Do I Keep My CE Knowledge Current?

There are many sources of information on medical technology, and it can often be overwhelming to keep up with changes and innovations. CEs usually subscribe to several journal publications, such as ECRI Institute's *Health Devices*, the Association for the Advancement of Medical Instrumentation (AAMI)'s *Biomedical*

Instrumentation and Technology (*BI&T*), and the *Journal of Clinical Engineering*. The Healthcare Information and Management Systems Society (HIMSS) produces the *Journal of Healthcare Information Management and Healthcare IT News* (published by MedTech). HIMSS, AAMI, ACCE, and CMBES all provide education courses during their conferences, as well as periodic webinars, teleconferences, special publications, etc.

Getting involved in the various societies is a great way to network with other CEs and learn from their experiences. ACCE conducts advanced clinical engineering workshops all over the developing world, and gives experienced CEs an opportunity to share their expertise in areas where there may not be any formal CE education available. It is an extremely rewarding experience, and a wonderful chance to travel to other parts of the world and learn about the challenges in managing medical equipment in resource-limited countries. (I have taught at workshops in Jamaica and Kenya.) Professional societies are always looking for volunteers to be on committees and boards, and to share knowledge with other members. The ACCE Healthcare Technology Foundation is involved in numerous interesting projects and there are several standards panels and advisory groups you can get involved in depending on your interests. Some other examples of groups you may become involved with include: The Hospital Bed Safety Workgroup, Integrating the Healthcare Enterprise (IHE) (developing communication standards for medical devices), and the FDA (or Health Canada) expert advisory committees and panels. Submitting papers and articles to conferences and journals is also a way to share knowledge and help you get motivated to keep current.

Concluding Remarks

Regardless of whether you focus on one area of clinical engineering or manage to juggle them all in a hospital environment, it is a very rewarding profession. CEs help improve patient safety, help healthcare organizations save money on the purchase and maintenance of medical equipment, and help solve problems so that clinicians can provide the best care possible. You have the chance to work with incredibly talented people from all disciplines, including physicians, nurses, medical researchers, technicians, information technology specialists, hospital systems engineers, and administrators, along with the medical equipment manufacturers and vendors. Anyone who is fascinated with medical technology and likes to work in an interdisciplinary, problem-solving, and multi-tasking environment will enjoy clinical engineering.

Suggested Readings

1. *Clinical Engineering Handbook* by Joseph Dyro (Academic Press, 2004).
2. *Clinical Engineering – Principles and Applications in Engineering* by Yadin David, Wolf W. von Maltzahn, Michael R. Neuman, Joseph D. Bronzino (CRC Press, 2003).

13

Entrepreneurship in Medical Device Technologies

Dany Bérubé, PhD

MicroCube, LLC, Fremont, California, USA

At the outset, I wish to candidly state that I am not an entrepreneur in the league of Richard Ferrari or Thomas Fogarty, who are considered to be prolific giants in the realm of medical device entrepreneurship. However, I am sharing my story and passion for the field, presenting some basic methodologies to select and run a project, and to maybe, somehow, influence talented individuals to join the field of bioengineering that I chose and love.

My Journey

I was a young teenager when I started to think about what I wanted to become in life. Like many other kids, I did not really know what I wanted to do, and I was not exposed to many professionals with some kind of a special success story in the business world. Indeed, there was not much diversity in my little town of Chibougamau, located at the geographic center of the province of Quebec, Canada. My home town was very far from everything. In fact, the place where I grew up could almost be considered the northern frontier of civilization in North America. Apart from all of the natural beauties, there is basically nothing much north of the city. Just to give you an idea, from my home town you would need to drive two hours south to reach the next city, and six hours to reach Quebec City. For me, taking a vacation in the south meant going to Quebec City or Montreal! There was not much opportunity either: the population was about ten thousand, and everybody was working in mining, forestry, or the service industry. Growing up in such an isolated community, my vision of the world was very limited.

I always loved science. I do not really know where this came from. I had no role model with a scientific background to follow in my family or my close surroundings. Discussion related to scientific subjects was almost nonexistent. However, every time I had the chance to read or watch something about science,

G. Madhavan et al. (eds.), *Career Development in Bioengineering and Biotechnology*, DOI: 10.1007/978-0-387-76495-5_13, © Springer Science+Business Media, LLC 2008

my mind would be totally absorbed by the subject. I was doing pretty well in school, but especially well in mathematics and science.

One day, when I was about fifteen years old, I had to meet a career advisor for young students. After talking for about ten minutes, it became obvious to him that I should become an engineer. I had absolutely no clue what engineering was, but the very fact that an engineer had to have a scientific background sounded appealing to me. However, when I heard that engineers were some of the most productive professionals for a country, in term of gross domestic product, I was sold. I thought that using an engineering background to produce useful goods was a very good way to make a living. So, on that day, career selection was a done deal for me. I decided that I would become an engineer (a decision that would later eliminate about 90% of the entire female population as potential girlfriends, but that is subject matter for another kind of book . . .). I was totally committed to my new endeavor, and never looked back or second guessed my decision, not even for a split second. If I had to go back in time, and choose again, I would do the exact same thing. It was right for me.

The next question for me was which field in engineering would better suit my interests. My problem was that I was interested in all engineering fields, as well as physics, which meant I had a hard time deciding which program I should apply for. One day, I came across a scientific article explaining different imaging modalities used in medicine. I was exposed to then relatively new imaging systems such as magnetic resonance, computed tomography, and positron emission tomography. It became evident to me that applying technologies in medicine would be very interesting and quite beneficial to society in general. I finally figured the whole thing out: I would be a biomedical engineer. On an emotional level, it felt right to use my talents to develop technologies used to treat or cure people, and to improve the quality of life. It gave a higher level, an almost spiritual sense, to my endeavor. I was at the beginning of my life, but felt that later on, when I was closer to the end, I would look back on what I did and just be satisfied to have used my natural talents to somehow treat sick people and give them better lives. It gave a real sense of pride and peace to the whole thing. It still does. Today, when I face difficult situations, I can still go back to those earlier days and remember why I am in this field. Knowing why you do what you do, and having an emotional connection with the field you choose, can give you a solid moral foundation from which you can grow as a professional and, more importantly, as an individual.

Biomedical engineer I would be! However, there was another problem: there was only one biomedical engineering program in the province of Quebec, and it was available only in graduate school. I still had to decide which undergraduate engineering program I would select. Since biomedical engineering is a multidisciplinary field, I decided to choose one of the broadest engineering programs available at Université Laval, in Quebec City, called "Physic Engineering." The program was basically an integration of electrical, mechanical, and material engineering, as well as physics. I still consider this choice to be one of the best decisions I made in my professional life, since it gave me a very broad engineering and scientific background that is very useful today to develop, analyze, or select technologies for medical applications. I completed my first degree, moved to Montreal, and started a master's program in biomedical engineering. When I joined, my director, Dr. Pierre Savard, presented several project options. The first was a project to create 3D rendering of images obtained with standard fluoroscopy. Another project was to use a mathematical technique to improve the contrast ratio of images created by an imaging modality. However, one project immediately caught my attention. My director explained that a new catheter-based ablation modality had been recently introduced in clinical practice to treat and cure irregular cardiac rhythms. With this new technique, certain arrhythmias that could not be cured without a major cardiac surgery before could now be cured safely, in a minimally invasive way, with a greater than 95% success rate. However, the new technique was not too effective in treating other kinds of arrhythmias. The ablations were too small and too shallow to treat rhythm abnormalities such as ventricular tachycardia and atrial fibrillation. The project was to develop a microwave-based ablation system to improve the current radiofrequency technology. I vividly remember the conversation. It soon became clear to me that there was an enormous opportunity to not only make a career out of it, but to start a company; the whole thing was in direct line with my dream to develop technologies to cure people.

Looking back, that is exactly what happened. I later joined a small start-up company called AFx, Inc., in California, where we developed a minimally invasive microwave-based surgical ablation system used to treat atrial fibrillation. The company was then acquired by Guidant in 2004 for about $110 million. There is a point to be made here. Opportunities are everywhere! At one point, you will find a good opportunity, or a good opportunity will find you. To be successful, you need to have a clear vision of what you want, why you want it, and put all of the elements of your life in place in such a way that, when the opportunity comes, you can see it, take it, and run with it. In any success story, there are also elements of luck. To me, luck is related to the level of your natural talents, and the potential of the opportunities that will come to you. For the rest, it is up to you. Note that there is an element of pro-activeness in "put all elements of your life in place." Some of those elements are associated with your mental and physical health, well-being, education, skills and empathy. Do not wait for the elements of your life to be put in place for you. If you take the latter approach, opportunities will still be present, and will still come to you, but you will not see them, nor be capable of exploiting them.

During my time at AFx, I had the chance to be exposed to all sides of the business early in my career. After we sold the company to Guidant, we decided to work on new start-up companies, where I now serve as a project president. My role is to identify business opportunities, define mission and strategies, and deliver measurable results. During these years, I have had the opportunity to get to know business leaders with a variety of leadership styles. I will later discuss methodologies to identify an opportunity, select a mission, define a plan, and execute the plan. Those are based on my personal experience, which has been forged and influenced by my own personality and way of performing, as well as my multiple interactions with leaders in various fields.

Medical Device Industry

The healthcare industry is generally considered a stable sector of the economy. It is a very broad industry, which includes sectors such as healthcare plan providers, professional services, health facilities, pharmaceuticals, and devices. The market for medical devices is "one of the largest and most stable industries in healthcare," according to Charles Whalen, a senior analyst with Frost & Sullivan. When you think about it, it makes a lot of sense: the demand is relatively independent of the economy. In good or bad times, people need to receive medical treatments when they are confronted with medical conditions, and physicians generally need to use medical devices to treat those conditions. It does not matter if the economy is in expansion or recession, or in a bull or a bear market. A sick patient has to be treated, period. I personally did consider this aspect when I chose to dedicate my career to this sector of the healthcare industry. There are plenty of opportunities to design new devices and procedures that could save or increase a patient's quality of life. The generation of baby boomers is getting older. This means that, beyond simple stability, the industry's future is being driven by favorable demographics.

The global market for medical products and hospital supplies is over $220 billion. The medical supply industry has had a consistent growth rate of over 10% for the last several years. In the US only, the medical device market is estimated at about $60 billion in revenue, with an 8 percent compound annual growth rate. The US market alone produces half of the medical devices and consumes about 40% of all medical devices. In fact, nine of the ten largest medical device companies are based in the US.

There are three major areas where medical devices are designed and produced in the US: California, mainly in the Silicon Valley, as well as the urban area surrounding Minneapolis/St-Paul in Minnesota and Boston in Massachusetts. Large companies produce most of the medical devices. There are about twenty medical device companies with revenues exceeding $500 million. Those twenty companies alone account for about 65% of the total revenue. There are about 6,000 medical companies in the US employing a total of roughly 400,000 people. Regarding the different sectors within the medical device industry, cardiovascular certainly is the largest, with about $15 billion in annual revenues. The sector itself is mainly divided into pacemakers, stents, defibrillators, and angiography products. The main players are Johnson & Johnson, Boston Scientific, Medtronic, and St Jude Medical.

Innovation – The Life and Blood of the Industry

Innovation is key to the industry: about 30% of leading medical company revenues come from products introduced within the last three years. This represents an enormous opportunity for entrepreneurs in the field. Large medical companies usually find it difficult to innovate. They without a doubt have bright and committed people, but their complex hierarchical structure usually gets in their way when it comes to innovation. In addition, the focus of large companies is often to protect their assets, a situation which does not necessarily promote innovation. Imagine for a moment that you are working for a large medical company and have an idea for a new product, procedure, or company. You can easily imagine how it would take quite some time to convince the management team to start a new project. In addition, if a project needs to take a new direction, the response will generally be slow, assuming there is a response. There are often many parties involved to start or change the direction, and the decision process usually takes time. It takes time and a lot of energy to steer a big boat. In a fast evolving environment, by the time the decision is made, the agreed upon solution is often already obsolete. It is usually easier for small companies to innovate. They are just more agile. Those last statements are general in nature; there are large companies that are pretty good at innovating, and small companies that are structured as if they were billion dollar companies. But those are the exceptions, not the rule. In general, large medical companies often rely on acquisition of smaller companies to access new products. In this model, innovation is created externally, and then acquired. For people with an entrepreneurial spirit, this represents an enormous opportunity! In a simplistic way, if you can identify an opportunity, find an innovative solution to address the opportunity, and demonstrate to a larger entity the merit of your solution, you can generate significant value. One needs to develop a method to identify those opportunities and, as soon as they are identified to design and execute a plan in order to take advantage of them.

Culture and Structure

To innovate, it is imperative to develop a culture designed to promote innovation. Trust and openness are crucial elements for innovating; they derive from a state of mind. If you are interested in entrepreneurship or innovation, surround yourself with people whom you can trust. Be trustworthy yourself. Innovation is a mental process involving bouncing concepts and ideas back and forth. It is an iterative process. It is about analyzing, thinking, deciding, polishing, challenging, realizing, rethinking, and restarting. How could anyone successfully go through the process if trust is not there? Of equal importance, surround yourself with open-minded people, and be open-minded yourself. You need to be in this state of mind to be capable of identifying opportunities and solutions. You do not necessarily need to find all of them by yourself. Believe me; finding all solutions yourself is a good thing for your ego, but not necessarily a good thing for the project. You might be a smart person and might find a lot of solutions without any help. However, you need to realize that your solutions are tainted by your personality, your emotions, your background, your comfort zones, and limited by your fears and blind spots. Whoever you are, your solutions and ideas are not always the best for the project. This is regardless of your age, experience, status, or title. Make sure that nobody on your team, including yourself, considers his or her own self as being bigger than the project. To innovate, the project requirements and needs always have to come first. Openness and freedom to think and share are important. Everyone needs to be in an open-minded state to successfully go through the iterative process and freely share ideas. The best solutions usually come when people can freely give the best of themselves. Then, when trust, openness, and freedom are there, the most straightforward aspects such as creativity, competence, intelligence, skills, and personal commitment are necessary.

All of those aspects are important parts of a culture designed to facilitate and encourage innovation. However, they are not sufficient. The decisions to be made can be very complex, and the best solution is not always obvious. Sometimes there is not even enough time to analyze, and decisions have to be made on the spot with incomplete information. That is where the leader comes in. Someone needs to be in charge, and it needs to be clear to everybody who this person is. Someone needs to consider all aspects, and make the right calls, which are sometimes based on inexplicable intuition. The leader needs to earn the trust of the team, endorse and promote all aspects of the culture, be committed to the project, and be capable of making those calls.

At this point, you might ask if having the right culture and the right leaders is enough to be successful. The answer is no. You also need to have a structured process to identify the opportunity, define a plan to take advantage of the opportunity, and execute the plan. Based on their experiences and personalities, every entrepreneur has his or her own recipes to define and design the process.

At MicroCube, we developed and use a process where broad business aspects are analyzed in a disciplined way. To briefly summarize the process, we start by identifying a clinical need, and analyze the overall market status. We then identify as many business opportunities as possible, with the associated restrictions, constraints, and limitations. When completed, we brainstorm to define different ways to take advantage of the opportunities within the known constraints or, in other words, we define several "missions." We then compare the risks and opportunities associated with the missions, select the one with the most attractive risk/opportunity ratio, and select a mission. Then, strategies are put in place to achieve the mission, and projects to implement the strategy.

Data Collection

This all sounds well and good, but it is not very specific. So, let me try to explain in more detail. The first step is to select a field, and start with an overall market analysis. The goal of this phase is to collect enough information to acquire a deep knowledge of the current status of a field. This is like a research phase. For those of you who went to post-graduate school, this is like performing a literature review before starting the research project. Understanding the clinical aspects of the disease, with its associated symptoms, consequences, mechanisms, and anatomy, is of primary importance. I probably do not have to work too hard to convince you that if you want to be an entrepreneur in medical device technologies, you have to understand the medical conditions you are trying to treat! A piece of advice – spend a lot of time defining what the current clinical approaches, procedures, and limitations are, and never underestimate the importance of knowing the relevant aspects of the anatomy. Identify the type of physicians that are treating the patients, and understand their background. For example, if you are considering an application in the cardiovascular field, will the patients be treated by a cardiologist, an electrophysiologist, a cardiac surgeon, or an interventional cardiologist? Also, identify the patient referral flow and the decision makers. As you can imagine, it is easier to manage a situation where the users of your technology are the decision makers who can control the patient flow. If not, you need to define strategies to address the situation, and find a win-win situation for all parties involved. Identify the reimbursement options for a procedure. This will have direct implications on the overall business value. Like it or not, in many of the capitalistic nations, physicians and hospitals need to make money when a patient is treated. Also, more traditional market research aspects need to be done. What is the incidence and prevalence of the disease? What are the demographics associated with the patients? Who are the main competitors? Which technology are they using? What are the advantages and disadvantages of their technologies? What is the price range of the competitive products? What are the new research trends? Who are the smaller start-up companies in the field? What are they doing? Who are their investors? Also, knowing the regulatory path taken by your competitors is tremendously useful. If you need to start a long PMA[1] process with the FDA[2] to get your regulatory approval, the overall risk, timeframe, and financial resources needed to complete your project will be quite different from a straightforward 510(k)[3] path with no clinical data required. You will also need to map the intellectual property related to your field. Identify the main patents that have been issued, and have an understanding of the currently submitted patent applications. You will need this information to later evaluate the protectability of your inventions, and to design around patents to make sure that you are not infringing on any device claims or inducing infringement of any method claims. This will help you in defining what is usually called "freedom to operate," in the jargon of the business. You also need to understand where the money will come from if you start a business. Consider financial aspects related to the field you are interested in. Is there investment money available? How many new companies in your field have been funded in the last few years? Where did the money come from? Regardless of your decision to start with a seeding round from "angels" (basically wealthy individuals), or venture capitalists, it is valuable to have an understanding of the financial aspects at the beginning of a project.

There is nothing magical or new in what I just explained. This is plain common sense. Just make sure that you have enough information so that you can consider the situation globally in order to make informed business decisions. Analyzing the information is more important than collecting the information. If you do not have

---

[1] Pre-Market Approval

[2] Food and Drug Administration

[3] A 510(k) is a term commonly used to refer to a pre-market application sent to the FDA. The information in this application documents the safety and efficacy of the finished medical device. If the FDA deems the device to be substantially equivalent to an adequate legally marketed medical device that is currently approved for marketing in the United States, clearance is granted to market the product. Visit www.fda.gov/cdrh/510khome.html for more information.

much business experience, make sure you team up with people having a solid business background in order to make not only informed, but wise, business decisions.

Opportunities and Constraints

As previously mentioned, the second step is to identify the opportunities and constraints. To perform this phase successfully, I would advise you to divide the field into sub-categories and identify related opportunities and constraints. For example, related to clinical aspects, a clinical solution might not be available for a disease or a segment of the patient population. There would therefore be an opportunity to develop such a clinical solution. The cost of existing technology could be too high to access an international market, and the situation could be considered an opportunity to develop lower cost technology to access those markets. There might be innovative ways to take a simpler regulatory path. For example, start with a relatively straightforward and simple 510(k) path and then go through a PMA route, develop a more effective technology, solve the technical challenges from a competitor and let them acquire the technology, license an existing patent, acquire an existing company, cover a broad area related to the intellectual property and then license your rights, change the referral path, create a new reimbursement pattern, etc. Other opportunities may lie in a change of status. For example, a broad patent might expire or a major competitor might disappear.

For me, this is like creating a list of hypotheses in a scientific process. The important thing here is to have an open mind, and think globally. If you have a technical background, do not limit yourself to finding technical opportunities only. The opportunities someone can identify are very often in direct line with their background and experience. Think broadly and surround yourself with people of different backgrounds to maximize the opportunities you will consider. Identifying constraints is of equal importance. For example, the technical skills of some physicians might be limited, so if a new technology is developed, it would have to be extremely simple. The medical condition you are trying to treat might not be life threatening. Therefore, the proposed solution will have to be very safe to have an attractive cost/benefit ratio. A broad patent might exist, which would limit the solution or embodiment you can propose. Reimbursements might be low or constantly decreasing. Or there might be several large companies already involved in the field. The goal here is to identify all kinds of opportunities and constraints in order to have a broad and complete perspective.

Missions Identification

The first two phases are the easy ones. Then things become slightly more difficult. The goal of our third phase is to identify missions to take advantage of an identified opportunity, (within the known constraints), and to develop strategies to achieve the mission. In this phase, you need experience, vision, and intuition. You might wonder what a mission is. Basically, a mission is a reason for the existence of a project, a company, or an entity. It defines the essence of what you are trying to accomplish. It embodies its goals and philosophies. It is not a catchy sentence hanging on the wall; it has deep meaning and reasons. It is basically a broad and overall objective. For example, at AFx, we identified as an opportunity the fact that ablation procedures for atrial fibrillation were long and technically difficult. We then defined our mission as "pioneering a two-hour cure for atrial fibrillation." The work "pioneering" carried the sense of innovation. Copying existing technologies (commonly known as "knock off") was not an option. The time aspect "two-hour" had deep implications related to ease of use, and therefore adoptability. Then, "cure for atrial fibrillation" defined precisely which type of arrhythmia we were going after, and the word "cure" was describing the desired clinical outcome. As a strategy to achieve the mission, we decided to develop minimally invasive surgical procedures using a microwave ablation technology. The mission was simple, crisp, inspiring, and the strategy was directly connected to the mission.

A good mission defines a common objective for all involved parties, including patients, physicians, hospitals, employees, and investors. When the mission is achieved, it creates a win-win situation for everybody. Without a mission, a company runs the risk of wandering through the project without having the ability to verify that it is on its intended course. With an improper mission, the company runs the risk of going in the wrong direction. A mission is somehow subjective in nature, it could be almost anything. But after it is defined, the mission directly influences all decisions: everything has to be rationally connected to it. Defining the right mission is a difficult and challenging task! Imagine a situation when you have a superb execution strategy, a flawless execution, a committed staff, but the wrong mission. What would happen? Your organization will be like a superb well-oiled machine going very effectively to the wrong place! Defining the mission of a project or a company is critical.

For example, if you consider the lack of a therapy for a sub-group of patients as an opportunity, the mission could be to provide treatment for those patients, and the strategy could be to use a new technology to accomplish the treatment. If the spread of a technology is limited to the domestic market because of its high cost, the mission could be to provide a cure for patients in foreign countries, and the strategy could be to outsource the production overseas. If a new field is emerging, the mission could be to cover a broad area of intellectual property, and the strategy could be to seek licensing agreements with interested third parties. The same concept applies in relation to the near field: a broad perspective is applied to define different missions, and their associated strategies.

Selection of "The" Mission

The next step is to select the right mission and decide whether you want to enter whether the field. Several sophisticated methods can be used to select the proper mission. In general, you need to identify the risks and benefits associated with the missions, give proper weight ratio to the most important aspects, and select the most attractive mission. Again, an effective way is to sub-divide the field into different categories. In technology, for example, is your mission based on a new or existing technology? What are the technical chances of success? What are the risks? What is the timeframe to develop the technology? Does your team master the technology? If not, can you put a knowledgeable team together? For the clinical aspects, is there an existing gold standard to treat the disease? If not, how much research needs to be done, and what are the risks? This last is an important aspect, since a project that would require significant clinical research would necessitate more time and resources, in addition to being more risky. There is a clear difference in the scope of the task between developing a cure for AIDS versus developing tools to perform an existing proven surgery in a minimally invasive way.

To continue our list, is the intellectual property associated with the proposed concept protectable? Will you have freedom to operate? Consider all other aspects such as reimbursement, patient flow, regulations, competition, etc. Analyze all aspects and identify the best possible solution. In general, make sure that the selected mission is connected to a real clinical need. You will need to demonstrate that the selected mission and strategy is capable of producing sustainable value and offering competitive advantages. You will find a multitude of books about the subject; everybody has their own way to analyze and make decisions. But at the end of the day, keep in mind that common sense has to prevail: you need to emotionally and rationally connect with the mission, and develop a team capable of implementing the strategy to achieve the mission. You might also realize that there is no attractive mission and not enter the field. It is better to make those hard decisions and face the reality early on, rather than start a project without proper global thinking and realize the presence of a major flaw years after the project has begun. The global exercise is also useful to identify the major risks and weaknesses in your plan so they can be addressed upfront. For example, if you are aware of an existing patent you might infringe, face the problem right away and find a way to design around it, by seeking a licensing agreement or perhaps modifying the strategy. If a design is not manufacturable, you had better find out right away. Face reality and be honest with yourself. At any cost, avoid falling in the trap of ignoring data and thinking that you can deal with things down the road. This is one of the most difficult phases, but is also one of the most challenging, exciting, and intellectually rewarding.

When the mission and main strategies are selected, then it all boils down to execution.

Define, Commit and Deliver

Human and Financial Resources

If you are not experienced in business or in the medical device field, I would strongly advise you to find an experienced partner or mentor. Find someone who can complement your strengths and fill in your weaknesses. Whatever your mission and strategy might be, your best asset is yourself and your team. One of the first challenges you will face is finding money to finance your project. You will have to prepare a business plan to present your mission, strategies, planning, risk management plan, and timeframe. Make sure your business plan and presentations are clear and easy to understand. The last thing you want is to have a potential investor trying to figure out what you are really trying to do after listening to your plan. If you are not particularly gifted in your communication skills, seek help and advice from experienced professionals, or find someone else to do it for you. Having a good simple business plan is necessary, but the truth of the matter is that private investors and venture capitalists usually do not read business plans. They do not invest in business plans, they invest in people! Specifically, they invest in you and your team. Surround yourself with a trustworthy team of people who have experience, credibility, and success stories. A good business plan is necessary, but not sufficient. It is a tool to clarify your overall strategy and show that proper due diligence has been done. For an early investor, however, when time comes to sign the big check, it is basically all about you and your team. They need to have the confidence that you can commit and deliver. They need to know that you are experienced, knowledgeable, and wise enough to put mechanisms in place in order to constantly monitor your field in its entirety, and assess the validity of your mission and strategies. They need to know that your organization will be agile enough to change the strategy when necessary. They need to know that money will be spent wisely, and staffing will be done effectively. They will invest in you, not in your business plan. Make sure you are surrounded by an "A+" team to increase your chance of getting funded and to maximize your overall chances of success.

If you have a good plan and a good team, you will most likely be able to raise the money you need. You might think that the provenance of money is irrelevant. Be careful! Smart money is better than any money. When you accept money, you basically give away part of the control of the company. The first investors will basically define the genes of the organization. This could later make or break you. As much as possible, only accept "smart money." Smart money comes from knowledgeable and trustworthy investors who can add net value to your organization. Look for investors capable of bringing expertise that you do not have, and that you need. For example, smart money could bring a deep market understanding to your organization. It could bring management expertise, credibility, clarity, perspective, and connections. Before accepting any money, you should perform due diligence and verify the potential investor reputation. Find out about their earlier investments and assess whether and how they added net value to the organization. In addition, figure out whether they have a repeatable and predictable history of high valuation successes, or whether they usually tend to liquidate companies at relatively low valuation. Furthermore, find out how the founders, management team, and employees of their earlier companies were treated at exit, and how the negotiation took place. Were stock options distributed fairly? Were the companies sold at a fair market value? Were the financial outcomes satisfactory for everybody? The best scenario is to find investors committed to finding the best "win-win" situation for all parties. If you have a good plan and a good team, your negotiation power will be increased, and you will be in better position to accept smart money only. Smart money is available, but you need to find it! It is like finding your soul mate: you know he or she is there somewhere, but sometimes you need to look hard in order to find him or her!

Leadership and Execution

When your mission and strategies are in place, and financial resources are available, then it all boils down to execution. You could have the best plan in the world, but what is the value of it if you cannot execute? You need to build a team of individuals who can commit to a task and deliver on time. This has to be the basis of your culture. Empower people who can deliver. Establish crystal clear goals, responsibilities, and due dates. If you want to manage a project, make sure you constantly monitor what is happening. Be connected with reality. The worst thing you can do is to assign ambiguous tasks and then disappear for a few weeks, leaving the team without a leader and hoping that they will all figure it out. If you take that path as an entrepreneur, you are heading for failure. Instead, be present, communicate with your staff, know what is happening, what is working and what is not, the challenges, and the requirements. Ask yourself how you can further clarify the objectives, optimally re-assign resources, reduce risks, confirm hypotheses, and improve strategies. Successful entrepreneurs are not passive – they take charge. To use an all-American example, they act as owner, general manager, coach, and quarter-back, all at the same time. They define, commit, and deliver, consistently, predictably, and effectively. They add net value. They have a vision, and transform resources into high-value products or concepts. You also need to teach, train, and mentor your staff, so that the responsibility is not entirely on your shoulders. This may sound totally crazy and counter-intuitive, but my goal as a leader is to make myself not needed in the project. If I can succeed in training people, and if they can become leaders themselves, then the team's capabilities will expand and reach the next quantum level, and the chances of success will be maximized.

An entrepreneur should ensure that the organizational chart is horizontal, especially during the early phases of a company. Do not try to imitate large companies and play the big guys by hiring several vice-presidents and directors, with just a few people actually producing value. Start-up companies with vertical organization chart are usually ineffective and highly political. This is not the culture you want. Make sure everybody is directly adding value.

For the technology development process, use common sense. Look for a *clinical pull* approach, rather than a *technology push*. In a clinical pull approach, the development is the consequence of a clinical market need. You need to identify the need, produce prototypes, prove your concept, and then manufacture. In such an approach, the clinical need drives the innovation, and the technology selection is made in a way to optimize the solution. The opposite approach may not give you satisfactory results. You do not want to develop a new technology and then figure out how it could be used to address a clinical need. It might just work in some instances, but, with such an approach, you are basically relying on luck to succeed. If that's the case, you would be better off going to Las Vegas – it takes less time, and it is a lot more fun!

Market Introduction

There are many critical steps between technology development and market release that I will not cover. When your technology is cleared by the relevant authorities to be marketed, you need to define sales and marketing strategies. This usually marks the beginning of a new era for your company. It is important to use a step-by-step approach. If you have an R&D (research and development) background, you can compare it to the early manufacturing phase. After you build your first prototype, can you say that you can manufacture thousands of units? Of course not. Similarly, you are not necessarily ready for full market release after you sell your first device. And selling ten devices to an account does not mean that the account adopted your technology. They are still just getting familiar with it!

A technology adoption is a commitment to your technology. It is a change of behavior, a change of choice. When your technology has been truly adopted by an account, the sales are consistent and predictable. For example, you will consistently see an order for twenty devices a month coming from the same account over and over again. The consequences of adoptability are predictability and measurability. You need to figure out and truly understand patterns that will lead to adoptability for your technology. Then you can go to full market release with an effective roll-out plan.

It is not necessarily a wise strategy to develop a nation-wide sales force right after, or just before, your regulatory clearance is obtained. During the research and development phase, you need to perform a design verification and validation with a pilot build before you can truly talk about manufacturability. Likewise, you need to do a smaller scale roll-out and understand buying patterns before talking about large scale market introduction. Such a disciplined step-by-step approach to understanding the path to adoption is also very useful if you are considering an exit by acquisition. Usually, the large medical companies do not care much about your absolute sales number. They are more interested in understanding the patterns and path to adoptability. Even if you only have the information on a limited number of accounts, they can easily take the information, use it as inputs, and scale-it-up in their own models. If they have the right pattern information, they will figure out their sales projection quite precisely. If you can demonstrate the conditions necessary to minimize the duration and resources to reach adoption, the potential acquiring companies will notice, and you will have created significant valuation and opportunities for exit. However, as for any other phase, not executing your early sales and marketing effort in a disciplined fashion can cost you a great deal of time and money. An inadequate sales and marketing strategy can end up costing a substantial amount to generate some revenues, without significantly increasing the overall valuation of your company. It is not always advisable to "play the big guys" and go for the home run early on. It might be fun and good for the ego, but it is not necessarily good for the company. Again, having a good team is important in analyzing the options and making wise business decisions. Figure out your key metrics early in the process, and then track them down in a rigorous fashion. In a sense, this is not different from any other phase or activity related to a company or project.

After you have completed that phase successfully, you are probably ready for the exit. That is where having smart money is critical. You want to have the right people on your team to negotiate the best outcome. When you have a clear mission, a committed team of trustworthy people with a sense of urgency, a culture that allows and forces people to face reality and take corrective actions, and effective execution skills, then the sky is the limit: you can achieve great accomplishments. So, take charge and go for it!

Conclusion

Life as an entrepreneur is demanding and sometimes difficult. It is also challenging and rewarding. There are many opportunities out there, and, if you can use common sense with the right thinking processes to make decisions, you could make a tremendous contribution for yourself and for society. The ride itself can be as enjoyable as reaching the final destination. What a beautiful and successful career it could be. Enjoy!

14 Entrepreneurship in Pharmaceutical and Biological Drug Discovery and Development

Rabbi Robert G.L. Shorr, PhD, DIC

Cornerstone Pharmaceuticals, Cranbury, New Jersey and Stony Brook, New York; Altira Capital and Consulting, LLC., Edison, New Jersey; Center for Biotechnology, Stony Brook, New York; Metabolic Research, Inc., Montgomery, Texas, USA; and BrainStorm Cell Therapeutics, Tel Aviv, Israel

I am delighted that you are considering a career in pharmaceutical development and biological drug discovery. I wish you every success because with your success, you will have helped to make the world a healthier place. To save a life is to save a world.

As you begin considering a career choice in the pharmaceutical development or drug discovery life sciences, or in related engineering and information technology, the possibilities will seem overwhelming. Even so, there are many overlaps in methods, equipment, basic technology, and training requirements. A strong background in one area can easily serve as a launching point for a career in another. Unique methods, approaches, and applications of clinical and scientific knowledge in one area can cross fertilize ideas and push the traditional envelope. Academia and industry, while not mutually exclusive from a technology perspective, have very different objectives, although more and more frequently discoveries made in academia serve as the basis for new pharmaceuticals. Both academia and industry have primary and subcultures to contend with, even from one campus or department to another or from company to company.

In this chapter, I intend to describe the possibilities available within the domain of biotechnology and biological drug discovery research and development from the perspective of my own life experiences. I hope to give you an idea of how personal self and career paths might evolve. I would especially like to address the entrepreneurial culture within biotechnology and drug

G. Madhavan et al. (eds.), *Career Development in Bioengineering and Biotechnology*,
DOI: 10.1007/978-0-387-76495-5_14, © Springer Science+Business Media, LLC 2008

development, which ultimately offers "a sky is the limit" path for scientific and career development. Some of the areas of responsibilities are outlined below.

- *Basic Discovery Research:* Basic discovery research typically seeks to understand the mechanisms of physiological function both in normal and pathological states – that is, the *hows* and *whats* of life. This is the place to go where no one has gone before, and to experience the thrill of discovery, as did the explorers of old. A simple question might be: how do muscles and nerve cells communicate, and what controls the interactions? Discoveries emerging from this type of question include the structure of the synapse, the biochemistry of neurotransmitter production and release, receptor interactions, and trophic factors that maintain health. Practical outcomes can include novel materials used in anesthesia, or drugs for treatment of peripheral and central nervous system disease. In my own experience, I began with a very strong interest in neurobiology as an undergraduate at the State University of New York at Buffalo, and participated in a work study program that gave me lab experience isolating and characterizing acetylcholine receptors. These experiences have proven invaluable throughout my career, as my mentors not only saw to it that I learned experimental techniques, but also how to plan and prioritize experimental details. I am grateful to Professor E.A. Barnard and Professor J.O. Dolly for those gifts. In choosing an internship or work study program, be certain that you will be guided and not left meandering, wondering what to do.
- *Product Development, Manufacturing (chemical synthesis, analytical chemistry, biotechnology, bulk product, final fill and finish, packaging), Quality Assurance and Control, Clinical Investigations, Technology Transfer, and Translational Research:* Each of these job functions relate to the particular details of the production of ethical drugs and testing for pharmaceutical use. The industry is heavily regulated by the FDA and other worldwide regulatory agencies to protect the public from mistakes or deceptions. A great deal of skill and discipline is required in each of these functions, as the products will ultimately be used to treat patients. Aside from the scientific aspects of the work, a great deal of paperwork is required to document and validate each step of the drug production development process. Those choosing these specialties in preparing drugs, as well as those choosing an FDA career path, can take pride in a system that assures the highest quality products for those who need them. Are you a person who enjoys task completion and checking off boxes as products are developed and manufactured? These areas may then be for you.
- *Strategic Planning and Project Management at Corporate/Institutional/Departmental Levels:* The most important questions ever asked in a business or institution are where the money going to come from and how it is going to be used. In a for-profit business, investors may purchase equity shares before there is any income. Many institutions rely on government or philanthropic grants.

How are these grants going to be obtained and money managed so that important milestones are met? It is here that success is defined, along with milestones that demonstrate progress. It is also here that progress against plan, timelines, and budgets are measured. Those choosing this career path will have the responsibility for making decisions that will affect the entire organization. The higher salaries here are accompanied by higher risk. It is extremely important to define success and outline the critical path of activities to achieve it. Anything else, although of interest, may be a distraction of time and money. To define success, look to groups that have gone before you. What made them successful?

- *Discovery, Formulation, Marketing and Sales of Ethical Pharmaceuticals or Nutritional Supplements/Business Development (buying, selling, mergers and acquisitions):* It is here that interactions with other people outside your own company or institution will be extensive and important. At the end of the day, everybody is selling something, either to a public clientele or business to business. Business is a contact sport. If you like people interactions and negotiations, including closing the deal, this may be for you. On a side note, there are those who are better at engaging interest in a deal than actually closing one, and the converse. In negotiating, think of the bigger picture. Win-win situations make for better long term relationships and future business. Build bridges.
- *Private or Public Financing, Investment Banking, Venture Capital, Investor/Public Relations, and Training and License as a Broker Dealer:* This is where you are likely to review companies at various stages of development, from simple technology start-ups to more mature organizations. These may be private or public companies and subject to numerous rules and regulations. It is here where a great deal of money can be made quickly. However, there is also risk. The fastest way to make a small fortune is to begin with a big one.

All of these specialties have subspecialties, niches, and distinct work environments and cultures, as well as career paths with differing levels and types of compensation. All require interaction, cooperation, and collaboration with your forerunners, colleagues, and mentors. If your interpersonal skills are poor, your career, simply put, will suffer. If people like you, they will forgive you anything (in most cases). If they do not, they will forgive you nothing. At the same time, a lack of sincerity will be obvious to all. You never have the right to be arrogant.

Does a career choice seem overwhelming? It should. You are stepping into a journey with an inability to predict the future. As you travel, remember this: no one has any power over you unless you give it to them. In addition, everyone needs an angel. Seek out mentors. As you grow and learn, you will change. Your life choices appear to be entirely in your hands. They are not, and you will find on reflection that you made choices but were guided in one way or another by events and circumstances beyond your control. The choice you do have is in deciding who you are and what you want to be. Much will depend on your ability to focus and finish. Define finish, and draw strength and confidence from each baby step forward. Do not be a person who quits at the first difficulty. Be strong and determined. Everyone gets knocked down. Will you get up and begin anew, or melt into the dust?

So, what is step one? What are you excited about? Ask yourself why. Talk to people who are doing what you consider your passion. Intern in different career and industrial settings to get a feel for how you would fit. Determine what tools you need to be in the game, then get them. Pay the price of admission. Did you have a chemistry set, telescope, or microscope as a child? I did. I knew that I wanted to have a lab and to conduct applied medical research. I also knew the price of entry would be a PhD. Beyond that, things were rather vague. As an undergraduate, until I was able to begin a work study program to pay tuition and expenses, I worked as a janitor, grocery clerk, bar tender, and guitarist in a rock and roll band. I was absolutely determined. Even so, in my early undergraduate years I nearly flunked out, and some professors advised me that I had no future in science despite my ambitions. How did I solve my inability to focus my efforts? I got married. At the time of this writing, nearly 34 years later, it is still one of the smartest decisions I ever made.

Credentials

As above, each career path will have a "ticket price." What you intend to do will dictate the credentials you will need. The world is also changing, and a typical bachelor's degree at a well known university may not have the same value as a certificate or degree from a program that trains you in a skill set. Check career forums for the areas that are of interest and stimulation to you. What are the requirements? For example: if you want to do research in someone else's research laboratory, and see yourself as a lab bench researcher or instrument specialist, then a bachelors, masters, or training certificate and license may open the door. Your hands-on skill set, ability to design studies, implement them, interpret results, and communicate in written report format or in presentations will be critical to placement and compensation level. If in industry, development knowledge of federal regulatory guidelines is extremely important. If you like getting things done and checking off boxes, then development may be a better choice than research, which may be more open ended.

If you want to do research in *your own* lab, then a PhD, MD or the combination, will be the price of entry. In the latter case, keep in mind that it is better to emerge from a highly respected lab in a lesser known school than from a lesser-known lab in a well-known school. Further, build up your interpersonal network and establish *sincere and honest* friendships and collaborations. Smile, and never be anything less than intellectually honest. As you move from an academic lab into industry, you will find a dramatic change in work culture, regulations, and even dress codes. Are you comfortable in a regulated environment with work schedules and timelines? Make sure you know what is expected of you. It may be different than what appears on the surface as the company or lab goals.

To Be an Entrepreneur: Uncertainty

It is difficult to know if an entrepreneur is born or made. I would describe such a person as having a "fire in the belly" to bring change into the world via the creation of new applied technologies that can ignite entire new industries. The drive cannot be stopped, and such a person (often eclectic) is often on the cutting edge of his or her field. To quote my dear friend Tia Hutchins "*An entrepreneur who is not living on the edge is taking up too much room.*" Certain events may bring out the entrepreneur in an individual. Job loss, a bad boss experience, or a creative, encouraging environment have all been historically known to awaken the beast. What does it take to be successful? Obviously, a vision and technology to expand or create new markets, and the ability to get others to believe in you and help you. The latter is best achieved by having apprenticed and established a successful track record. Even more, you will need to be able to tolerate life on the high wire, and risk.

While job security is an illusion, not everyone has the ability to withstand situations of higher degrees of uncertainty. Do you have the fire in the belly to be an entrepreneur with all of the risks involved? Would you mortgage your home to self finance? Do you have a substantial track record and references from having worked for someone else to establish a credible track record with investors other than friends and family? Is the market willing to take substantial risks with investment capital in a field so new that a beach head is sought even with the youngest of specialists (For example biotech in the 1970s, or the Internet in the 1980s). Are you willing to sign up with a start up biotech or pharmaceutical company, or do you need the security of an established infrastructure and a large company? Like the Lone Ranger and Tonto, keep your ears to the ground listening for what is coming down the road, but not so closely without paying attention to what is happening now that the stage coach runs over your head. Develop the skills to cope with uncertainty. These will be assets for the rest of your life. If you have had a challenging or difficult life, do what you need to get over it and always move onward and upward. The fastest way to become an entrepreneur is to join such a company or group.

Core Values and Self Identity

When I was an undergraduate student at the State University of New York at Buffalo, I was thrilled to be studying neurobiology with a research group directed by Sir John Eccles (Nobel Prize Laureate in Physiology or Medicine, 1963). I sought him out and read his books extensively. In each and every one of your experiences, take the opportunity to learn about people, decision making, and behavior. Seek out the leaders, movers, and shakers. To learn from your own mistakes is smart, and to learn from those of others is even smarter. I paid attention to what worked and what did not for others. I tried on different personality aspects of people I admired to see what fit me. Each aspect was an experiment to learn more about myself, and what felt comfortable to me. I accordingly modified that aspect to suit my personality. I have been working in pharmaceutical/biological drug discovery research and development for more than thirty years. I am still learning, and expanding into corporate financing, business development, and the full gamut of the pharmaceutical industry from concept to standard of care. More than thirty years later, I have been privileged to lead or be a contributor to teams that have brought to market life-saving and quality-of-life improving drugs that are considered blockbusters (sales > $1 billion), along with nearly 200 patents issued and pending worldwide. I have helped finance companies with more than $150 million which today are considered among the most successful on the pharmaceutical/biotech landscape. I hope these words do not sound arrogant or boasting. I have been very fortunate to be blessed with numerous life gifts, including exceptional teachers and mentors. I do like to think that I am just getting warmed up.

As an undergraduate in one elective class, "The Individual and Health Care," I found that the course focus was on self and interpersonal communication. The primary text was "I'm OK, You're OK." This was an extremely important course. The ability to communicate, and I mean listening and understanding as well as being understood, is perhaps the most important skill to work on, no matter the career choice. In my case, every day I came home from school my mother would ask what I learned. I did not shrug off the question but answered, usually about science, until we understood the subject. The first languages in my home were Portuguese, Yiddish, and English. English was a new language for my mother. I even took my mother with me to some of my university classes. She taught me how to speak science to a lay audience, and I am forever grateful.

As to the "I'm OK you're OK" course, I still recall the first day when about 15 to 20 students were asked to sit in a circle, introduce, and identify themselves. Each person gave their name, year, and major. That is until the young woman next to me announced her name and her identification as a lesbian. Needless to say, that was a tough act to follow. So I did as the others had done. On reflection, I realized that the young woman had answered the question of self identity far more deeply and honestly than the rest of us. I realized that the Who We Are question takes a great deal of self exploration and discovery that only life in combination with guidance and experience can be answered. For me the honest answer was to be found in rabbinic studies. What we choose to do is NOT who we are, but "who we are" will dramatically impact what we decide to do, and HOW we do what we do, and the moral and ethical choices we make as individuals, families, communities, and nations.

Who am I?

Aside from the genes you came into the world with, you will have been taught and guided by your extended family, friends, and surrounding cultures, religions, and environment. If you want to know who you are, ask these questions:

1. What do you value? Honesty, integrity, money? Rank, power, prestige, and honor? Fame? How deep are these values? Do you take home office supplies for

personal use? What will you compromise for fame or fortune? Whatever you choose, know that your money will leave you at the gate to the cemetery and your friends will depart at the grave site. Only your acts of kindness will accompany you farther.
2. What do you do with your spare time? Pursue activities that interest you? Why do you choose them?
3. Are you smiling and friendly at work? What *drives* you toward what you would like to accomplish at work?
4. How do you treat your family? Is your home a house of horrors where your children and spouse fear your temper? If you are angry, ask yourself why. Anger is one of the most powerful of teachers.

Does Your Career Choice Suit Who You Are?

Below, I have tried to summarize guidelines for finding and succeeding in a career within the realm of life sciences engineering and technology.†

1. **The Quest:** The status of the economy both in the United States and elsewhere will have a profound effect on the availability of jobs. Tough times, cutbacks, recession, and unemployment can create intense competition for jobs. A growing economy and expansion in diverse fields can create opportunities for jobs that did not exist before, and a shortage for people with the right skills for the field you have selected. In today's economy, what is the career outlook? Check into the placements of recent graduates in your field. How did they do, and where did they go?
2. **The Approach:** You will need to prepare a resume that highlights your accomplishments made relevant to the field you wish to enter. Do not underestimate the importance of networking with friends and colleagues regarding your job search. In your written materials and personal interviews, project confidence and enthusiasm. Never negotiate salary. Negotiate the job. Companies will have salary guidelines that are linked to job descriptions. Impress your interviewer that you can do more than the job you are interviewing for. Who knows what may be available. Does your job choice provide a steady income? Opportunities for advancement? Increased training?
3. **The Qualifications:** We have already discussed credentials. What is even more relevant is experience. Do not minimize your contributions.
4. **The Motivation:** Are you looking for a 9–5 job and then going home, Or are you enthusiastic about what you are doing? You should love your work. Interviewers and colleagues will know the difference, and it will affect your career path. Even if you hate work, push yourself to have a positive attitude. Fake it until you make it (this is truly a viable philosophy). Are you loyal to your employer? Is the loyalty returned?

† Developed from M. Goldberger, *Bitachon Guidelines for Finding or Succeeding at One's Job* (1992), Staten Island, New York.

5. **Persistence:** Keep pushing until it gives. Follow your dream, even if you have to do something else to pay the bills. Keep in mind that dark clouds come with silver linings. Select the job that matches your personality. It will come your way.
6. **Foresight:** Train yourself to pay attention to the effect your words and actions have on other people. Are you communicating effectively? Prepare for future possibilities. If you have experienced failure, what can you learn from it? About yourself? Others? Have realistic goals and timelines for what you define as success. Is your personality type compatible with the corporate culture you are looking to work in?
7. **Determination:** Are you prepared to stay the course and pay the dues to get where you want to go? If you fall, will you pick yourself up and go at it again? Do your work honestly.
8. **Patience:** It can take time for the right job to be available at the right time and place to meet your needs. Be patient and prepare ahead.

Do You Have What it Takes to be an Entrepreneur?

What you will require is an idea that you have refined in discussion with others and believe in wholeheartedly, a willingness to work long and hard, the support of your family, and angel investors, grants, or other sources of funds to build your business. Make sure your objectives and plans are realistic. The more money you need, the more you will need to convince others to invest. Start with proof of principle. Make sure you understand patents and intellectual property law. Most importantly, seek out a mentor! Do not neglect government and university support systems. You will need a great deal of confidence. It is OK to be afraid. Courage is what you do when you are scared.

As you grow in your career, be careful not to define your identity as your job. If you do, then who will you be when you retire?

Part III

Innovative Alternative Careers in Bioengineering and Biotechnology

15 Human Implantable Technologies

Joseph H. Schulman, PhD

Alfred E. Mann Foundation for Biomedical Engineering, Valencia, California; Department of Biomedical Engineering, Viterbi School of Engineering, University of Southern California, Los Angeles, California, USA

If you like technology and want to make better gadgets that make a difference in the lives of people, this is the field for you! When I was 16, I enjoyed building electronic gadgets. If you are into making electronic, mechanical, or chemical things for the fun of it, a career in human implantable technologies will satisfy your puzzle solving skills and give you great pride in your work. If you like near future science fiction, you can make some of it come true.

I have been working in the field of implantable medical technology since 1969. It was my first job out of school, and I am still in that same job. In March 1969, I went to work for a prolific individual named Alfred Mann, who was starting a pacemaker company. Each implantable device led to another. Over a dozen companies were initiated since that time, and the process has not stopped yet.

I was asked about five years ago to list all of the items I have to take into consideration when we design a new implantable device. I came up with a list of over a hundred items. I randomly start the list with the initial considerations of technical feasibility, need, usage, and market size. Some items are repeated in different categories. Each person will usually start from the area he or she likes best, or the area that he or she thinks will give the most trouble. This would also depend on the expertise of the person. For example, whether the individual's expertise is mechanical, electronic, chemical, materials, packaging, business, finance, sales, or some other area will determine how a project is started. Everybody will tell you why their part of the project is important, and that without taking their part into account, the whole project will fail. The technical person will tell you that if it does not work, the project will fail. The financial person will tell you that if you cannot finance it, it will fail. The sales person will tell you that if you cannot sell it, it will fail. The lawyer will tell you that if you cannot guarantee ownership and government approval of the project, it will fail. The truth is that you must look at the total project and make sure that all aspects of the program can be made successful.

G. Madhavan et al. (eds.), *Career Development in Bioengineering and Biotechnology*,
DOI: 10.1007/978-0-387-76495-5_15, © Springer Science+Business Media, LLC 2008

Two last comments before I start the list:

A. A major change (improvement?) in all implantable devices over time is reduction in size, (i.e., miniaturization), and addition of complexity in function. We went from 11 transistor pacemakers in 1973 to multi-thousand and probably many million transistor gates in modern pacemakers.
B. The pacemakers might last about ten years until the primary battery wears down (or until a rechargeable pacemaker has outlasted its patient), but in two to three years all pacemaker models always become obsolete with regard to new sales, and we need to make better, smaller, reliable, and more functional pacemakers to stay ahead of our competitors.

Considerations for Implantable Technology Development

My boss, Alfred Mann, stated in a presentation to the US Congress that the initial items for a successful business include the following:

1. Finance
2. Finance
3. Finance
4. Finance
5. Finance
6. Finance
7. Finance
8. Finance
9. A sound market
10. A good product

That's a terrific list. But below is an abbreviated version of the various items I also like to take into account when designing a new implantable device.

1. **Application:** What disease(s) or illness(s) are you treating? For example, a cardiac pacer treats heart block, a cochlear implant treats deafness, and an insulin pump treats diabetes.
2. **Novelty:** What improvement are you making over the competition with this application? Is the device smaller, more user friendly, easier to implant surgically, avoids a major problem?
3. **Significance:** Can the customer justify the change to your new technology? The increased price? The additional complexity in programming? The convenience to the patient?
4. **Selecting clinician:** What type of clinician will be involved in selecting the treatment? For example, a cardiologist determines that the patient needs a pacemaker. An audiologist determines what type of cochlear implant is being used.
5. **Implanting clinician:** What type of clinician will be performing the implantation? A cardiac surgeon actually implants the cardiac pacemaker. An ear-nose-throat (ENT) surgeon implants the cochlear implant. Are there other doctors who can fill one or both of these roles?

6. **Market size:** How many potential patients are there? How many more patients are developing the problem each year who will be treated by this technology?
7. **Product cost and reasonability:** How much will the implantable device cost? Is this cost reasonable for the obtained improvement? There will be a need to know the R&D (research and development) cost, the manufacturing cost, and the final production costs.
8. **Who will pay for it?** There are many possibilities. An insurance company, the patient, the government, or charities are among the first things that come to mind.
9. **Who is the customer?** This is not such a simple question. For a cardiac pacemaker, the doctor selects the implanted device, the hospital orders it, inventories it, has to pay for it, and to get somebody else to reimburse them. The reimburser is the patient, an insurance company, or the government through the insurance agencies. Also, which doctor selects the pacemaker, the cardiologist or the cardiac surgeon? Sometimes different groups of doctors have competed for this function.
10. **Technical feasibility:** Technology is moving rapidly forward. (See Figure 15.1) One must make sure it is possible to make the desired device. If you make a giant leap and are too far ahead of your time, it will most likely take longer, but if you complete the giant leap, you will most likely have a monopoly. If you make a small leap forward, it is possible that others are out there trying to do the same thing. This has to be weighed carefully.
11. **Unique product versus better mousetrap:** If you make a unique product, you need to educate your customer in what your product does, how to use it, and why it is so wonderful. This is a major effort if you have a hundred thousand customers. If you make a better mousetrap, you don't need to educate your customers. The old mousetrap salesman had to do that job. You simply start selling to your competitors customers, all one hundred thousand of them. It is a much easier job. I mention this because I have heard many company presidents lament that they were not first in selling a certain product, and had to compete. However, when I gave them a unique product, they had a more difficult time trying to educate every single customer as to why this was a great product, and had to train them in how to use it. Education became a major budget issue.

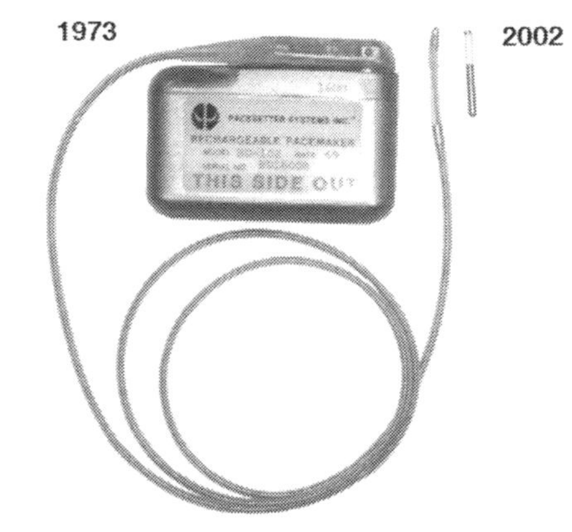

Figure 15.1 Fixed rate rechargeable cardiac pacemaker, released in 1973, next to a programmable rechargeable injectable microstimulator, released in 2002. The microstimulator can be programmed to the same stimulator parameters as the pacemaker.

12. **Device life time:** There are several factors that determine the lifetime of the implant, such as (a) the internal battery lifetime, (b) hermaticity (the time it takes water to leak in and create a bubble drop), (c) time for corrosion to destroy the ceramic case, (d) time for corrosion to damage the metal leads and/or electrodes, and (e) other potential failure modes such as purple plague, (a degradation of wire bonding between aluminium and gold).
13. **Biocompatibility issues:** For an implantable device, there are two kinds of biocompatibility issues. The first is the potential damage to the patient by the device. Obviously this must be zero or negligible. Second is the potential damage to the device by the patient. The body has mechanisms to prevent foreign items from being inside the body, and to destroy or seal them off if they get into the body. The potential damage to the device or to its functionality must also be zero or negligible.
14. **Package designing:** There are many aspects to the designing of the package. Note that there are two worlds here, the world outside the hermetic seal, and the world inside the hermetic seal. Outside the hermetic seal, one must assume that salt water will pervade all areas. Inside the hermetic seal everything will be very dry. If there is a device filled with liquid inside the hermetic seal, then the liquid container must also be hermetically sealed. For example, this would be the case for a wet electrolyte battery inside the hermetic seal. Besides galvanic corrosion, there are other dangerous mechanisms that can occur. Migration is the tendency for certain metals to move along insulated paths as if they are flowing. Sometimes this flow is stimulated by electric currents, and is called electrolytic deposition.
15. **Sterilization methods:** There are many ways to sterilize a device. Some of the methods will damage a complex enzyme chemical, for example, glucose oxidase and some will damage an integrated circuit. What is hard to find is a sterilization method that will sterilize both an integrated circuit and a complex enzyme without damaging either of them. Radiation such as x-rays or gamma-rays normally damage a chip. Poisons such as ethylene oxide normally damage a protein.
16. **Surgical implantation:** The method of surgical implantation is dependent on the size, shape, and design of the implant. If the method is very minor, such as a simple injection, it is possible to perform the entire procedure in an office by a single skilled doctor in a few minutes, and the wound can be closed with a Band-Aid. If the method is complex and time consuming, such as connecting to the wall of the heart with a large device and several wires, it could require a hospital operating room, a team of operating room personnel, and require many stitches to close the wound. Today, many surgical procedures are being converted to minimally invasive procedures. For example, the micro-stimulator in Figure 1, which can be injected, can replace the cardiac pacemaker, which required a deep surgical cut to locate the blood vesicle and the making of a surgical skin pocket to store the pacemaker. One needs to design the implant for a specific implant procedure.
17. **Power sources:** Wires through the skin to an external source, implanted batteries, and transformer coupling have been used in cardiac implants since 1958. Primary, rechargeable, nuclear, and bio-galvanic batteries were used at different times. Today, lithium-ion batteries are used in most commercial pacemakers. These require surgical replacement about every ten years. Experiments using piezoelectric generators powered by the mechanical motion of the heart, and

solar power using external light through the skin, have demonstrated that these other methods are feasible. Recently, rechargeable lithium-ion battery stimulators for spinal cord stimulators and microstimulators have become available.

18. **Communication methods:** Today, most biotelemetry communication methods use transformer coupling. Recently, a radio channel at about 403 MHz has been made available and has begun to be used for medical biotelemetry. Light, sound, and electrical conduction have also been used experimentally to communicate with implantable devices.
19. **Shipping and storage:** Implantable devices must be boxed up for shipping in a sufficiently strong container that the tumbling during shipping cannot damage the device. Most major shipping companies have a testing method that involves shipping the package several times back and forth across the country. The greatest danger is a large temperature change during shipping. Most batteries cannot tolerate temperatures above 50 °C. A simple disposable temperature monitor, such as a dot that changes color above a certain temperature, can detect this potential problem.
20. **Accessories:** Practically every implantable device needs some type of accessory for special or normal situations. For example, a rechargeable implant is normally shipped with a wall charger. However, if the patient happens to be a truck driver, he may need a cigarette lighter-powered charger. For devices with leads that are too long or too short for a specific application, the implanting physician may need a kit to safely lengthen or shorten the lead. Also, if the device is a better mousetrap, it would be convenient to be able to safely attach the device to a competitors lead.
21. **Clinician's tester:** The initial implanting physician can run into a problem that at first glance is indistinguishable from a failed device. The best situation is for the initial implanting physician is to have an automatic implant device tester, so he can immediately determine if the problem is in the implant or something outside of the implant that he is in control of. If you did a good job, most of the problems will not be with the implant.
22. **Animal and lab testing:** Testing in the lab and in animals might reveal a dangerous situation. Dangerous situations like the circuit breaking into an oscillation when the battery runs down, or at the beginning of charge up, must be determined. A high frequency oscillation out of a cardiac pacer can be deadly, just as direct current in the body.
23. **Brainstorming:** The most valuable team activity is to review all of the dangerous situations that can occur and how to test for them.
24. **Documentation:** From the first day, every meeting that is held should have an assigned person taking notes, and also the chairperson should take notes. The two sets of notes should be combined and reviewed at the next meeting to make sure that nothing was left out.
25. **Patents:** Before any paper is given or submitted, a provisional paper with that information should be sent to the patent office. The initial simple concept is the most valuable one to patent. To evaluate a patent attorney, ask him questions about previous patents that he wrote. The attorney should understand the patents with respect to an understanding of your field. If the patent lawyer asks you many questions, that is good. If the attorney asks you the same question four times related to something he or she should know, look elsewhere.

Strategies for Success

Pathways to enter the field: Since nearly all major engineering and science fields, as well as business, accounting, and legal fields are involved, one can enter bioengineering from an existing field. The most difficult place from which to enter this field is an education in bioengineering without a specialization in something else. Bioengineering by its very name is a multi-disciplinary field.

A wide open frontier: Most medicine, and a great deal of biology, is more an art than a science. Training in a specific area is very important. In this field, a Nobel Prize was given simply for developing the voltage versus current curve of a nerve membrane. Many, if not most, Nobel Prizes in physiology or medicine were given to non-biologists.

Helping mankind: Unfortunately, in many cases, people still die before we figure out that something is dangerous. Simple things like insulin pumps, pacemakers, and intravenous systems had to wait decades for somebody to come along and do it. If you are familiar with present day technology and then walk through a hospital, you will find a great many things being done with antiquated technology. The medical field is so burdened with maintenance of life that very little improvements come directly from the field. Bioengineering is a ripe field to grow new technology. This is one field that will make you feel good just to see it in operation and to help people.

Concluding Remarks

The rewards of a career in implantable technologies are many, and are received at unlikely places. Each time I accidentally come across one of these rewards, I am uplifted for days at a time. It makes me feel good just writing about it. Here are a few incidents.

The first time I received one of these rewards was when I was walking through the large lobby at Stanford University Medical Center on some unrelated business. I saw a cart being pushed along with one of our pacemaker programmers on it. My pacemaker was being used at Stanford!

Once while walking through the airport in Chicago, I saw a meeting going on of parents with children who used our cochlear implants. I stood quietly and watched for a few minutes, then walked on with a tear in my eye.

Another time, my wife and I were invited to a family birthday party of a neighbor who had been our daughter's teacher. When we walked in everybody clapped because three of her relatives were wearing insulin pumps or cochlear implants that I had helped develop. I had no idea.

A few months back I saw a picture on the front page of the *Los Angeles Times* of a stretcher coming off a military plane carrying a woman reporter who had been hurt in a bomb explosion in Iraq. On the stretcher piled on top of her and around her were several of the triple channel computerized intravenous pumps we had developed. I realized that they could not easily insert the needles into a vein on a vibrating airplane, so they installed all the needles that they required for the long flight, and the caretaker simply had to turn on each pump as needed for each drug or solution.

Today I am working on making the "Star Wars" hand!

16 Specialized Careers in Healthcare

Charles H. Kachmarik, Jr., MS

Salem Solutions Corporation, Salem, New Hampshire; Paradigm Physician Partners, LLC, Fairfield, Connecticut, USA

To address the question of why you should choose the healthcare sector for your career, let me begin by sharing my own career path. I started my career in public accounting, having an undergraduate degree with an accounting major and a graduate degree in management science. Healthcare was not on my list of careers to pursue, although I had spent my first few years in many different industries, including healthcare. During those early years, I was exposed to manufacturing, wholesale, retail, broadcasting and journalism, and numerous service industries. As I gained an understanding of the business issues involved in each industry, I became more and more fascinated with the complexities surrounding healthcare systems. The business issues were significant, and at that time it was an industry on the verge of a growth trend that would outpace most sectors of the economy. The year was 1969. The Medicare[1] and Medicaid[2] programs, governed by the United States Department of Health and Human Services, were only a few years old. The industry at that stage had limited business talent. That would change dramatically over the years.

I saw all this as an opportunity that would professionally challenge me and provide significant career growth. The initial needs of the consumer in the sector were just starting to be met and expectations were growing faster than the industry could handle. Cost of care was accelerating. New technology development was occurring at a rate much faster than it had in decades. I felt that with my strong financial background, I could make a difference. But to be successful I needed to understand the "business," that is, help people and make a difference in their lives. People may try to isolate themselves from others, but with that type of mindset, it is not likely that they would have a rewarding or enjoyable career in the healthcare sector.

The industry went from being approximately 7% of the United States Gross Domestic Product (GDP) in 1970 to 16% in 2006, equaling $2.1 trillion

[1] A US federal health insurance program for people aged 65 and over and for individuals with disabilities.

[2] A program sponsored by the US federal government and administered by states that is intended to provide health care and health-related services to low-income individuals.

G. Madhavan et al. (eds.), *Career Development in Bioengineering and Biotechnology*,
DOI: 10.1007/978-0-387-76495-5_16, © Springer Science+Business Media, LLC 2008

in expenditures. Current estimates are that spending by 2016 will total $4.1 trillion and be almost 20% of the GDP. Healthcare has had the largest growth in jobs over the last decade, more than any other industry segment.

Currently, the United States has 46 million people without healthcare coverage, and ranks 17th of the industrial nations in clinical outcomes. Healthcare spending in other parts of the world is significantly less, with much better outcomes. Currently, healthcare is the most frequently debated major political issue in the United States. The population is aging, and the need for care will grow. Talented developing nations like India and China, among many others, see the opportunity to train healthcare workers, such as information technologists and nurses, for the US market.

The US healthcare industry, which I describe as a $2.1 trillion "cottage industry," is composed of many facets, which can be categorized into five sub-sectors: patient care, suppliers, service/support, new technology development, and regulatory affairs, each of them having numerous career branches, as described below

1. ***Patient Care***, including:

   - Acute care hospitals
   - Specialty hospitals (rehabilitation, psychiatric, etc.)
   - Long-term care hospitals
   - Nursing homes
   - Assisted living facilities
   - Ambulatory care centers
   - Ambulatory surgical centers
   - Physicians offices
   - Diagnostic centers

Careers in patient care are seemingly unlimited. There are shortages in most professional care areas. From primary care physicians to nurses to laboratory technicians to pharmacists, the industry is constantly in need of more people. The industry does require licensing for most positions that "touch the patient." This creates a defined career path that includes specific education requirements, training, experience, continuing professional education, and licensing. Select your area and there will be a professional organization that represents it. Information on what you need to know related to that area is readily available. It is also very easy to understand your potential financial rewards since compensation and benefits are tracked by many organizations. Also, there may be a union that represents the area you are interested in. Healthcare has more union representation for professionals than other industries.

Typically, acute care hospitals have the most sophisticated services, and large academic acute care facilities are known for dealing with the most difficult cases. From a career prospective, it becomes important to focus on what work environment interests you. Do you want to work for the "biggest and/or the best"? Do you have a social agenda focused on improving access and quality? Do you feel more comfortable in a smaller organization where you want to be an agent of change? Healthcare has all of these settings and more. You should not go into patient care service unless you have the mind set and compassion to deal with people at some of the best and worst times in their lives. If you are only driven by money, patient care services is not for you.

Healthcare is like many other professions. The major academic medical centers are aggressively trying to attract the top of the class. Other service providers are trying to attract those with a mission. As I indicated before, there are shortages of medical personnel in virtually all areas of patient care. The ambulatory care area has been growing faster than the inpatient sector, and has become very sophisticated in its service offerings. Medical centers and hospitals have seen their revenue base go from inpatient-dominated to one split between in- and outpatient business.

Do you have the feeling that this career path might not be for you because you have a strong independent business spirit? You need to know that healthcare is also known for people setting up their own businesses to provide these services. Independent contractors and specialty companies have thrived in this regulated environment, much like entrepreneurship itself, which is alive and vibrant in healthcare.

2. ***Suppliers to the Industry***, including:

- Pharmaceuticals
- Equipment manufacturers and distributors
- Medical supply and distributors
- Bankers and investment bankers

The supply side of the industry may be more inviting than the direct patient care space. This area has multiple aspects. Pharmaceuticals are the most profitable, "deep pocket" side of the industry. They have been able to expand from basic clinical support drugs to "life-style" drugs and personalized health solutions. They have developed their own products and have also been very aggressive in buying startup companies with creative products and processes. They have also become the target for how to fund expansion of healthcare coverage for the needy, and to cover the growing profitability gap that providers in some parts of the United States suffer. Many in government and in the provider sector want to force pharmaceutical companies to share profits so that more can access care. It would be a major mistake to curtail the pharmaceutical industry profits since innovation and research for new solutions would suffer. From a career prospective, the pharmaceutical companies are known for hiring the best and the brightest. This includes not only researchers and production staff, but marketing, sales, and finance people. An emerging area is the protection of products and markets. The industry is faced with an epidemic of counterfeiting and diversion of product to black market dealers. Creative solutions for these problems can create new and exciting career paths.

Equipment vendors include not only high tech products like computed tomography scans and robotic lab equipment, but implants from pacemakers to hips. This area is growing significantly as the baby boomer generation needs and wants to be made over. Employment opportunities in this sector should accelerate in the next decade. In many cases, the clinical and educational background needed could rival physicians. You need to know the product and the way it affects the person using it. Profitability will be tied to creative product development and quality.

Medical supplies could be called the legacy products of the industry. They are in high demand, but at the lowest possible price. By providing quality and dependability, a company can dominate a market. The concept of volume is meaningful in this sector. Strong marketing skills coupled with creativity can drive a successful career. An adjunct to the supply side is the group purchasing organization. These organizations started out primarily as cooperatives to obtain better supply deals by leveraging the buying power of their members. They have expanded in to multibillion dollar businesses that can buy nearly anything a provider needs or uses.

What are financial and investment bankers doing here? Healthcare has an insatiable and perennial need for capital, both with long term investment and short term liquidity. If you look back at the areas we have talked about so far, the question that underlies what is being said is, how can the industry financially support all of this growth? The opportunities for careers in healthcare investment banking are enormous. My advice to anyone with technical skills, business understanding of the industry, and a financial background is look into healthcare investment banking. There will be hundreds of billions of dollars needing to be infused in to the industry in the next decade. There are opportunities with both banks and investment banking firms. The ability to understand whether the organization's strategy makes sense and will position it as a survivor in the future is critical to a successful lending decision. Keep in mind that although we are in a growth market, there will still be failures because not everyone understands how to survive and grow.

3. ***Service/Support***, including:

   - Outsourcing of support services
   - Insurance/actuaries/managed care
   - Consulting and legal services
   - Computer systems and information technology

What is meant by outsourcing of support services? This is centered on the non-clinical delivery of services. A growth part of healthcare has been the recognition by many providers that non-core services are best performed by professionals with a specific focus. This includes clinical engineering technology upkeep and information technology services. All these services are expanding into the revenue cycle, which is the most complicated business system of a provider. Physicians have been outsourcing collections for some time. Now hospitals and other providers are looking to do it. The revenue cycle has become more clinical in recent years, and requires a significant knowledge base. The business opportunities are enormous. The concept of core businesses has been looked at many different ways over the last few decades. Some have embraced it, others have rejected it. It has its successes and failures. It will continue to grow, and will expand into the clinical sector.

The insurance and managed care side of healthcare is another *very* profitable side of the healthcare business. This is another area that the provider sector believes has excessive profits at their expense. Yet there are multiple dimensions to this sector. Let us discuss the easier part first. Malpractice insurance that the providers purchase, self insurance, or a combination of both is affected by litigation. Providers have proven that quality driven organizations can limit risk. Malpractice issues have stabilized in recent years, yet there are still some provider failures linked to malpractice.

The other side is the health insurance that companies and individuals buy. The managed care company becomes the negotiator for the insured. This is part of the provider's revenue cycle, and in a perfect world would pay the providers an equitable amount, but we do not have the kind of insurance environment that existed twenty years ago where the providers were in more control of their pricing. The ultimate risk of receiving the proper payment for services has been shifted to the providers. Just because you provided the service does not necessarily mean you get paid. Many insurers refuse to pay for services for a variety of reasons, such as that treatment was not necessary or that advanced approval was not received. The upshot is, there are many interesting positions with managed care insurers. Managed care companies have become the advocates for the buyer of the insurance by negotiating better pricing coverage and refusing to pay for what they consider unnecessary services. The fairness of this relationship can be debated, but this is the current environment. The top three or four publicly-owned managed care companies have been reporting excellent earnings, and top executives have earned hundreds of millions of dollars. There are many lawsuits challenging the practices of the managed care companies, but so far they have suffered no catastrophic set backs.

Consulting and legal services have grown significantly over the years. Consulting firms are faced with a shortage of qualified people. They are some of the biggest recruiters from the college campuses in terms of number of people and compensation. They are also aggressively hiring from industry. Consulting services cover all aspects and sectors of healthcare. Typically, hospitals and pharmaceuticals are the sectors that buy large multi-million dollar projects. The other sectors buy smaller projects and tend to use small boutique firms with specific skills. Services provided include revenue cycle, operations improvement, service delivery, and compliance support. Revenue cycle has become so complex that it is not unusual for a hospital to have six to ten different boutique firms providing services. Of course, there are a few large firms that offer all of the services, but they often have to share engagements with the boutique firms. Operations improvement and service delivery services overlap. There are firms that provide clinical people to restructure and/or reengineer how departments operate. There are firms that provide elaborate benchmarking as drivers for change. Compliance support covers two areas: those needed to maintain accreditations, and those to deal with government compliance. The former services are commonly used by providers to maintain industry licenses and accreditations. The latter is used to deal with government agencies, and many times overlaps with the revenue cycle services.

Law firms overall are busy and are recruiting aggressively. Healthcare is growing at a faster pace than many industries, and is attracting people with not only law degrees, but clinical experience. Lawyers, much like consultants, touch all aspects of healthcare. Representative areas include civil investigations and litigations, anti-kickback issues, payment claims and appeals, clinical trials, non-compliance matters, credit recovery, mergers and acquisitions, healthcare governance, anti-trust, trade regulations, bankruptcy, and organizational restructuring. Any license or compliance issue should have legal involvement. Some law firms are starting their own consulting groups to support their practices and hire the same type of people as management consulting firms.

Computer systems and software are shown as a separate category because of the enormous opportunity they present for healthcare employment, both in the provider setting and in the consulting area. Healthcare, as an industry, has not invested in computer and information technology. Typical spending has been at 2% to 3% percent of operating expenses, while other industries have spent 5% and higher. Providers that have invested in technology view it as a competitive edge. Medicare and Medicaid programs are stimulating more investment in this area by grants and other incentives. They are also encouraging providers to share successful technology, which is not being embraced by many in the industry. The next decade may have computer and technology growth in healthcare that has not been seen in decades.

4. ***New Product Development***, including:

- Industry-based research and development
- Small business development and entrepreneurship
- Government
- Academic research

Healthcare has significant investments being made in all the above aspects within the industry. Some of it is in the area of new product development, while others are new approaches to delivering services. I have separated new product development from the suppliers of the industry because it drives many new and unique career paths. We are dealing with many of the same companies that supply the industry now. These giants of the healthcare industry are creating new products by developing their own research and development models that run outside of the normal corporate setting. (In the case of pharmaceuticals, there is a well established model.) They are recruiting creative people to lead them forward in innovation. They are also buying up new companies that are starting up to take advantage of the market opportunities.

New healthcare companies are typically focused on a single product or process, and are private companies funded by venture capital firms. Their goal is to discover a successful product or process and sell it to the big players, or to go public. Many of these companies have their roots in the academic world. Universities and academic medical centers are creating their own "incubators" to encourage breakthrough research that will provide them with not only leading-edge medicine, but future economic benefits. State and federal government are investing in future innovations. The states especially see it as a way to attract business and industry.

5. ***Regulatory Affairs***, including:

   – Federal regulations
   – Self-review

Healthcare has been called the most regulated unregulated industry in the United States. This may be more of a United States phenomenon than what is seen in the rest of the world. All of the efforts to deregulate it have lead to more regulation. The Health Insurance Portability and Accountability Act (HIPAA) has created more reporting and exposure than the industry has ever had. What it has also done is to create more job opportunities. If you have the mind set and the personality, regulatory enforcement can be a rewarding career path. This can be with the government or working for a provider. Also, enforcement of the regulations around the revenue cycle is big business. The federal and state governments budget billions of dollars to be recovered from providers for abusing the payment system.

In conclusion, we have covered a significant number of areas that have promising career opportunities. Some are similar to one another, others are unique and specialized, and some are dealt with in greater detail elsewhere in this book. All have effects, in varying degrees, on the quality of people's lives. No matter whether it is direct patient care, creating new cures, or managing the business, people are impacted. Healthcare can be an attractive career for those who have the motivation, energy, and dedication to help. It can also attract those because of the financial economics of the business. Those that believe they can make a difference will have a successful career. Those that are in it for other reasons may or may not be successful. Whatever career track you may choose in healthcare, always remember the foundational principle of medicine – *do no harm*!

Finance and Investment Industry

Kristi A. Tange, MS

Derivatives Operations, Goldman, Sachs & Co., New York, New York, USA

I wanted to work in Japan after graduation. With a bachelor's degree in finance and international business, and a master's degree in foreign service with proficiency in the Japanese language from Georgetown University in Washington, District of Columbia, what better entry point than a global investment bank? Goldman, Sachs & Co. took a chance on me in 1997, hiring me into their Treasury Operations team for a year of training in New York, followed by a year-long assignment in Tokyo. Ten years later, I have come to enjoy building a career in this firm and industry, spending time as an operations analyst in treasury in two countries, a project manager on a cross-product team, and finally as a team manager in derivatives operations, managing at turns the firm's collateral team, credit derivatives middle office, interest rate products, documentation and settlements, and, most recently, global derivatives prime brokerage middle office.

Introduction to the Finance and Investment Industry

The financial services industry often brings to mind bankers in navy Brooks Brothers suits or academics looking to be the next Robert C. Merton, (who popularized the Black-Scholes options pricing model). The reality is that with the explosive growth over the past decade, the financial industry has diversified its hiring, and engineers of all academic persuasions have been targeted, along with traditional business majors and increasingly more diverse liberal arts, mathematics, and economics majors.

What exactly is the finance and investment industry? It is a fairly broad ranging group of institutions that provide various financial services. Examples of such institutions include commercial banks, merchant (or investment) banks, insurance companies, brokerage firms, asset managers, and an assortment of other institutions. Services provided range from money management for private individuals to debt and equity underwriting for corporations to issuance of insurance policies, and many other services. In today's global markets, this industry is the engine of the global economy, enabling corporations and governments to grow and expand.

G. Madhavan et al. (eds.), *Career Development in Bioengineering and Biotechnology*,
DOI: 10.1007/978-0-387-76495-5_17, © Springer Science+Business Media, LLC 2008

Why Engineers in Financial Services?

Much as with those who have business, military, and technology backgrounds, students of engineering are trained in problem-solving, process design, and modeling. They also have solid quantitative skills. There is a constant theme of "how" embedded in the curriculum – how to perform calculations, how to organize information, and, ultimately, how to solve problems. The required courses of an undergraduate program include classes that range from basic to relatively sophisticated calculus, classic physics, thermodynamics, and other topics in which classes are structured around problems to solve, and the steps on how to solve them. Most of the basic course work involves a familiar theme: here is what you know, here is what you don't know, and this is what you're looking for – now lay out the steps on how to find what you're looking for within a given time frame. Usually there are several choices on how to get the answer, but, generally speaking, there is one preferred method to solve the problem. This theme is carried through to the majority of classes, and after a few years of this mental conditioning, students will say in jest that modeling, calculus, and physics all looks like the same material, since many of these courses are governed by the principles of what has been published as Newton's work (i.e., calculus). For each class, students are constantly solving problems using these methods for homework, tests, and finals.

A common observation that students make is that it becomes easy for some to lose sight of what is actually occurring when these problems are being solved. Usually, engineering degrees require a large amount of course work with heavy calculation (one can usually spot the engineering student in the room by the model of his or her Texas Instruments calculator). While the answer to a question on a complicated senior level geotechnical engineering class final exam might be "4" of a given unit, it becomes easy to be caught up in *how* to arrive at the answer, and not take a step back and ask *why* the calculation is being done in the first place. The course work can be demanding, and before a student can reflect on why something was done, figuring out how to solve the next problem becomes the priority.

Upon completing a program, there are usually many different career choices for students of engineering. Some believe that the ability to break down business problems the same way engineering students are taught to break down their engineering problems is a benefit to a business career, since the generic engineering approach calls for information to be organized into three boxes: known, unknown, and solution. This skill is thought to be a benefit, since in business, there are often many unknowns to any given problem.

In addition to traditional engineering career paths, the financial services industry offers a unique opportunity for students of engineering to apply their technical background, and really begin to see *why* solving problems is significant by leveraging all the years of learning *how* solving problems is best facilitated. In particular, the growth in derivatives over the past 10 years has created opportunities for students of mathematics and engineering to use the techniques they learned in school to create value for their respective firms, and for themselves. Strategists who build models, traders who price securities, operations professionals who create and manage the systems that house the securities, structurers who customize client needs, and bankers who put together financial deals all represent areas that would benefit from an individual with an engineering background who also has a drive for business. Those who know how to perform functions such as Gaussian Copulas will likely find it satisfying when they apply the copulas to a correlation trading client's needs, thus adding value to the firm. Engineers can add insight to banking teams by understanding the details of a technology company's core products, and adding a unique view to the analysis.

By applying the logic that is set in the foundation of an engineering education to a fast-paced, demanding atmosphere such as financial services, many opportunities exist to apply the specific technical skills that one masters in school to a broad array of opportunities. Additional general skills are also important – organization, communication, teamwork, and a desire for excellence. These assets are attractive to several divisions in particular within the financial industry:

- *Operations* – The operations division drives the efficiency and scale of businesses, manages and controls operational risk, processes transactions executed by sales, trading, and clients, and enhances a client's experience through good service. What does all this mean? Operations professionals need in depth product, process, and industry knowledge. When the firm decides to enter a new market, such as Russia or Brazil, operations understands the local regulations and client expectations and sets up the trade processing. When a client approaches Goldman, Sachs & Co with a request, for example, to automate a legal confirmation process, operations works with technology and legal to determine how best to meet the client's needs. When sales and trading decide to introduce a new product, operations partners to help book, process, clear, and settle those trades, including any legal confirmations, cash flows, and margin requirements. When a derivative product grows from 10 to 1000 trades per day, operations works internally with trading and technology, and externally with industry groups, to design solutions to build scale.
- *Technology* – Engineers have training in processes and logic, often with computer programming skills. Certain roles require specific languages, for example, Java or web programming languages, but other roles work with a proprietary technology language that can only be learned at the firm, so on-the-job training is required. Many engineers have the academic training to succeed in such a role, and can conceptualize and design the elaborate technology needed to process the firm's transactions or keep custody of the firm's positions.
- *Quantitative strategies* – Strategists use a combination of quantitative and technological techniques to develop models for derivatives pricing, risk modeling, and creating insights into financial markets. A graduate degree in financial engineering is a good foundation, but anyone with a mathematical

or other computational background is desirable. A typical project that a strategist could work through includes designing a pricing model for an interest rate derivative. Another example is constructing a portfolio margining scheme that takes into effect the diversity of the client's portfolio and the various risks that the portfolio faces, including market shocks across equities, foreign exchange, and interest rates.

Other areas may also be interested in candidates with an engineering background. However, applicants generally need at least a basic understanding of markets and services to get through the interview process. Such services would include sales and trading, that trades stocks, bonds, currencies, commodities, and their derivatives in the world's markets; and investment banking, that provides advice and financing solutions to clients on topics such as mergers, acquisitions, securities issuance, and risk managing transactions.

Engineering as a field is broadly attractive to financial firms. Bioengineers, in particular may be well suited to quantitative strategies modeling work, given their familiarity with biological behaviors and simulations that could also apply to global financial markets. Research teams that cover the pharmaceutical field may also be interested in the educational training of a bioengineer.

As with all jobs, financial experience is not usually required, although there should be a willingness to learn. Reading a few relevant industry books should spark your interest as well as provide good conversation during interviews.

Getting Hired: How to Get Started

As a bioengineer, how can you be sure that your resume will be attractive to a potential financial services recruiter or hiring manager? Know that you are not at a disadvantage in relation to other career paths. Potential employers in the financial services industry enjoy diverse backgrounds, both academically and professionally, as they know that potential hires should be both qualified and interesting to converse with. Case in point – a group from the prime brokerage desk at our firm took a client team out to dinner. At one point during cocktails, I overheard a technologist from our firm having an engaging conversation about aeronautics with a senior person on our client's team. The technologist had an engineering and computer science background and formerly worked at NASA, and the contact with our client has a brother who worked at Los Alamos National Laboratory. It is often interesting points in one's background that help form connections with employers, colleagues, and clients.

On your resume, emphasize the academic qualifications that may be relevant to the job, for example, mathematics, economics, computer science, management/business courses, and advanced sciences. In regards to work experience, describe not only the functions you performed, but what you learned from the experience. Generally, it is better to focus on a few key growth points or contributions than to create a laundry list of less meaningful tasks. Be cautious about including an objectives section on your resume. If your objective is too targeted, you may be passed over in the process if the employer does not have an exact match. If your school offers business classes for non-business majors, consider taking a course or two, such as "math in finance" or "accounting for non-business majors." Such courses on your resume demonstrate an interest in the field.

In an interview, again emphasize your academic qualifications, particularly if the job has a specific requirement such as basic computer programming or modeling. Describe a project or two, and the relevant skills/experiences that you honed through the process – What thought processes led you down the path to the conclusion? How did you overcome any obstacles? Global financial services firms enjoy a diversity of backgrounds. No one is more interesting than someone who has the base qualifications for a job, and has an interesting accomplishment in their background. For example, we recently hired an engineer who worked for several years in the construction industry, and had actually worked on buildings in the surrounding neighborhood. The candidate had relevant project management skills and a technical background that would translate well to a credit derivatives middle office role in operations, and was hired based on several interesting interviews.

One idea to gain insight into the financial services industry on a trial basis is to apply for a summer internship, which is generally available to both undergraduate and graduate level students. Summer internships are typically 10-week programs that give the intern exposure to a specific division within a commercial or investment bank. Interns spend time on a desk or with a team to experience day-to-day life, and will often also have a project to deliver by summer's end to demonstrate their capabilities. This opportunity will enable you to obtain some perspective before making a bigger decision regarding full-time employment. Note that resume deadlines and interviews for major firms are typically early in the year (January/February), so you will want to research opportunities as early as possible.

A Taste for Daily Life in the Industry

What is daily life like at a financial services firm? This question is depends on the point of entry into the field. For example, in common areas of an investment bank:

- Operations' daily work is also correlated with the markets supported, starting the day when trading begins or clients arrive. Most operations roles interact with either external clients or internal business partners. Any new, high volume, or unusual activity will drive operations' work on a given day. For instance, a record volume trading day may present a spike in trade-related problems that need sorting with the firms counterparts, or a new type of trade may need urgent addressing by operations, technology, credit, legal, and/or compliance to be able to book in the systems and settle the transaction. Project managers within operations, similar to their counterparts in technology, have a schedule that rotates around their deliverables.
- Technology can differ based on the role. Technologists who support trading activities may work similar hours to market-facing jobs, ensuring that technology is running as needed by sales, trading, and operations. New products or clients will

likely generate development work. Technologists who code projects as a primary responsibility will have schedules that revolve around their project work, busy at delivery times and less so off project cycle. Starting time may be closer to 9:00am.

- Quantitative strategists often work closely with trading desks and technologists to construct pricing models, and tweak the models based on real-time trading needs. As such, their work hours will coincide with trading desks as described below.
- Trading and sales have a fairly predictable experience (although there are always exceptions). The day starts early (7:00 – 7:30am) with calls on the overseas markets and gearing up for the trading day. Market hours are incredibly busy, with client orders and trades flooding in throughout the day. When the market closes, after reviewing PnL (profit and loss calculations) and ensuring risks are managed, the day is done.
- Investment banking is project-oriented, typically with long hours centered around preparing for a client presentation. Bankers will typically work on teams consisting of an analyst, associate, vice president, and managing director, and may interact with trading, credit, compliance, legal, and structuring.
- Research is on a cyclical schedule, with peak hours required during earnings announcement seasons. Work may include conference calls, due diligence visits, and writing briefs. Research works closely with sales and trading in support of our client base. Given recent heightened focus on "Chinese wall" considerations, research must treat confidential information quite sensitively until it is made public.

Training is typically provided by the firm. New graduates who join a sales and trading role will typically participate in several weeks of classroom training to prepare for their future jobs, as well as any required industry licenses (typically National Association of Securities Dealers' Series 7 and 63 examinations). In other divisions, such as technology, new graduates will attend a two-month training program introducing the firm's technology and coding practices. The classroom training may be taught by a combination of local professors, professional training instructors, and experienced staff at the firm. The intention is to ensure that all participants have a common foundation. These classes are especially beneficial for non-business majors. The classroom training is typically followed by assignment to a desk, where you will be paired with senior, experienced staff who shows you the ropes. In other divisions, such as operations, new graduates participate in two weeks of training on the organization and key principles, but most training comes on the job. This tends to be the case for mid-career lateral hires, who often will receive on the job training experience supplemented by training courses on products, industry, and personal development skills.

Career Path

Where can these careers lead? Take the example of one engineer who worked in the energy industry building gas stations. He was hired as a project manager in derivative operations. An opportunity to manage a team of analysts supporting interest rate derivatives client confirmations arose as a result of previous successful project initiatives. That role soon expanded into credit derivatives, which has been a continuously booming field over the past few years. This person developed an interest in the industry, and through internal networking, found a role in the investment banking division, where he currently pitches products to municipal clients.

Another engineer with a graduate degree joined our firm as an analyst on the valuations team in operations, which is responsible for delivering trade pricing to the firm's clients. After a successful year in that role, an internal mobility opportunity arose to build a derivatives team in the prime brokerage area. The chance to build a team from the start is an interesting experience not to be passed up, and this person is now an associate on the desk, specializing in prime brokerage for several derivatives products.

Others choose to move within the financial services industry, particularly to the growing hedge fund industry. Hedge funds tend to hire experienced candidates, as their training infrastructure is less than that of a commercial or investment bank. The hedge fund experience is typically a smaller, more intimate work environment that appeals to people who enjoy the hedge fund's culture or who are looking for flexibility that bigger firms cannot easily offer. Some hedge funds are even becoming as large as investment banks, with similarly diverse investments. The latest trend is to raise capital in traditional fashion. For example, look to a leading hedge fund, whose debt offering of $500 million five-year notes in 2006 was the first-ever sale of bonds by a hedge fund to raise permanent capital.

If you take one thought away from this chapter, let it be this – give finance a chance. If you have some curiosity about the industry, rest assured that the industry is interested in your background. With a nicely put together resume, you can soon have someone with an interview that may open up an opportunity for you in the world of finance.

Recommended Reading

1. Bernstein PL. *Against the Gods: The Remarkable Story of Risk.* New Jersey: Wiley, John and Sons, Incorporated; 1998.
2. Burrough B, Helyar J. *Barbarians at the Gate: The Rise and Fall of RJR Nabisco.* HarperCollins Publishers; 2003.
3. Endlich L. *Goldman Sachs: The Culture of Success.* New Jersey: Simon & Schuster Adult Publishing Group; 2000.
4. Fabozzi F (ed), Mann SV (ed). *The Handbook of Fixed Income Securities.* New York: The McGraw-Hill Companies; 2005.
5. Knee J. *The Accidental Investment Banker: Inside the Decade that Transformed Wall Street.* New York: Oxford University Press, USA; 2006.
6. Levitt S, Dubner S. *Freakonomics: A Rogue Economist Explores the Hidden Side of Everything.* New York: HarperCollins Publishers; 2006.
7. Lewis M. *Liar's Poker: Rising Through the Wreckage on Wall Street.* New York: Penguin Group, USA; 1990.
8. Lowenstein R. *When Genius Failed: The Rise and Fall of Long Term Capital Management.* New York: Random House Incorporated; 2001.
9. Weiss D. *After the Trade is Made: Processing Securities Transactions.* New York: Penguin Group, USA; 2006.

18 Regulatory Affairs

Ronald A. Guido, MS

Global Research and Development, Pfizer, Inc., New York, New York, USA

Alan V. McEmber, MS

Global Research and Development, Pfizer, Inc., New York, New York, USA

When exploring alternative career paths for life scientists, discussions of the regulatory affairs role are met with what may be best described as uninformed interest. After lecturing on topics in regulatory affairs at the graduate level, hallway discussions eventually gravitate towards the type of questions typically heard at a job fair. Highly trained and scientifically astute students approach us looking for insight into pharmaceutical career options. Peppered among the queries of "how do I break in" and "do you have a business card," most will say they often heard that regulatory affairs is a great career – but are unsure what it is. We will address these and other related issues in this brief chapter.

Describing careers in regulatory affairs should be an easy task for someone with years of experience in the field – right? Not exactly. The constant evolution and differences in the practice of regulatory affairs make it difficult to succinctly describe the field. Moreover, descriptions are always colored key individual experiences, and will reflect variations in regulatory practices within a global geographic area, time in the discipline, and the practice within a company. Confusing? Let's just say that a bit of exploration may dispel the hearsay and reveal one of the most attractive life sciences career options that you've ever heard about. At least, that is what we believe and hope to convey.

Regulatory affairs is a career that touches every facet of the biologic and pharmacological drug development process. That is not an overstated selling point. Regulatory affairs is a career whose definition has evolved greatly over a relatively short period of time, and continues to change rapidly. No boredom here. Rote practitioners need not apply. In fact, writing a chapter such as this would have been far easier a couple of decades back. Back then, regulatory affairs was a function, but could hardly be called a viable career. It was a post often occupied by quality assurance or research and development folks in the twilight of their

G. Madhavan et al. (eds.), *Career Development in Bioengineering and Biotechnology*,
DOI: 10.1007/978-0-387-76495-5_18, © Springer Science+Business Media, LLC 2008

careers. Often there was no department *per se*. Individuals contemplating a move from a science position to regulatory affairs in the early 1980s, would be advised against joining the profession just yet because they were too young and too up to date on their scientific knowledge to switch. However, even then, the winds of change were blowing. These gathered intensity, and brought us to the transformed field we have today.

In that not-too-distant past, the regulatory affairs role seemed to only be comfortable with the application and interpretation of black-and-white rules. The regulatory practitioner was typically seen as the policeman or overseer, ensuring that applications for product approval met with all applicable sections of the US Code of Federal Regulations detailing drug approval applications. In many cases, the designation of the regulatory affairs function as *registrations* signaled the attitude that this was solely a mail stop that received completed product (that is, documentation) for subsequent transmission to regulators for approval. The registration officer presided over the multitude of government guidelines for format and content of applications. The completion of proper forms was a mainstay – a quagmire of poly covers, binder thickness, pagination, tables of contents, and margins. This is no longer the case. The role now reflects a highly-technical member of a product development team. As you might imagine, unless you are speaking to a current practitioner, you may receive a slanted view of this discipline based on past practice – one of filling out forms, and execution by the book. But time has moved on, and that image of the role could not be farther from the truth.

As the world changed, the field of view of the regulatory professional shifted from the microscopic to the telescopic. Perspective has broadened. Focus has switched from regional to global. The involvement of the regulatory professional in the drug development process moved from managing the tail end (application submission) to one of weighing in on the earliest decisions on a suitable target for a candidate drug [1–10]. Regulatory insight is now sought through all phases of the development process, and even following registration. The career has escaped its earlier limits which related to intense scrutiny of minute detail, and now enjoys a diverse role in the ongoing assessment of complex research and drug development programs.

How so? Let us be more specific. Broadly, the regulatory professional is charged with ensuring compliance with established regulations and current standards for approval, and that the development and approval programs for medically necessary products are well-designed and well-executed. This function promotes bringing safe and medically important products to market as quickly as possible and with an appropriate product label.

How is this accomplished? Briefly, a regulatory professional applies expertise, which encompasses a synthesis of current regulatory knowledge, the ability to discern and apply scientific rigor, and the flexibility to use these talents in a creative manner in the development of new biopharmaceuticals/pharmaceuticals. In this respect, the regulatory professional is unique and set apart in this growing and changing field.

Why Regulatory Affairs?

The reason for regulatory affairs in certain industries can be seen almost daily on the front page of the newspaper or leading the evening news. It is a sad fact of life that tragic things happen when something goes wrong in a complex industry. Airplanes crash, and there is monumental loss of life. Food is tainted, and there is rampant illness. Wall Street and board room scandals lead to losses of a hardworking individual's life savings. Quack cures harm the unsuspecting public. Luckily, these events happen far less frequently than they could, given the scope and reach of these industries. The airlines, food safety, banking and finance, and medical product industries are controlled by government oversight – translated, that means pages and pages of regulations. Inspection of aircraft and their airworthiness, food sanitation, financial transactions, and medical products' effectiveness and safety are among the industries and processes subject to government regulations. That is, they are regulated.

The development, approval for marketing, manufacturing and ongoing compliance of pharmaceutical products, medical devices, and biotechnology products are among the most regulated of any industry. While varying in complexity and stringency, all countries have a framework of legislation to control the manufacturing and distribution of pharmaceutical products. Typically these are complex systems of interrelating rules – that is, regulations. These rules cover a broad range of activities, and have been modified (or amended) and supplemented over time in response to changes in science, medical practice, politics, or, many times, following adverse circumstance. These rules assure, among other things, that the products distributed will do no harm (be safe) and will in fact do some good in the treatment of disease (be effective). These rules impinge on all aspects of the processes and procedures that are required in the development and subsequent marketing of a pharmaceutical product. Noncompliance or even unfamiliarity with these requirements during development and following approval of the product carries with it a tangible cost that can be measured in both time and money. These costs can encompass such things as unusable data from clinical trials, delayed time to approval, delayed approval of manufacturing sites, or fees associated with compliance actions. It is the role of the regulatory professional to navigate these complex waters, keeping the product in the best position to achieve and maintain marketing approval.

The evolution of the regulatory role has been driven in large measure by a business need. The process for developing a drug has become extremely complex, expensive, and uncertain. This is borne out by the recent spate of newspaper headlines declaring spectacular program failures following huge investment. Often, seemingly small errors or missed signals precipitate these failures. What this means is that getting that new and promising target through clinical research into the approval process and ultimately to market has proven to be no sure thing – and it is getting more and more uncertain. The road to the pharmacy shelf is littered with great compounds that did not make it. Some of these fell away not because of product flaws or poor efficacy; but rather due to bad regulatory processes that stalled review or created an impasse between regulatory and application sponsor. Regardless of outcome, even making the journey can be costly. According to the Tufts Center for the Study of Drug Development, if we account for the products that fail testing and associated time costs, the actual cost to develop a biotechnology product is $1.2 billion. Of this, $615 million are sunk into preclinical and $626 million into clinical testing. Biopharmaceuticals have only 30.2% success rate [10]. Given this, it is clear that a misstep resulting in a regulatory impasse at an early clinical stage – end of phase II is a typical danger point – will result in a significant loss of revenue. Any actions a company can take to avoid costly missteps, hasten approval, or add some measure of certainty to an approval process, are critical. This potential benefit is enough to drive the need for the regulatory role. While this will not prevent all program failures, there is enough at stake here to warrant the creation of this specialized discipline, and to justify the salaries of our colleagues.

So, What Does the Regulatory Affairs Role Look Like in the Pharmaceutical Sector?

Regulatory affairs is a specialized profession found within regulated industries. While this is a role that may be found in any regulated industry, our focus in this article is this function within the pharmaceutical/biotechnology sector. The regulatory affairs professional oversees company compliance with all of the regulations and laws pertaining to the manufacture, marketing, and development of regulated medical products. Regulatory affairs professionals act as point of contact between the company, its products, and regulatory authorities The regulatory professional interacts with worldwide, federal, state, and local regulatory agencies (e.g., FDA (US), EMEA (EU), Health Canada, and BfARM (Germany)) to ensure the licensing, registration, development, manufacturing, marketing, and labeling of pharmaceutical and medical products are conducted in compliance with all applicable established rules.

While this may sound one-dimensional, this level of oversight requires expertise and training that demands facility with the scientific, legal, and business issues that encompass the pharmaceutical development process. This specialized expertise assures that products, at the end of their development, are good candidates for marketing approval, and that the products that are developed, manufactured, or distributed continue to comply with established regulatory mandates. In the pharmaceutical/biotechnology sector, the regulatory affairs professional acts as a *de facto* internal company advisor in the approval and registration of pharmaceuticals, medical devices, nutritional, consumer healthcare, and other related products.

So, If I Were to Pursue This Role, What Might I Be Doing Day-To-Day?

Typical daily activities might include but not be limited to:

- Determining whether a new product licensing candidate meets applicable worldwide regulations, and thereby is a suitable candidate for worldwide marketing.

- Rapidly developing clear responses to regulator queries, keeping a product in compliance and on the market.
- Participating in the development of clinical development programs to develop adequate data to support product registration applications.
- Reviewing complex data to develop product communications.
- Developing clear, comprehensive product labels and associated patient information.
- Overseeing the initiation of clinical trials.
- Reviewing clinical trials data.
- Advising product development and manufacturing operations teams on issues of regulatory requirements.
- Preparing product approval application strategies, applications, and associated documentation.
- Advising teams on issues of timing and milestones associated with the development of new products.
- Developing and establishing processes and procedures to support product quality assurance systems.
- Managing regulatory inspections and customer audits.
- Communications with, and developing presentations to, regulatory authorities;
- Submitting registration and clinical trial support applications to regulatory authorities.
- Negotiating with regulatory authorities for marketing authorization approvals.

I Have Expertise in Science. Will My Success in This Role Benefit from My Background? What other Skills Will I Need?

A regulatory affairs position is multi-faceted, and touches on many diverse aspects of drug development. It draws upon a diverse set of skills. The regulatory professional needs to be recognized as credible by both the external regulator and internal company specialists; therefore, the most critical foundational attribute would be scientific expertise (favoring broad experience rather than specialization). Because of this, companies typically prefer candidates for regulatory affairs positions to have an degree in a scientific or medical health discipline. The minimum standard is typically a bachelor's degree, while most companies prefer master's, PhD, or an advanced professional degree. This is not the end of it, however, and additional mandatory skills and areas of expertise include:

- Interpersonal communication and facilitation
- Writing, presenting, and speaking
- Ability to simplify complex data sets
- Negotiation
- Interpretation and data analysis
- Team dynamics
- Project management
- Information organization

Of all of these, the ability to clearly communicate, negotiate, and build consensus using complex data is most pivotal to the successful regulatory professional.

This might be a good time to look more deeply at some of the areas this role interacts with in developing and managing new medical products.

Early Regulatory Advice

Rather than acting as an archive, mail drop, or repository of completed documents, regulatory affairs has now been recognized as a critical player in the design and execution of a successful program. In the past, regulatory received completed documentation, and needed to make do with it in developing a submission. At that time, the approval process seemed to be comprised of an endless string of regulator queries, and associated, additional data submissions. In the worst case, the agencies determined that these submissions were unsuitable for filing or approval. In some cases, clinical programs encompassing thousands of patients were rejected outright as unsupportive of a proposed product's intended clinical indication. Therefore, one of the key roles that emerged for the regulatory professional was to advise on the early development and clinical development program to assure that it was designed to meet standards for regulatory approval.

Approval Standards

The regulatory affairs professional is charged with familiarity and currency with all established standards for approval for any specific program. While there are base approval standards related to proof of safety and efficacy, specific drugs have their own unique approval standards that need to be researched. Further, general and specific approval standards are subject to changes in the regulatory environment and in the evolving practice of medicine. In this regard the regulatory professional must stay current and keep internal company teams apprised of change.

Clinical Program Development

The regulatory affairs professional is a key participant in the clinical development program. This program must be designed to support the approvability of the product as well as the company's desired label indication for use. In this regard, the studies must be well-designed, well-controlled, with sufficient subjects, and conducted using the appropriate population. Beyond this, the regulatory professional must ensure that the clinical assessment methods used are recognized, validated (that is, measure what they purport to measure), and demonstrate a positive effect on clinically important aspect of the disease, to a degree judged to be medically significant.

From a compliance standpoint, the regulatory professional oversees many aspects of the clinical study, including assurance that the application to permit initiation of human trial has been submitted (IND, CTA), the regulated waiting time (regulator review) has elapsed, study physicians are properly documented, required ethics committee reviews have been conducted, consent forms are adequate, and systems for protection of human subjects have been established. Further, the regulatory professional will often ensure that any trial-related adverse event is expeditiously reported to regulator, investigator, and ethics committees as required.

Label Development

Label format, content, and associated negotiation with any interested regulator all falls into the jurisdiction of the regulatory professional. It is quite easy to believe that the final label a product carries is a direct reflection of the sum total of the data collected in support of it. But this is not so. What, beyond the minimally compliant information, will be included and how things are said (or depicted) are open to discussion and negotiation, and these negotiations are some of the most intense that any regulatory professional may undertake. Limited by the data collected, the regulatory professional must gain regulator concurrence that provides the best information to the prescribing physician. Product use by the healthcare practitioner, determination of reporting requirements for safety events, as well as any claims made in subsequent advertising and promotion campaigns, will be based on the established and approved label.

The established label must also be kept current as to the latest safety information, any changes associated with the drug class, or any changes to the product or its approved clinical uses. This is overseen by the regulatory professional, who coordinates submission and review of post-approval label revisions with regulatory bodies worldwide.

Application Design

Notwithstanding all the requirements for a complete, comprehensive and approvable application, the regulatory professional is now called upon to assist in the design of the approval application and its subparts. The ultimate intent of this exercise is to render an application that is reviewer friendly. In the past, this ended at good copies, complete tables of contents, tabs, and reviewer guides. Now, with the advent of electronic applications, the sponsor may creatively use electronic commentary, hyperlinks, and application schematics to ease review. All of this translates into an efficient and potentially more rapid review, which is one of the cardinal aims.

Regulator Negotiation – Relationship Management

For the regulatory professional, discussions with regulators span the continuum of complexity. One moment, these discussions may be gaining concurrence on the most intricate issues – product quality issues, the design of clinical development program, acceptable study endpoints, the design or update of product labeling. At other times, it may be as simple as updating a project manager, inquiring as to review status, or requesting an extension on a deadline for a query. Complex or simple, all of these negotiations require excellent communication and interpersonal skills. Miscommunications, misinterpretations, incorrect responses, perceived lack of immediacy, or absence of trustworthiness can all help slow – or stall – in any negotiation.

Regulatory Communication – Ongoing Transparency

The regulatory professional must be seen as competent, trustworthy, and as a true partner to the regulator. This relationship must be built, nurtured, and maintained to assure good communication and review practice. It is built upon an *a priori* assumption that the regulatory professional is technically astute, sensitive to the reviewer's concerns, and established as (ever and always) a guardian of patient safety.

Advertising and Promotion Oversight and Review

The advertising and promotion of medical products is likewise regulated, and therefore under the oversight of the regulatory professional. The regulatory professional is called to ensure that the piece is factually correct, and all claims, both explicit and implied, are based in direct data collected on the drug, and fall under the agreed agency labeling. This extends not only to the printed matter, but also to television, radio, and, for that matter, all electronic media. This oversight even extends to educational presentations, seminars, and promotional items provided for professionals.

Safety Reporting

The regulatory professional is advisor as to the ongoing reporting of safety events to regulators worldwide. Virtually all regulatory bodies have established programs to collect this information with the intent of assuring patient safety. The regulatory professional ensures that events are reported in a timely fashion, and that the label reflects the most up-to-date safety data to permit the physician and patient to make an informed risk/benefit assessment. The regulator also ensures the continued safe use of the product.

cGMP Compliance, Preapproval Inspection/Routine Inspectional Support

In many cases, the regulatory professional will play a pivotal role in developing and maintaining the quality systems that will ensure that a drug product will be produced day-after-day in a manner that is compliant with current standards for good manufacturing practice (cGMP). Associated with this role, the regulatory professional will be called to host and manage regulator audits of company facilities in cases of "for cause," routine, or pre-approval audits. The regulatory professional is trained to guide the inspection process and provide succinct, accurate responses to onsite regulator queries.

The regulatory professional must establish a process to ensure that inspections can be entertained at any time. What this means is that staff are trained to calmly answer inspector queries, a support team is established in advance to provide supportive data as required, and individuals are designated to manage the inspection at the production line level. The inspector will likely follow all parts of a planned inspection, and will "wait to observe" if required. One can always anticipate the arrival of an inspector at the most inopportune time.

Manufacturing & Controls

For all approved pharmaceutical and biotechnology products, the site of manufacturing, the manufacturing process, and the processes for testing and product release are all part of the basis for approval, are fixed by regulation, and, in most cases, will require submission and/or approval prior to any modification. Regulatory professionals participate in the establishment of the chemical test parameters that establish the identity of the active substance – commonly referred to as *characterization*. Regulatory professionals weigh in on the establishment of specifications (control parameters) for the active ingredient (drug substance), in process checks and finished product. By this means, the process is kept in control and the product is consistently of the same quality. These same specifications are applied to determine the continued quality of the product as it sits on a shelf for its expiry period. This ensures the patient taking the medicine always gets the expected dose of medicine.

Following approval, the regulatory professional is there to field a seemingly endless stream of manufacturing changes. All these changes are assessed for impact on the established product quality and risk to the patient. On this basis, the regulatory professional decides the manner and content of submission, and, related to this, the timing for implementation. It is safe to say that application holders continually look to modify these parameters to optimize production, increase yield, increase efficiency, and to decrease costs of production and maintenance. Some changes even focus on the product itself, to perhaps optimize performance or provide a new product size. All these changes need early assessment by the regulatory professional as to anticipated costs (testing, validation, stability assessment) and timing for approval. This information will permit the anticipated benefit to be offset by the true cost, permitting a valid cost/benefit assessment. If positive, the regulatory professional will assess and determine the best route for approval, the submission contents, and ultimately, the worldwide impact. Further, the regulatory professional will submit a suitable support package, manage its review, and oversee the post-approval implementation. This oversight is required for all post approval manufacturing changes, product changes, and changes in established specification.

Recall / Field-Correction

Just the word "recall" is enough to strike fear into the heart of the most seasoned regulatory professional. It means that the sponsor has discovered something seriously wrong with a product that has been distributed (is "in the field"), and that product must be brought back. Perhaps it has been adulterated, mislabeled, or misbranded. In any event, the immediacy and scope of the recall action is commensurate with the imminent hazard to public health conveyed by the product in the field. The most severe recall, category I, demands immediate, broad communication to end users, expeditious return of the product, tracking of product returns, reconciliation of all distributed product, and regulator supervised destruction of the adulterated product. This usually includes negotiation of the recall category, communication with the local field office of the FDA, ongoing recall status correspondence, and a formal close out on completion. Beyond this, it is important to note that the majority of recall actions trigger regulatory inspection.

Worldwide Strategy

The drug development process is truly global in scope. In the development of a plan to achieve the broadest collection of approvals possible, the regulatory professional must design development programs and applications for approval, educated assessment of best options for issue resolution, globally recognized clinical trials, worldwide labeling, planned global manufacturing, and appropriate indications. This must, of course, be accomplished in the least time, at the lowest cost, benefiting from facilitated review, with robust database, and providing optimal patient safety.

Is there any Formal Training?

Historically, regulatory affairs has been learned at the kneecap of the master. There are now formal, degreed university programs, and training courses sponsored by professional societies. This is a career where keeping current is elevated to an art form. The landscape changes so quickly that this chapter will be out of date by the time you finish reading this book. The best training is often engendered by establishing a good network of practitioners who are aware of the latest changes, and what the latest concerns might be. Trading war stories of travails in the "grey" area often proves eminently more helpful than poring through pages of black and white regulation.

Concluding Remarks

It is our sincere hope that this short chapter has afforded you a brief glimpse of what regulatory affairs offers as an interesting and attractive career option. We would, however, be remiss not to touch on the dark side. Stay with us here – it is not really all that bad.

Above all, one must remember that in this role you may often need to advise on the best regulatory strategy. This strategy may divert from the most popular company strategy, or even, on rare occasions (believe it or not), the most scientifically advanced strategy. This could – at times – make the regulatory professional as popular as a visit from a plague infested rodent. The role requires scientific and interpersonal expertise, engendered trust, and a touch of intestinal fortitude. As one of us is so fond of saying: . . . *when you work in regulatory affairs, you should really consider buying a dog so that on those occasionally trying days, there is at least one living thing who is thrilled to see you.* Should you choose this complex and challenging, but very rewarding career path, we promise that you will never be bored, you will likely learn at least one thing new each day, and you will be at the heartbeat of discovery and developmental activity. In the end, after the dust settles, perhaps you will be staring down at a newly approved pharmaceutical/biopharmaceutical product that will treat a previously untreatable serious medical condition. In this role you will – without question – have played an instrumental part in bringing it to market.

Bibliography

1. Miller, L. L., Cocchetto, David M., "The Regulatory Affairs Professional in the Hot Seat in the Drug Development Process," *Drug Information Journal* 31: 805–823, 1997.
2. Omenn, G. S. "Regulatory Frameworks and Decisions Matter to the Development of Biotechnology and the Approvals for Biotechnology Products," *Current Opinion in Biotechnology* 10: 287–288, 1999.
3. Evers, P. "Pharmaceutical Regulatory Affairs Outlook," *Business Insights.*, 2002.
4. Marasco, C. A. "Careers for 2002 and Beyond: Regulatory Affairs," *Chemical and Engineering News* 80(48): 75–80, 2002.
5. Keramidas, S. "The State of Affairs: An Overview of the Regulatory Affairs Profession," *Science Careers.* Novemeber, 2003.
6. Turner, P. "A Career in Product Registration and Regulatory Affairs," *Careers with the Pharmaceutical Industry (second edition)*: 203–212, 2003.
7. Mantus, D. S. "The Practice of Regulatory Affairs." *FDA Regulatory Affairs: A Guide for Prescription Drugs, Medical Devices, and Biologics*, The CRC Press, 2004.
8. Carroll, J. "Global Deals Grow; So Do Regulatory Issues," *Biotechnology Healthcare*: 9–10. April 2006.
9. Hoag, H. "Rules rule," *Nature* 441, 1022–1023, June 2006.
10. The Tufts Center for the Study of Drug Development,"Outlook 2007," *Impact*, April 2007.

Clinical Research Careers

Kathi G. Durdon, MA

Welch Allyn, Inc., Skaneateles Falls, New York, and The Society of Clinical Research Associates, Chalfont, Pennsylvania, USA

Clinical research, in part, involves evaluation of medical therapeutics and diagnostics to determine their safety and efficacy for improving public health and augmenting medical knowledge. Most importantly, clinical research involves using humans as willing and informed study participants. Those involved in conducting clinical research can essentially be grouped into one main career category: "Clinical Research Professional (CRP)." A CRP may have a more definitive title that is aligned with the job site, which may be a private practice physician site, a Clinical Research Organization (CRO), an Academic Medical Center (AMC), a pharmaceutical, biotechnology, or medical device manufacturer ("industry"), an Institutional Review Board (IRB), a consultant, or a training organization. However, the clinical research profession is not standardized, and positions with similar job qualifications will more than likely have diverse titles. There are also many clinical, scientific, administrative, legal, and regulatory routes that can be taken within the CRP field that could also dictate the job description and title.

Along with having diverse job titles, there is currently no well-defined educational path or formal process to enter into the clinical research profession. Some of us were lucky and found ourselves at the right place at the right time, and often came with little to no CRP skills. Thus, "on-the-job" training was critical. For the most part, positions require a two-year college degree, while some specifically require a clinical (nursing or medical) degree. Senior-level CRPs typically have a track-record of experience, and possibly certification. As the growth and need for CRPs expands, the clinical research profession is making its mark as a top job placement category, with educational institutions and training organizations moving quickly to fill the openings. Due to the critically responsible nature of the underlying goal of all CRPs—that is, to protect human subjects in the conduct of clinical research—training and education are absolutely essential, and it behooves us to continue to expand our knowledge.

To become a CRP, you need to talk to and network with current CRPs and let them know you are interested in the discipline. Start to become familiar with the terminology, regulations, standards, and clinical research ethical history, for

G. Madhavan et al. (eds.), *Career Development in Bioengineering and Biotechnology*,
DOI: 10.1007/978-0-387-76495-5_19, © Springer Science+Business Media, LLC 2008

instance, the Nuremburg Code, Declaration of Helsinki, and Belmont Report. As the profession becomes a more mature and well traveled path, you will see increases in offerings of professional degrees, internships, and certifications for CRPs.

Some representative pathways of how you can become a CRP include the following:

1. As a student, if you are interested in research and associated logistics and a physician/faculty member at your academic medical institution is doing research as a principal investigator (PI), you could become a clinical research coordinator ("CRC," or a similar title) and schedule research subjects, attend investigator meetings, become proficient at IRB application submissions, negotiate study budgets, and help write grants. This will require you to constantly update and renew your training knowledge by means of attending many clinical research conferences and gaining necessary regulatory and Good Clinical Practice (GCP) training.
2. If you are already in a Clinical Research Organization (CRO; that is, a medical device, biotechnology, or pharmaceutical company), as a Clinical Research Associate (CRA), you may be able to manage clinical study at any sites that are contracted to conduct the sponsors' study. Your duties may include monitoring the qualified study sites for regulatory compliance, following the study protocol (investigational plan), using GCP to provide scientifically sound data to the sponsor and/or to the FDA, and ensuring that the rights and safety of the study participants are protected. The candidates for CRA positions are usually experienced research coordinators and research nurses. Unfortunately, there is often difficulty in moving as a study coordinator from the site (private physician office, hospital, academic medical center) to industry CRA (pharmaceutical, biotechnical, device company or CRO), as the ability to gain requisite monitoring experience may not be available. How to bridge this gap? The best option is to find an investigator at your facility who is conducting studies as a sponsor-investigator. Another option is to work with a private site that is willing to bring you on as a lower level associate and teach you the ropes. Once you gain monitoring experience, watch out, you will be besieged with job opportunities. Also, be prepared, some CRA positions require up to 80% travel, coupled with opportunities to work from home.

Both of the above sample pathways can lead you to positions in project management, another tier on the clinical research rung. This position incorporates all you have learned to date as a CRC and CRA, and is a highly detail-oriented position involving heightened organizational, supervisory, and management skills. As a project manager you will typically be managing all phases of a project. You will be responsible for making sure that the goals of the project are met within scope, budget, and timeliness. This includes developing resources to streamline process (such as creation of Standard Operation Procedures), delegation of project tasks, assessment of and plans to alleviate risks to the project, promoting optimal communication, and, ultimately, being accountable for project success or failure. Not necessarily a stress-free career, but certainly a growth-oriented one.

One point I want to emphasize is that to become a CRP, you do not need to have a clinical background. CRPs should excel at administrative and management tasks, be detail-oriented, and well-organized. Clinical research is conducted in a regulatory environment that is absolutely necessary, but involves a great deal of documentation and oversight to ensure protection of study participants and to provide necessary support of safe and effective product through clean viable data.

Another piece of advice is to never consider yourself an "expert." Regulations try to keep up with the advances in medical science and technology through guidance documents and revisions to current regulations. To keep abreast of these advances, you need to constantly interact with fellow professionals and experts in the field, share your own knowledge, and gain access to updated resources and tools. I highly encourage attending prominent clinical research-oriented annual conferences and more specialized workshops. Additionally, in regards to professional certification, most organizations require a prescribed number of continuing education credits to retain your certification. So consider enrolling as a member in some professional research societies, and actively update your knowledge.

Finally, to give you a flavor of the CRP today, I asked several of my colleagues to share how they became involved in clinical research, and to describe their current roles and responsibilities. This is just a small sampling, but I think by reading these, you will begin to see how happenstance and luck helped many of us find what has become a fulfilling and rewarding experience.

Associate Project Coordinator

"When I began working in clinical research as an oncology site coordinator about 15 years ago, I knew immediately that clinical research was the career for me. However, I did not seek this profession out. It found me. I was working as a secretary on a medical floor where the majority of our census was cancer patients. I was approached by the director of oncology who had an opportunity in research for a part-time site coordinator. The opportunity sparked my interest, gave me a sense of doing something important for others, and it worked into my busy family life. I now oversee international studies in neurological disorders, involving nearly 90 research sites, including development of protocol, model consent form, and data collection tools; preparation of materials for grant and other funding applications; writing operations manuals; conducting orientation meetings; maintaining regulatory documentation; daily liaison with site investigators and coordinators, and working with data management processes specific to the conduct of clinical trials."

—Elaine Julian-Baros, University of Rochester, New York, USA

Clinical Operations Associate

"I have a bachelor's degree in anthropology and English, a master's in business policy, and had amassed ten years of experience in the health administration field (I needed to feed my children, and digging at an archaeological site or writing for the community newspaper was not going to cut it). Fortunately, my varied education and experience met many of the qualifications requested for directorship of a new central clinical trials office being developed at the academic medical center where I worked. Even more fortunately, they hired me. I knew immediately that this was the place for me, but I had a lot to learn, as this was one of the few central clinical trials offices in existence. So it was trial and error for a while, but colleagues in clinical research came to my rescue often, and I thank them for sharing their expertise. After five years, industry called with a new opportunity. The position being offered was challenging, I had met my current objectives and could not really move any further, and, let's face it, industry paid lots better. So I am now a clinical operations associate at a leading medical device manufacturer. I fill many roles in a highly supportive clinical affairs department, and although I did not think I could. I am enjoying this job even more than the one before. It only gets better."

—Kathi Durdon, MA, Welch Allyn, Inc., New York, USA

Clinical Research Associate

"I have a master's in exercise physiology, and the program I attended was heavily research-based. During a graduate school internship, I assisted with some industry-sponsored trials, but mostly was involved in investigator initiated studies. I liked the job of the nurse coordinator there, although at the time I did not fully understand the job title or responsibilities. After graduate school, I was recruited to work with a scientist who was starting a large NIH trial. I got the job as coordinator for that trial, and when that concluded, I took a job as clinical research associate in the division of cardiology, where I am now. I am the only coordinator for a large volume of open industry-sponsored clinical trials. I maintain subject and other study-related records, and ensure smooth, accurate progress of clinical trial from planning stage and budget negotiations through trial completion and closure. My duties also include maintaining regulatory documents on all studies, coordinating all participant visits, attending study meetings, training, and all aspects of trial administration. This includes correspondence with our local IRB and sponsors, including applications, continuing reviews, adverse event reporting, and clinical research form completion."

—Annie Pennella, MS, SUNY Upstate Medical University, New York, USA

Clinical Research Manager

"I began my career as a registered nurse in the emergency room (ER). After about five years in the ER, I transferred to the cardiac catheterization lab (CCL) and began working with interventional cardiologists; this is where I was introduced to research. The director of cardiology was very involved in research trials, and many of them took place in the CCL. A friend took a position in cardiology research, and as I began to hear more about this area, my interest grew. After about 2 1/2 years in the CCL I transferred to cardiology research and began to coordinate clinical trials. I had worked as a CRC for 2 1/2 years when a company approached me about working for them as a regional clinical site manager. I took the job and have found my niche. I have been in industry as a regional site manager for about 5 years now. In the course of this position, for various companies I have contributed to and designed case record forms (CRF), source documents, worksheets, pocket folders, and handouts to support a variety of trials. I have evaluated sites for participation in trials, reviewed budgets and consent forms, conducted site initiation visits (training research staff, patient care staff, and physicians on the protocol and device use), monitored CRFs and regulatory binders, and closed out sites. Basically, I have overseen all phases of the clinical trial from start to finish. I have also helped plan coordinator meetings held at central locations, and I have been a speaker at both company and outside meetings. I have had many opportunities to travel in Germany, Poland, and the Netherlands to initiate trials and proctor physicians. Recently, I have begun to take on some project management duties. I love my job! To be on the cutting edge of discovering new treatments, having such autonomy, and to travel to places I may have never seen keep it constantly new and challenging."

—Carla Terrill, RN, Radiant Medical, California, USA

Clinical Trial Support Unit Manager

"I have been in clinical research for more than twenty-five years, prior to that I worked at a university, coordinating programs for a medical school. I essentially was at the right place at the right time. I was looking to get back into the work force after an absence to have children. I knew a lot of people in the hospital setting through my previous work at the medical school. In the early days of hiring research coordinators, academic sites had little money and no idea what they were looking for. I believe it was my tenacity, maturity, and common sense that led my interviewers to take a chance on me. I have not looked back once. I now manage a team of a dozen plus dedicated, amazing research coordinators and research nurses who facilitate clinical trials in pediatric hematology/oncology and bone marrow transplant. Our studies originate from cooperative trial groups, industry, and investigator-initiated research carried out locally and at multiple sites. We are similar to an in-house CRO managing all aspects of the trials from study initiation through close out. There is no better job than this – an opportunity to make a difference."

—Susan Devine, Hospital for Sick Children, Ontario, Canada

Consultant

"My degrees (BS, MS, PhD) are all in animal science, with graduate training focused on physiology/endocrinology. My coursework also involved a lot of chemistry and nutrition. My professional career started as a staff physiologist at a small research facility attempting to establish a breeding colony of nine-banded armadillos (an animal model for leprosy research). I also supported a toxicology program—the National Cancer Institute's carcinogenicity bioassay program—and general toxicology programs for agricultural and pharmaceutical clients. The bulk of my research at my first company was government sponsored, so salaries were low. Some of my colleagues went to work for industry for much higher salaries, so I answered an advertisement to work in the pharmaceutical industry. I was hired to be a medical writer and worked on various products in clinical research and development. After a few years, I became a manager in a rapidly growing department in a changing regulatory environment. Because I was managing medical writers, CRAs, and safety analysts, I had to learn the Investigational New Drug (IND) regulations well enough to be compliant with new regulations. This then led to presidency of a consultant business. We have three primary components: medical writing (protocols, study reports, clinical sections of non-disclosure agreements (NDA)), regulatory services (filing and maintaining INDs), and training classes (GCPs, drug development, preparing study reports)."

—George D'Addamio, PhD, PharmConsult, Inc., Georgia, USA

Consumer Safety Officer

"At the Food and Drug Administration (FDA), I perform audits of clinical data containing pre-market approvals and some 510(k) submissions; perform audits of Investigational Device Exemptions (IDE) submissions; analyze and audit research misconduct and fraud complaints associate with medical device research; perform inspections of Institutional Review Boards that monitor investigational device studies; provide education, training and guidance to regulated industry; and implement FDA federal rules and regulations to ensure safe and effective decisions related to devices."

—Donna Headlee, RN, FDA Center for Devices and Radiological Health, Maryland, USA

"I lucked into clinical research. My first job was in the proposals department, and from there I became involved with contracts, finance, regulatory quality assurance, and, finally, my current role, where I develop and manage partnerships with high volume sites, including collaborating on installing infrastructure necessary to fully leverage the sites' patient populations."

—Adam Chasse, Quintiles, Inc., Missouri, USA

"I went to graduate school to learn clinical pharmacy. My career objective was a joint academic/hospital pharmacy position. Clinical pharmacy had just become the hottest pharmacy specialty. I chose Baxter Labs over the University of Cincinnati College of Pharmacy. I have never regretted that decision. My career started at Baxter Laboratories as a senior clinical research associate. The responsibilities were mostly writing clinical protocols, overseeing clinical studies, and writing and compiling IND's and NDA's. This led to learning regulatory affairs. From there, my career included both clinical research and regulatory affairs. Eventually I picked up quality assurance and product development. Currently, my responsibilities include: review of IND and IDE regulations, education regarding obligations of a sponsor-investigator, product classification, assessment of IND or IDE applicability, contents of IND or IDE, submitting IND and IDE applications to FDA, developing regulatory strategy, template documents for clinical monitoring, protocol development, annual reports, product disposition, scheduling, coordinating, preparing, attending meetings with FDA, interaction with IRB regarding research protocols, reminders of regulatory obligations, periodic document reviews, offer assistance provided by the clinical trial monitoring service, assistance during FDA inspections of clinical trials, and registering clinical trials in federal registries."

—Harvey Arbit, PhD, MBA, University of Minnesota, Minnesota, USA

IRB Coordinator

"I was an administrative and legal assistant before I became involved in clinical research. A friend of mine that was working for an IRB asked me if I was interested in a new career and a position that they were looking to fill. That was six years ago and I am happy that I decided on a new career path. I am now responsible for the processing of all studies that meet the criteria for expedited review in human clinical research."

—Pricilla Gage, CIM, IntegReview, Austin, Texas, USA

Research Compliance Officer

"My background is in nursing. I worked briefly as a staff nurse in the hospital. I then worked as research coordinator over several years. As a research compliance officer, I am now responsible for overseeing all research activities involving human subjects, including compliance with state and federal regulations, developing and implementing a human subjects protection program, and the administration of the IRB and the Institutional Biosafety Committee."

—Marti Benedict, RN, SUNY Upstate Medical University, New York, USA

Principal Investigator

"I first obtained my PhD in basic sensory physiology before going to medical school—I was intending on a research career from the beginning. Clinical research offers the chance to perform meaningful studies on the same patients one cares for, which for me was a powerful incentive. I started as a trial investigator, and my activities over time have helped me reach a stage where I could assume a leadership role in multicenter trials. I spend about 40% time on departmental administration, 30% seeing patients, and 30% on research. I am co-chair of a clinical trials consortium, and we coordinate training and certification of sites on the outcome measures used, as well as supervision of study monitoring."

—Jeremy Shefner, MD, PhD, Department of Neurology, SUNY Upstate Medical University, New York, USA

Program Manager

"In my final year of graduate school, I was a research assistant. It was very satisfying to help find answers or solutions to research problems. At the same time, as clinical nursing supervisor on the night shift, my director would ask me to design and conduct audits. I did a technology assessment study that showed how some technology in healthcare easily becomes mainstream and overused, even in settings where they are not required. Afterwards, I worked *per diem* for the cardiology division, coordinating in-patient clinical trials. Our then vice-president for nursing "volunteered" my name to a family practice physician who wanted to be a principal investigator (PI) in a clinical trial for in-patients with community acquired pneumonia. I enrolled patients following my night duty tour. As it turned out, our site was one of the top ten enrollers. Who knew! As a result, our site was audited by the sponsor, and we got top marks. Since then, the hospitals supported the creation of our current office, centralizing all sponsored clinical research with that PI as its founding director. I am mostly concerned with administrative functions, but focus on leadership, growth, active search for long-term research alliances/ partners, actual medical grant writing, research design and methodology, budget preparation, contract negotiation, project implementation, execution, quality improvement, risk management activities, and support for medical education research activities in the hospitals in terms of lectures related to the responsible conduct of research.

—Martha Nelson, RN, MS, MPA, United Health Services Hospitals, New York, USA

Clinical Research Coordinator

"I started out as a registered nurse on a patient floor. I will be the first to admit I fell into research. I had had enough of the clinical floor and a nurse's aide telling me how to be a nurse. During a little meltdown at the nurse's station, two physicians overheard me say that I would rather work in a grocery store. They said they were looking for a nurse to do research studies. The rest is history. Now I pretty much do everything: prepare IRB applications, regulatory documents, budgets, contracts, consent patients, perform study specific visits/ procedures document data and case report forms, and track money."

—Martha Castle, RN, Crouse Memorial Hospital, New York, USA

Clinical Research Associate

"Getting involved in clinical research was a stroke of luck. My first "real" job out of college was as a cytogenetic technologist. I was working in a diagnostic lab investigating karyotypic abnormalities for about three years. I then joined a pain control clinic as a clinical research data collector. I have since worked through the ranks to study coordinator, assistant project manager, project manager, and now research associate (and am up for a promotion to senior research associate). I am the coordinating center regulatory monitor for a 77-site international, government-funded, multi-center trial focusing on type 2 diabetes and cardiovascular disease. I keep track of all local IRB submissions/approvals and help guide study management (that is, the PI and sponsor) in regulatory and IRB issues. I also serve on the IRB, though that is not a paid position."

—Angela Kimel, MBA, Wake Forest University School of Medicine, North Carolina, USA

Site Study Coordinator

> "I was not involved in clinical research initially as a career. I was a medical records analyst/correspondent for a family practice health maintenance organization (HMO). A friend told me about the position, I applied, and here I am, nine years later. I am now responsible for large-scale respiratory studies within oncology, dealing with physician-initiated and industrial clinical trials. My responsibilities include data capturing, IRB submission, correspondence, billing, labs, site visits, monitoring visits, and so much more."
>
> —*Linda Spillett, SUNY Upstate Medical University, New York, USA*

To find additional positions and descriptions, visit the sites referenced in this chapter, as well as industry, academic medical centers, private clinical research facilities, and vendor job opportunity web pages. Clinical research is a vibrant and growing field with many opportunities. However, a formalized educational system needs to be promoted to aggressively market this exciting and worthwhile profession.

Useful Resources:

Human subjects protection web-based training:

- Belmont Report (can be accessed from the FDA website: www.fda.gov)
- National Cancer Institute's Human Participant Protections Education for Research Teams: www.cme.cancer.gov/clinicaltrials/learning/humanparticipant-protections.asp
- The CITI Program for the Protection of Human Research Subjects: www.citiprogram.org/

Certification opportunities available through various industry, institution and clinical research educational non profit organizations:

- Drug Information Association, or DIA, offers an Investigator Course.
- Society for Clinical Research Associates (SoCRA) hosts both an Investigator Program as well as a week-long research seminar, and also hosts a Clinical Research Professional Certification (CCRP) Prep Course and Exam for those who meet the required qualifications.
- The Association of Clinical Research Professionals (ACRP) holds two certification exams, one for Clinical Research Coordinator (CRC), and another for Clinical Research Associate (CRA).
- The Society of Research Associates (SRA) and the National Council of University Research Administrators (NCURA) provide outlets aligned for clinical research for site and academic medical institutions.
- There are additional certifications, post-graduate, and training programs available through a variety of vendor companies, just search on "clinical research" and see what results.

Websites to Visit

1. Association of American Medical Colleges: www.aamc.org/research/clinicalresearch/training/start.htm
2. Advamed: www.advamed.org
3. Applied Clinical Trials: www.actmagazine.com/appliedclinicaltrials/
4. Biotechnology Industry Organization: www.bio.org/
5. CenterWatch: www.centerwatch.com
6. CenterWatch Books and Publications: www.centerwatch.com/bookstore/pubs_profsindex.html
7. Clinicaltrials.gov: www.clinicaltrials.gov
8. Department of Health and Human Services: www.hhs.gov
9. Food and Drug Administration: www.fda.gov
10. NIH Introduction to Principles and Practice of Clinical Research (IPPCR): www.clinicalcenter.nih.gov/researchers/training/ippcr.shtml
11. NIH Roadmap for Medical Research: www.nihroadmap.nih.gov/
12. PhRMA: www.phrma.org/
13. Regulatory Affairs Professionals Society: www.raps.org/s_raps/index.asp

20 Science and Technology Policy

Luis G. Kun, PhD

Information Resources Management College, National Defense University; American Institute of Medical and Biological Engineering; Institute of Electrical and Electronic Engineers—USA, Washington, District of Columbia, USA

In June 2007, I was listening to a discussion on science, courts, and public policy on National Public Radio. They noted: "*Most lawmakers and judges did not major in biology, chemistry, or statistics. Yet they are increasingly expected to draft and enforce laws directly influencing the cutting-edge worlds of science and technology, ranging from forensic evidence in the courtroom to the contentious politics of global warming.*" Part of the discussion was about how major policy level decisions are often in the hands of individuals without a scientific education. Such individuals do not necessarily understand the implications of the science and technology that they are dealing with. In rebuttal, one of the guests opined that in most cases, lawmakers hire very smart staffers to deal with those types of issues.

Later that year, I went with three colleagues from the American Institute of Medical and Biological Engineering (AIMBE) to visit the offices of three senators and three state representatives from both the Republican and Democratic parties. The titles of the individuals we met included legislative counsel, counsel, senior legislative assistant, chief of staff, floor director, legislative correspondent, and press secretary. We discussed many contemporary issues related to medical and biological engineering, with a special focus on the role of the bioengineering and biotechnology in national science and technology policy. In a follow-up presentation, we emphasized the potential of biosensors and biometrics for homeland security. We also discussed how advancements in bioengineering, particularly involving prosthetic limbs and maxillofacial or plastic surgery, can be useful in helping wounded soldiers.

One member of our group always asked about staffers' backgrounds. To our amazement, we discovered that not a single staffer we met had a technical background in science, math, or engineering. Most, in fact, had majored in journalism—a few others had majored in political science or English. It was a shocking surprise. But this surprise also helped us realize that for bioengineering and biotechnology to

G. Madhavan et al. (eds.), *Career Development in Bioengineering and Biotechnology*,
DOI: 10.1007/978-0-387-76495-5_20, © Springer Science+Business Media, LLC 2008

contribute to society, it is important to help educate staffers about these critical subjects. Playing the role of educator for political staffers is actually a key activity for bioengineers and biotechnologists. Such activities help ensure continuous flow of funds from the government to institutions and laboratories for research and development. Within this context, there are two key issues facing us:

1. How can projects related to technology be funded if those that make the decisions do not understand the concepts or the implications?
2. How can we educate ourselves and effectively explain the science and technology of bioengineering and biotechnology to lawmakers and staffers?

Addressing these two issues is where an educated bioengineer or biotechnologist can make a significant impact. By *educated*, I mean someone who, regardless of age, stays on top of current trends, emerging technologies, and is willing to analyze, synthesize, and communicate the needs of other parts of the profession, even in areas where he or she does not have expertise. Also relevant is the fact that bioengineers may find themselves very good with their understanding of science and technology and even to be good communicators, but may not be sensitive to the needs and ideologies of the policy makers. It can take a frustratingly long time for engineers and technologists to understand and assimilate the dynamics of the legislative processes.

Unforeseen Opportunities: My Professional Trajectory

When I finished my doctorate in bioengineering at the University of California at Los Angeles in 1978, my expertise was mainly in cardiac rhythm management, (that is, pacemakers), along with biosignal processing, cardiovascular and neuromuscular physiology, and computer modeling. I went to work for IBM, and slowly became involved in the medical informatics field. Although I am not a programmer, I developed the first healthcare applications for personal computers, primarily to showcase how these devices could be used in this field. Further projects with IBM Healthcare in Dallas, Texas, led me to develop a semi-expert system for the operating room and intensive care environments, followed by some visualization and satellite transferable image processing algorithms for teleradiology, imaging, and archival applications.

While these experiences extracted the best of my technical know-how, they also led me to join various IBM committees on technology planning. This allowed me to better understand the future of medical imaging and to more clearly understand the business and decision-making side of industry. These experiences became useful later, when I began working with legislation, and in my role as member of various professional societies.

By the summer of 1986, I became the technical manager of a bedside computer system for nurses. I was also participating, together with colleagues from industry, academia, and government, in an IEEE Working Group that was trying to define standards for a healthcare information data transfer. Two friends and former presidents of the IEEE Engineering in Medicine and Biology Society

(EMBS) approached me with a proposal. There was an emerging committee at the IEEE-USA headquarters in Washington, District of Columbia, which had a charter of examining science, technology, and policy issues in the realm of healthcare and medicine. Both of my friends felt that, since I was one of the few EMBS members who was working in industry, rather than academia, I had a different perspective and set of skills to offer to the committee. As I recall this conversation, I asked myself: "Engineering *policy*? What do I know about it? Why should I be involved? What value can I bring to the table and what benefit could I receive?" As I now reflect, more than two decades later, the decision to participate in that voluntary committee changed both my professional and personal life forever.

Voluntary Professional Service: An Open Resume

People fresh from school rarely have the practical experience that an employer may be seeking. But as a career progresses, you will find that experiences gained from your participation on volunteer committees can sometimes be even more important than the experiences gained from your own salary-based job. Here is why:

- The volunteer committee environment is one that you choose—as such, you are more likely to do a good job that is noticed by influential individuals.
- A volunteer committee provides opportunities to meet and learn from others with similar interests. Of course, you will contribute as well.

Volunteer activities allow your fellow members and outsiders to observe you in action. Certain aspects that can be observed include:

- Your work ethic. Can you be trusted? Do you comply with deadlines?
- Are you self-motivated? Can you work alone? Are you a self-starter, or do you always need direction?
- How well do you work in groups?
- Your leadership. Can you delegate or be delegated to?
- Are you flexible? Are you a micromanager?
- How do you react to adversity and diversity?
- How do you deal with difficult colleagues and un-met deadlines?

After I joined the IEEE-USA Health Care Engineering Policy Committee, I helped found, and ultimately chaired, the Electronic Medical Record (EMR) and High Performance Computers and Communications (HPCC) Subcommittees. In the latter role, I was chosen to be an expert witness to Congress on the area of HPCC, and to craft a policy-level position statement. I also became a member of the Department of Health and Human Services Security team of the Health Insurance Portability and Accountability Act (HIPAA) of 1996, which developed the report to Congress on the security of health data and communications.

Besides explaining high-end computing applications to congress members through public presentations and publications, I was also involved in explaining implications of science and technology that congress members and staffers were often unaware of. This type of interaction deserves attention, as government agencies (example, the Food and Drug Administration) are responsible for creating a series of publications, called *blue books*, that accompany the President's budget for different areas such as health, defense, weather, space, intelligence, education, energy, and the Internet. Blue books are critical, because they are read by members of Congress to better understand a technology's applications and cost-effectiveness. In my interactions with various government agencies, I focused on electronic health records, computer decision support systems, and telemedicine applications, and helped translate scientific information into blue book format to help guide congressional decision-making.

My public speaking to increase government awareness about telehealth eventually led me to speak at the White House Office of Science and Technology Policy in July 1997, where I re-emphasized the criticality of global health information infrastructure. and its implications on homecare, elderly care, and chronic disease management. A bill containing many of the policies I advocated was proposed and was later signed by President Clinton as part of the Balanced Budget Act of 1997.

I was asked to give many talks both nationally and internationally that year. I was also asked to represent the Secretary of the Department of Health and Human Services in a major international convention. In the following years, I also became the acting Chief Information Technology Officer at the Office of the Director of the National Immunization Program (NIP) at the Centers for Disease Control (CDC), with the charge of creating a National Strategic Plan for Information Technology, and with developing the National Electronic Disease Surveillance System. All of these experiences provided extraordinary opportunities to use my science, technology, and engineering skills for the benefit of public health and allowed me to incorporate many of the lessons I had learned in public health to national security.

Pathways to Careers in Policy

My own entry into science and technology policy was through a less traveled, but not completely unusual, path. Professional society volunteerism has helped me in key ways, and can help you as well. By volunteering in numerous committees, you not only stay at the cutting-edge of addressing policy issues, but also get a chance to improve your cross-disciplinary growth. My advice to you is to become active in your professional society through volunteering your efforts and expertise.

With increasing awareness of the inter-connectedness of policy with science and technology, a number of fellowships are becoming available for students and professionals. The duration of these fellowships can range from a few months to several years. The most notable ones include programs offered by professional societies such as the American Association for Advancement of Science (AAAS) and IEEE-USA; government agencies such as the National Academies, National Institutes of Health, and Food and Drug Administration; private foundations such as the Robert Wood Johnson Foundation; and, lastly, fellowship programs available through many universities. These programs allow fellows to experience the inner workings of the government through collaborations with legislative law makers. They also allow fellows to formulate policy statements and write actual legislation. Fellowship experiences can be instrumental for a long-term career in policy-making.

In terms of training, there are increasing numbers of universities offering graduate degrees in policy. If you are considering a career in this field, I would advise you to review those programs and to consider securing an advanced degree related to policy. A background in bioengineering and biotechnology, coupled with advanced training in policy-making, can provide professional opportunities to tackle wide-ranging policy issues in science, technology, and public health.

Final Reflections

A career in science and technology policy can be a wonderful opportunity for bioengineers and biotechnologists interested in helping society in a different, yet important, way. Both the generalist and the specialist can have an opportunity to see how other people and organizations are tackling similar problems within the policy sector. This can be invaluable, because it prevents us from re-inventing the wheel, and can also allow others to see how the field you represent may use the same tools.

One of the benefits of knowing and understanding what others are trying to do with the same technology is that you may increase your chances of getting funded in research. Technologies that serve multiple applications are more likely to be funded and adopted by governments than those with single use applications.

Students and professionals entering a career in policy should learn to cultivate their listening skills. Policy is all about interactions and human relations, especially with people from varied perspectives, disciplines, backgrounds, and cultures. Individuals should also work toward keeping themselves sharp by reading broadly and staying aware of technology developments across the world. This is crucial when you are working under economic constraints, and are trying to bring in extant technologies as opposed to building new technologies for some particular application.

A career in policy can be simultaneously rewarding and frustrating. The frustration mainly emerges when personalities of different natures collide. For example, a policy committee typically has members with dissimilar view points. Diffusing personal differences stemming from religious, political, or personal perspectives can always be a challenge and a cause of frustration. However, the better you understand the perspectives of others, the more successful you will be. Moreover, working with a diverse set of individuals can help you with your personal growth and professional maturity. Chances are that you will meet individuals with similar interests who may like you and open doors for your career development. Hence, I wish you good luck in forming those sometimes unlikely, yet important, partnerships.

Representative Policy Fellowship Resources

- IEEE-USA Government Policy and Fellowships, www.ieeeusa.org/policy/
- AAAS Science and Technology Policy Fellowships, www.fellowships.aaas.org/
- Christine Mirzayan Graduate Science and Technology Policy Fellowship Program, The National Academies, www7.nationalacademies.org/policyfellows/
- U.S. Office of Science and Technology Policy, www.ostp.gov/
- *Guide to Graduate Education in Science, Engineering and Public Policy*, www.aaas.org/spp/sepp/sepslpc.htm

Forensic Psychology

Diana M. Falkenbach, PhD

Department of Psychology, John Jay College of Criminal Justice, The City University of New York, New York, New York, USA

I teach in a forensic psychology program for undergraduate and graduate students, and on the first day of class each semester I ask students what they want to do with their degree. Inevitably, they answer that they want to be like Clarice Starling from *Silence of the Lambs*. More recently, I have also been getting answers like "I want to be like the profilers from the TV show, *Criminal Minds*." I do get a few students who hope to be like the characters on *CSI*,[1] however, given that *CSI* is actually not psychology at all, I promptly send them to the forensic science department. The media, movies, and television have dramatized and popularized "forensics" in all fields. However, forensic psychologists are not psychics that can look at a crime scene and predict the criminal's every move. Whenever there is a big media-publicized killing, I immediately get phone calls from all the news channels asking me to tell them what made that person commit that crime – but we cannot really answer that question. Typically, forensic psychologists are scientists who apply psychological research to real life legal situations. If we are clinicians, we typically need to meet a person and gather data about them before we can make a diagnosis.

Forensic psychology is about the intersection of psychology and law. It includes the science of profiling, but is much broader than that. While the character of Hannibal is fascinating, and I can certainly see the allure of Agent Starling's character, I arrived in my career as a psychologist studying forensic issues via a slightly different pathway. I always knew I wanted to be a psychologist and I had the vague notion that I wanted to help people. I was a psychology major in college, but about halfway through my undergraduate degree, I realized how difficult becoming a psychologist actually is. I worked hard in my classes and decided to get a master's degree. I thought I wanted to do therapy with kids, but I soon realized that for me, working with children was extremely difficult and emotionally exhausting. Through a job, I became interested in the effect of hormones (in birth

[1] *Crime Scene Investigation:* A television series in the United States run by CBS Broadcasting, Inc.

G. Madhavan et al. (eds.), *Career Development in Bioengineering and Biotechnology*, DOI: 10.1007/978-0-387-76495-5_21,

© Springer Science+Business Media, LLC 2008

control pills) on sexuality and aggressive behavior of women. This interest, and very good grades in my master's program, led me to a doctoral program in clinical psychology. While there, I stumbled upon a research study a faculty member was conducting looking at aggressive behavior in psychopaths. My interest was piqued, and it never stopped being interesting. I followed through with a dissertation on psychopaths and aggression, and a clinical internship in forensics, and now that is my area of expertise.

History

Hugo Munsterberg, in 1908 [1], was the first to advocate the application of psychology to the legal system, suggesting that some people might be coerced into providing false confessions [2]. The law-psychology combination did not come into vogue until the 1960s, with the organization of the American Psychology Law Society (AP-LS) [3]. Just prior to that, in 1954, the *Brown v. Board of Education* [4] case indicating that separate but equal schools violated the constitutional rights of African American children, cited the works of social scientists, and "this is the high point in the periodic flirtation between law and social science" [5]. Shortly after that, *Jenkins v. United States* [6], established psychologists as experts in court [7].

There are now many different training options for forensic psychology, and the field has several scholarly journals. There are now three major career paths; those that are interested in the social science issues of law or research about how the law works, study of how the law impacts and regulates psychology in research and practice, and social policy research and connecting psychological data and research on legal issues [8].

General Overview of Profession and Employment Opportunities

The definition of forensic psychology continues to evolve. The broad definition remains the research and application of psychological knowledge to the civil and criminal justice systems. A more specific definition may be, "the development and application of psychological principles to the problems and administration of legal, judicial, correctional, and law enforcement systems . . . [it] is clearly rooted in the discipline of psychology and draws on other areas of the field including clinical, developmental, social, and experimental psychology" [9–10].

Forensic psychology is generally made up of clinical psychologists (e.g., clinical, counseling, school, and neuropsychologists) who have specialized training in the delivery of applied psychological services; and experimental psychologists (e.g., cognitive, personality, and social psychologists) who focus on research and theory, and the application of psychological concepts, but not the clinical application to patients. All types of *forensic psychologists* are most often employed by universities to do academic research and teaching.

Forensic Clinical Psychology

Clinical psychology is the evaluation, assessment, or treatment of people with mental illnesses, including everything from marital distress and fear of public speaking to major illnesses such as schizophrenia and bipolar disorder. Clinical psychologists are trained in the use of measures, interviews, and instruments to assess mental functioning – they learn about symptom patterns in order to diagnose illness. When clinical psychology is taken into the forensic area, it includes these activities in relation to any process or person in the legal system, such as risk assessment for potential dangerousness, criminal behavior, aggression, legal competency and insanity, corrections, probation, parole, domestic violence, and family law, including custody evaluations and child abuse investigation, victim services, prevention and treatment of psychopathic behavior, and juvenile delinquency.

A *forensic clinical psychologist* evaluates how someone's mental state may affect their behavior, and then give an expert opinion in court or in a report to a lawyer or judge. Some examples include competence, insanity, and sentencing. *Competence to stand trial* is a concept related to determining whether defendants are capable of participating in their trial, understanding what the trial is, who is participating, and what the consequences are. Criminal responsibility or insanity evaluations include assessing whether a person's mental illness contributed to or was the reason for committing a crime. In capital sentencing cases, forensic criminal psychologists may evaluate the mitigating sentencing issues at hand when the defendant committed the crime.

Forensic clinical psychologists may also get involved in civil cases such as personal lawsuits, or workman's compensation where mental health or personal suffering is part of the case. They are often involved in guardianship decisions, to help determine if person has a mental illness that inhibits their ability to make life decisions (i.e., activities of daily living, money management, and legal choices), custody cases where they may evaluate the children and/or the parents, and assessment of abuse or neglect and decisions about placement of these children.

Outside of court, forensic clinical psychologists work in prisons, treatment centers, or forensic hospitals to treat or manage people who have mental illness and have been involved in the legal system. Here, they may be involved in developing treatment programs and assessing outcomes, changing a person's behavior using therapy, anger management and social skills training, or developing rehabilitation programs for criminal behavior and substance abuse. They are also part of mental health courts, where first time offenders who have committed a misdemeanor crime and are mentally ill are taken through specialty court where those involved are educated and knowledgeable about mental health issues. *Police psychology* is also part of forensic clinical psychology that involves working for police, screening new recruits, counseling police about work issues, debriefing after traumatic events, training on dealing with mentally ill criminals, or hostage negotiations.

Profiling

While it is not the profiling of television, profiling may be about creating a psychological or behavioral picture of a criminal, based on information from the crime. Forensic clinical psychologists or *forensic experimental psychologists* may use crime scene variables, modus operandi, signature behaviors, or other behaviors that occur as part of crime but are not needed to do the crime, such as positioning, trophies, and wound patterns, to help point out who the criminal is psychologically and perhaps what purpose the crime held for them. These psychologists typically work for universities as researchers, and can also work as consultants.

Profiling as dramatized on television is typically associated with the Federal Bureau of Investigation (FBI) and the National Center for the Analysis of Violent Crime (NCAVC). If you are interested in this work, you are better off going into law enforcement rather than psychology; however, very few people become profilers for the FBI. To work for the profiling unit, you must apply and be trained as an agent in the FBI, including meeting the strict qualification standards. You then must work as an agent in a violent crime unit for a minimum of three years. Those in the NCAVC unit typically have an advanced degree in forensic science or psychology, because positions in this unit for research would require education, training, and experience in research.

Forensic Developmental Psychology

Developmental psychology considers all areas of psychology across a lifespan; child, adolescent, or elderly issues. Most often *developmental psychologists* do research at universities on issues such as child testimony, memory welfare issues after divorce, abuse, neglect, and child competence, or their psychological capacity to make decisions about their own wellbeing, or ability to understand the legal proceedings and the consequences for their legal choices. They may also work in areas such as development of diversion programs for juveniles in the justice system, or as consultants for court as experts on developmental issues in general. For instance, if someone is interested in having a child testify as a witness or victim, they may have a *forensic developmental psychologist* present research on whether that process might traumatize children, or whether children can accurately describe or remember their experiences. So this would not be an evaluation of that particular child, but rather an opinion on children in general.

Forensic Social & Cognitive Psychology

Social psychology considers the social influences affecting our lives; the impact of situational factors. Cognitive psychology is about our thoughts, perceptions, reasoning abilities, and memory. People believe their cognitive processes process the world around them perfectly, however, this is often not true. *Forensic social and cognitive psychologists* apply social or cognitive theory to legal process, and typically work in university positions or as independent consultants where they may testify about their area of study. They may research areas such as detection of deception, psychology of judge and jury decision making and problem solving, willingness to accept plea bargains, settlement processes, discrimination in the legal system, and sexual harassment.

Eyewitness memory is one large area of study. Problems can occur in each phase of memory. Some forensic psychologists study this area because "the single most important factor contributing to wrongful conviction is eyewitness misidentification" [12], and therefore eyewitness misidentification has profound implications for many court cases. Research looks at memory and retrieval, interferences, and the impact of types of questions on those processes.

Jury selection and the *voir dire* process is another area forensic psychologists study. *Voir dire* is the process where attorneys from both sides of a case question the jury to inform decisions about selection. This process was developed to ensure fair trials for those accused, but it is often considered strategically. Demographic and social information is often used to make guesses based on statistical tendencies of particular groups. Attitudinal questions assess what the jurors themselves admit to in terms of their beliefs on issues related to the case. Psychologists study this process and the ability to predict jury decisions.

False confessions are also studied in forensic psychology. Many crimes are "solved" based on confessions; however, there is documentation that people give false confessions. For instance, there have been cases where people have confessed, with intricate detail, to murders, but then the victim was found alive, or there is no evidence of that person at the scene. Forensic psychologists also study why someone would confess to a crime they did not commit.

Training Requirements

Training in forensic psychology is a difficult task. There are almost no jobs with just a bachelor's degree in forensic psychology. However, with an undergraduate training in bioengineering and biotechnology, one may embark onto graduate studies in forensic psychology. There are some positions for those with a master's degree in forensic psychology working under the supervision of doctoral level psychologists in a hospital, clinic, correctional setting, or research laboratory. A doctoral degree is required for most positions as a forensic psychologist. Doctoral programs are extremely competitive, with one study estimating acceptance into clinical programs to be 6% [13]. Each doctoral program has different requirements for entry and prerequisites. What course of study in psychology you take depends largely on your area of interest.

If you are interested in only providing client or patient care or services, such as evaluations for the court or treating sex offenders, and are not interested in doing research or teaching, then earning a Doctorate in Psychology (PsyD) will suffice. A Doctorate of Philosophy in Clinical Psychology (PhD) is a joint clinical and research degree, and so allows for working with patients and training in research. If you are interested in forensic psychology research and consultation, then a PhD in the specialty of psychology is recommended, with a research focus on forensic issues (e.g., clinical, cognitive, social, experimental or social). There are also a few programs that offer a specific PhD. For example, you can do a joint law and psychology degree (JD/PhD), which requires several additional years to meet the requirements of both degrees. You can also get a PhD in a more general psychology discipline, and work under someone who does research in forensics. So someone interested in jury perception might pursue a PhD in experimental or cognitive psychology, and work under someone who specifically researches that area, while someone interested in psychopaths, as I am, might do a PhD in clinical psychology, and focus their research on psychopaths or aggressive behavior. The key is that you apply to a program where there is already someone working in your particular area of interest who can be your mentor.

The timeframe for earning a PhD in psychology depends largely on the amount of time it takes you to complete the required research projects and dissertation. Typically, it requires four to five years of coursework and research. A PhD in clinical psychology and a PsyD require an additional final year of clinical internship and a license. Every state has slightly different licensure laws, but each state requires a license to practice. Typically, you have to pass the state exam and have a doctoral degree in clinical, forensic, or counseling psychology, plus two years of supervised clinical experience (internship plus one year of supervised post doctoral degree) to be a licensed psychologist. With further experience, you can become board certified as a forensic psychologist, which typically handled through the American Board of Forensic Examiners.

Advice for Preparing for Forensic Psychology Career

The best advice I can give to anyone interested in forensic psychology as a career is to study hard and start planning early. I have several recommendations if you feel that ultimately, you might like to get a doctorate. Doctoral programs are looking for candidates who excel in their school performance and test scores (numbers matter), demonstrate research interest (meaning – seek out opportunities for research experience), and have good letters of recommendation.

Demonstration of research interest is best provided through experience. Experience in your stated area of interest is best; however, any demonstration of dedication to research is important. To gain this experience, I would recommend volunteering to work with a professor on his or her projects. All schools have websites now that list the faculty and their research interests, so email someone working in your area of interest. (It helps a lot if you have read some of the faculty member's work and can write intelligently about it in your email. Spamming lots of professors with a generic email professing interest is a bad idea.) If your pre-doctoral program offers a thesis option, I would highly recommend completing one. This will show additional dedication to research, and provide doctoral programs with a sample of your writing and research abilities. Letters of recommendation can also be enhanced through work with an advisor on a thesis, or volunteering in a research laboratory. If your letters only come from professors in your classes, they will lack the personal connection that makes them stand out. Get to know your professors beyond the classroom, attend their office hours, participate in class, and talk to them after class so that they can comment on more than just your course grade.

Concluding Remarks

Working in forensics means you may have exposure to very difficult populations, so one important aspect of this career choice is that you need to get to know your limits and capabilities. For example, I began my career in psychology wanting to work with children, and I realized emotionally that was too difficult for me. However, I am able to interview someone about how they mutilated and dismembered a child with much more emotional distance; many people would not be able to stay impartial in such situations. Ask yourself whether you could sit in a room with a person who you know has committed a horrific crime and objectively evaluate their mental status. Or whether you can you sit in a room with women who are abused over and over, and help while not become too emotionally involved. Also, in forensic clinical work there is some degree of danger involved in interviewing the violently mentally ill. My father was a police officer, and I remember him saying that he would not want his daughters to work in law enforcement because of the danger. He was so proud and excited when I said I wanted to be a psychologist. Unfortunately, he now has a daughter who interviews psychopaths for a living.

Forensic psychology is a career that is heavily based in research, and any job in forensic psychology requires extreme meticulousness. Research is all about attention to detail, and any clinical work that is going to be submitted in court or that impacts the life of a human being requires checking and double checking of facts and details. The evaluations and research that I do are puzzles that need to be solved. Sadly, I must sometimes conclude that they are just not solvable. All forensic psychology also involves having multiple projects and responsibilities. You need to meticulously gather information from many sources, which means interacting with clerks, doctors, family members, victims, and perpetrators. Often people tease me that I always procrastinate on a project; however, what they do not realize is that as a university professor, I have never-ending deadlines regarding research, writing, teaching, mentoring, and committee meetings. As a clinician working on forensic evaluations, I have multiple cases going on at the same time, and deadlines change for the many people involved in the cases.

While overall this career requires a great deal of time and school commitment, with a rather small salary payoff, I love what I do. One of the best parts for me is the ability to work on research and mentorship in a college setting, but also do some clinical work. I have unlimited options and variety, as well as a great deal of flexibility in terms of how I schedule my time. In the end, whether through research, teaching, or clinical work, I have an impact on others.

Bibliography

1. Munsterberg, H. *On the Witness Stand: Essays on Psychology and Crime.* New York: Doubleday; 1908.
2. Ogloff, J.R.P, Beavers, D.J. & DeLeon, P.H. Psychology and the law: A shared vision for the 21st century. *Professional Psychology: Research and Practice.* 1999 30(4): 331–332.
3. Grisso, T. A developmental history of the American Psychology-Law society. *Law and Human Behavior.* 1991 15: 213–231.

4. *Brown v. Board of Education*, 347 U.S. 483 (1954).
5. *Jenkins v. United States*, 307 F.2d 637
6. Kalven, H. Some comments on the Law and Behavioral Science Project at the University of Pennsylvania. *Journal of Legal Education*. 1958 11: 94–99.
7. Ogloff, J. R. P. Law and psychology in Canada: The need for training and research. *Canadian Psychology*. 1990 31: 61–73.
8. Grisso, T., Sales, B. D., & Bayless, S. Law-related courses and programs in graduate psychology departments. *American Psychologist*. 1982 37: 267–278.
9. Otto R, Heilbrun K. The practice of forensic psychology: A look toward the future in light of the past. *American Psychologist*. 2002 57(1): 5–18.
10. John Jay College of Criminal Justice Psychology Webpage [www.jjay.cuny.edu/psychology/aboutforensicpsychology.asp]. New York, New York.
11. Howard L, Hilgendorf L. Psychologist as Expert Witness. In: Joanna Shapland, editor. Lawyers & Psychologists – The Forward, p 7–19, 1981.
12. Huff C, Rattner A. Convicted But Innocent: False Positives & the Criminal Justice Process. *Controversial Issues in Crime & Justice*. 1988 130–144.
13. Norcross J, Sayette M, Mayne T, Karg R, Turkson M. Selecting a doctoral program in professional psychology: Some comparisons among PhD counseling, PhD clinical, and PsyD clinical psychology programs. *Professional Psychology: Research and Practice*. 1998 29(6): 609–614.

Suggested Readings and Resources

1. American Psychology-Law Society, www.ap-ls.org/
2. Committee on Ethical Guidelines for Forensic Psychologists. (1991). Specialty guidelines for forensic psychologists. *Law and Human Behavior*, 15, 655–665.
3. Hafemeister, T. L., Ogloff, J. R. P., & Small, M. A. (1990). Training and careers in law and psychology: The perspectives of students and graduates of dual degree programs. *Behavioral Sciences and the Law*, 8, 263–283.
4. Hess, A. and Weiner, I. (Eds), *The Handbook of Forensic Psychology* (2nd ed.). New York: Wiley.
5. Kagehiro, D.K. & Laufer, W.S. (Eds.) *Handbook of Psychology and Law*. New York: Springer-Verlag.
6. Kuther, T. L. (2003). *Your Career in Psychology: Psychology and Law*. Wadsworth.
7. Otto, R.K., Heilbrun, K., Grisso, T. (1990). Training and credentialing in forensic psychology. *Behavioral Sciences and the Law*, 8, 217–232.
8. Roesch, R., Grisso, T., & Poythress, N. G. (1986). Training programs, courses, and workshops in psychology and law. In M. F.Kaplan (Ed.), *The impact of social psychology on procedural justice* (pp. 83–108). Springfield, Illinois: Charles C Thomas.
9. Roesch, R., Hart, S., & Ogloff, J. (Eds.). *Psychology and Law: The State of the Discipline*. New York: Kluwer/Plenum.
10. Tomkins, A. J., & Ogloff, J. R. P. (1990). Training and career options in psychology and law. *Behavioral Sciences and the Law*, 8, 205–216.
11. *Law and Human Behavior, Behavioral Sciences and the Law*, and *Psychology, Public Policy and the Law*.

Energy

Mary E. Reidy, EDM, PE

National Grid USA, Syracuse, New York, USA

Introduction

You may wonder how a bioengineering/biotechnology education can be helpful to work in the energy industry. I have a bioengineering background, and I hope to share with you the excitement and wonder I have experienced in my professional career within the energy industry.

The energy industry is a very active one, with many exciting and innovative sectors that encompass generating, transmitting, and distributing energy. This includes oil, electricity, and natural gas, as well as new renewable sources of energy such as wind, biomass, and biofuels. Increasing consumer demands for new and innovative energy sources will challenge the geopolitical landscape. Consumers are searching for improved transportation, coupled with faster and more efficient sources of communication – they expect the energy industry to serve as a foundation for acceptable solutions. These developments will continue to push toward a further "flattening" of the world [1].

At the center of this transition is the need for rapid response and adaptation. Engineers and technical professionals and skilled designers have long been expected to provide solutions to consumer needs. However, the critical difference in today's climate is that innovation is occurring so quickly. This requires a team that works together efficiently, while optimizing time and concentrating on solutions development. The technical team must understand their critical role in product development.

Insight on Change by Viewing the Past

The business model in the power industry was developed over 100 years ago, and was based on central station power plants that would serve the local needs of all consumers, including industrial, commercial, and residential. The model depended on a closed system where engineers and business owners controlled and managed the choice of generation, purchase of fuel, design of generation stations, and location of electric facilities—all under government oversight, management and rate setting.

G. Madhavan et al. (eds.), *Career Development in Bioengineering and Biotechnology*,
DOI: 10.1007/978-0-387-76495-5_22, © Springer Science+Business Media, LLC 2008

What are the Present Day Changes

Today, the energy industry is partially deregulated, meaning that the generation function is owned by entrepreneurial, regional, or global firms that may specialize in the design and delivery of generation plants. These firms often concentrate in renewables, such as wind power development, nuclear, photovoltaic, biomass, or natural gas. Still, traditional coal-fired generation held a 49% share of the total major sources of generation in 2006 [2]. Renewable energy sources during the same timeframe contributed a total of only 9% of overall power generation.

Transporting energy from the generating source to the market is now often managed by firms that are regulated by at least one regulatory agency. Industry consolidation is bringing increased innovation. Although careful management by regulators helps ensure that residential customers are provided cost-effective electricity, it is not likely that all aspects of the electric industry will be deregulated completely [3].

Relating Energy and Bioengineering

Let us investigate two areas of similarities:

1. Both fields benefit from interdisciplinary approaches. For example, in the energy industry, if a wind farm is proposed for a particular area, project managers work on scheduling the design, permitting, siting, and construction of the project [4]. The managers must often present the project to engineers and financial teams that may be working on completing an environmental review of, say, the avian path of an endangered bat. Project managers, engineers, and developers may also wish to construct and transport materials to the site.

   Failure to understand the ecological balance of nature and the protections in place for some species may not only delay the construction schedules, but may require re-routing of the windfarm. Such a relocation would not only increase costs and delay final acceptance of the design plans, but would also bring excessive public scrutiny into the proposed development. Ultimately, success requires an appreciation for the framework and timing, as well as the sociological aspect of the technology culture and its use and integration [5].
2. Both fields require the talents of people who enjoy opening up networks that were previously designed in very specialized ways. For example, the energy industry is currently investigating nanotechnology, whereby materials are deconstructed to the atomic level in a quest for materials and approaches that will allow infrastructure such as conductors, towers, poles, and insulators to be built to withstand the forces of nature [6].

   If conductors can be built with nanoparticles that will carry more current and withstand the extremes of summer and winter temperatures, and transmission towers and associated supporting structures could be designed to carry more conductors per structure, additional efficiencies will be possible. The concern here is to build and construct infrastructure that is environmentally neutral (that is, symbiotic with nature), acceptable to the regulatory agency, and visually acceptable for the local residents. Bioengineering and biotechnology involves the study of interdependencies of biological systems. In a similar fashion, the energy industry works to preserve existing species, using their knowledge of environmental needs, as well as electrical, mechanical, civil, and structural engineering disciplines, so that electrical infrastructure can be built in harmony with nature while also supplying consumer energy needs. It is these interdependencies that create fantastic opportunities for achievement even as they require broad interdisciplinary knowledge.

Innovation and Integration

Both energy and bioengineering fields are investigating new methods of analysis and new business opportunities. What began as technological momentum over 100 years ago (Frankfurt, Germany in 1891, Chicago in 1893, and Niagara Falls in 1895) has led to a framework upon which an energy industry has served and continues to successfully serve the diverse needs of customers, small businesses, and large industrial customer needs. Originating as demonstrations of the technical practicality of the polyphase universal electric supply system, the new concepts amazed, inspired, and served to create a wonderful new electrical transmission and distribution supply system [7]. Building the initial distribution lines, which were supported by wooden poles that ran along residential streets parallel to existing rail tracks, created an opportunity for engineers to test varying delivery voltages and frequencies. This evolving system transformed formerly labor-intensive operations into operations that facilitated more rapid movement of products and services to markets. Manufacturers and educational institutions cooperatively worked to train and employ engineers who understood electrical theory and could apply the theory to industry. As industry grew and provided faster delivery to customers, additional demand for skilled engineers created opportunities for advanced applications. Manufacturers adapted new methods of production, encouraging adoption by their suppliers. Successful employees adapted and learned how to apply the changes to the new and emerging systems. Customers benefited from lower prices and increased options [7].

As the frameworks of biotechnology and bioengineering continue to develop, cross-disciplinary opportunities will arise and create emerging industries that will need energy to power their business and research units. Understanding the needs of bioengineering-related firms will spur increased business opportunities for the power industry, and may create new fields of energy supply.

Project Management

When you are working on your first project after graduation, whether in research or in industry, considerable time will be spent on the negotiation process. It is important that any design you are working on takes into consideration the goal of the project and the needs of the client [8]. When does the client review the proposed design, and how often? How many changes will be considered after the first proposed design has been approved? When does the client have an opportunity to visualize and test the design in a simulated setting? Are there external parties, customers, regulatory agencies, neighbors, or suppliers and vendors who will rely on the final product as a potential input to their own work? Will the design impact other customers? How does your plan integrate the discussions and needs of all of these external audiences? Once the design has been completed, you will have to present it and its benefits to the finance team, project management, production team, sales and marketing team, and most often the customer team. It is important to take these various constituencies into account [9].

The design process is an extension of the course work that you have taken throughout your undergraduate and graduate work. You spent a considerable amount of time in lectures, listening and learning to the instructors teach engineering design. When you join a firm, you may understand theories and concepts, but it is very important for you to understand that firm's history – how it was established, and what distinguishes this firm from its competitors. Begin to ask yourself these questions and you will shortly find that your contribution can be much richer and deeper.

Connecting the Components

Bioengineering and biotechnology are rapidly growing applied fields with a solid foundation in biology and chemistry. The basic tenet is understanding how parts fit. When faced with a problem, many potential solutions exist. Development occurs as problems are disassembled and viewed from the component parts. Unlike a puzzle, the components do not come together in only one unique way. Consider a box of Legos[1], the components are many and the "creation" is determined by the designer. It is important for the designer to understand the relationships among the pieces; that is, where the connection is possible. The ultimate end product will be a solution based on the designer's understanding of the customer's unmet needs. The process that the designer uses to determine how the existing pieces interact will ultimately shape and define the end product. The answer is not the pieces, but rather the picture of the solution that rests in the designer's mind.

Here is your opportunity. Consider this approach, which allows you to shift from "forcing the solution" to "illuminating the possibilities." For example, as you continue to develop design solutions, consider the impact that the pre-existing design will have on your solution, as well as the impact your solution will have on the next level. Do you understand the relationships among the component and

[1] Legos is a brand name to describe a series of small, rectangular shapes that can be combined to build children's toys, bridges, and buildings.

supporting designs? If not, consider the impact of just such an instance during the Northeast Blackout of 2003. An electrical disturbance first noticed in the Midwest cascaded and impacted electrical operations throughout the Northeast USA and Canada, rendering some systems inoperable.

Approaches to Consider

Consider your engineering design as an amoeba, not an oak tree. Does your design adapt and support the system you are a part of and continue to respond and move forward? Is your design a system that will absorb the inputs and continue to stabilize? What does the environment around your proposed design need to continue to perform? Are there stand-alone systems, or components that need to communicate with the surrounding systems?

The Future Beckons

As you begin your journey, consider that success is a journey, measured individually, to be reconsidered, redesigned, and reshaped by you. In shaping your journey, it is important to determine your life's balance is [10]. Global developments and consumer needs will create a dynamic adventure for you. Determine what goal you are striving for – it is likely to be much different from the goals of your friends and associates. When considering possibilities for career growth, whether within your current firm, or outside, wonder "what if" rather than "when." Investigate each day and start each moment with an opportunity to truly *see.*

Developments, innovations, and possibilities will be there if you consider that the future is limitless, and that the world needs what you can develop, build, improve, and innovate. When a task appears to be extremely challenging and has no apparent solution, investigate your understanding of the interactions at play in these systems, look to your background in biology, and ask *how, why, what,* and *if.* Challenge yourself to look outside the present situation; it is not only important to remember past ways of solving the problems, but also to look for different methods, manners, configurations, and interactions.

Consider the origins needed to create a beautiful string of pearls. The magnificent presence of the pearl only occurs after the introduction of a new particle. The new particle changes and is changed by its environment, and emerges as a unique element. Managing your dream may be thought of as you holding many pearls in your hand [11]. Each may represent a design, a place, a goal but all are waiting for you to recognize them, and, at a time that is right for you, recognize and release each so that they may be realized and the world can see and appreciate and welcome them as well.

Now, Your Assignment

This is your task – to absorb knowledge, to watch, to observe nature for its resident beauty, sustainability, adaptive ability, and for the incredible wonderment and awe it engenders in humankind. Education is your stage. It creates, encourages, provides inspiration, and allows you, the actor, to bring a unique contribution to the part. In the words of William Shakespeare [12]:

> "All the world's a stage,
> And all the men and women merely players.
> They have their exits and their entrances,
> and one man in his time plays many parts . . ."

Consider how your part can be the protagonist. The *sine qua non* is the symphony that you create as a part of this world. I wish you great adventures and many challenges in your career.

Bibliography

1. Friedman, Thomas L., *The World is Flat, A Brief History of the Twenty-First Century*, Farrar, Straus and Giroux, 2005.
2. U.S. Government, Energy Information Administration, Department of Energy, Monthly Energy Review, May 2007, www.eia.doe.gov/emeu/mer.pdf.
3. Kuttner, Robert, *Everything for Sale The Virtues and Limits of Markets*, The University of Chicago Press, 1996.
4. Rose, Jonathan D., Hiskens Ian A, "Challenges of Integrating Large Amounts of Wind Power", *1st Annual IEEE Systems Conference*, 2007, Waikiki Beach, Honolulu, Hawaii, USA, April 9–12, 2007, pgs 259–265.
5. Bijker, Wiebe E., *Of Bicycles, Bakelites and Bulbs: Toward a Theory of Sociotechnical Change*, The Massachusetts Institute of Technology, 1995.
6. Sargent, Ted, *The Dance of Molecules: How Nanotechnology is Changing our Lives*, Thunder's Mouth Press, 2006.
7. Hughes, Thomas, P. *Networks of Power, Electrification in Western Society, 1880–1930*, The Johns Hopkins University Press, 1983.
8. Gil Nuno, Tommelein Iris D., Schruben, Lee W., "External Change in Large Engineering Design Projects: The Role of the Client", *IEEE Transactions on Engineering Management*, Vol. 53, No. 3 August 2006.
9. Bucciarelli, Louis L., *Designing Engineers*, Massachusetts Institute of Technology, 1994.
10. Lenehan, Anne E., *Story: The Way of Water*, The Maple-Vail Book Manufacturing Group, 2004.
11. Bennis, Warren, *An Invented Life: Reflections on Leadership and Change*, Addison and Wesley, 1993.
12. *The Portable Shakespeare*, The Viking Portable Library, *As You Like It, Scene VII.*

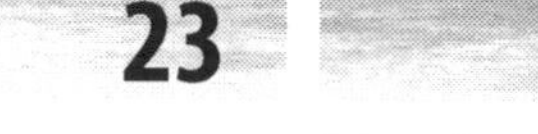

Technology Transfer

Eugene B. Krentsel, PhD

Office of Technology Transfer and Innovation Partnerships, State University of New York, Binghamton, New York, USA

"The value of an idea lies in the using of it."

—*Thomas A. Edison*

First of all, what does *technology transfer* mean, and why do people get involved in it? There are many somewhat varying definitions, but the Association of University Technology Managers (AUTM) defines technology transfer as, "the process of transferring scientific findings from one organization to another for the purpose of further development and commercialization." [1] If you take a deeper look at what the core business of an academic institution is, it is generation and dissemination of knowledge. Technology transfer is indeed a part of that general agenda, and thus becomes an intricate part of core business of research institutions all over the globe, and especially in the United States.

Technology transfer occurs in many ways, be it through research publications, conference presentations, or through increasingly more sophisticated concerted institutional efforts coordinated by academic technology transfer offices. Each institution that actively engages in technology transfer does that for a multitude of reasons. Here are some of them:

- It is in human nature to seek recognition for one's achievements. Credit for discoveries made at a university adds to that institution's reputation and visibility, both nationally and internationally.
- The passage of the Bayh-Dole Act, which was sponsored by two senators, Birch Bayh of Indiana and Robert Dole of Kansas a little more than a quarter of a century ago, dramatically changed the landscape of technology transfer from universities to industry. This had been a special area of contention when federal government funds had been used to develop the inventions. The legislation has stipulated that default ownership of the title to such inventions was to be given to non-profits, including universities and small businesses. Prior to that it had remained with the government. Academic institutions are now encouraged to partner with private companies to advance public and private use of inventions

G. Madhavan et al. (eds.), *Career Development in Bioengineering and Biotechnology*, DOI: 10.1007/978-0-387-76495-5_23, © Springer Science+Business Media, LLC 2008

stemming from federal funding. At the same time, the Bayh-Dole Act puts certain requirements on academic institutions receiving federal funding (e.g., disclosing each new invention to the federal funding agency within two months after the inventor discloses it in writing to the university; filing a patent application within one year upon election of the title; and providing the government, through a confirmatory license, a non-exclusive, non-transferable, irrevocable, paid-up right to practice or have practiced the invention on behalf of the US throughout the world). The government monitors compliance with those mandatory requirements.

- It is no secret that academic institutions compete fiercely for the best and brightest faculty. Institutional support in bringing faculty inventions to the market is an important tool in recruiting and retaining these highly sought-after researchers.
- The link between successful technology commercialization and local economic development is very apparent. Names like "Silicon Valley" and "North Carolina Research Triangle" have become associated with success in creating a working mechanism for creating new enterprises that take advantage of innovative research and technologies developed in academia. This innovation yields hundreds and thousands of patents and licenses, and produces tens of thousands high-paying jobs, dramatically impacting the local and national economy.
- Research funded by industry contributes a relatively small amount to the total volume of academic-sponsored research, but the amounts are growing, as are the importance of these amounts as a consequence of the stagnation, and even downward trend, in funding from governmental sources. Ability to attract industrial sponsors largely depends on an institution's abilities to be efficient and flexible in handling issues related to intellectual property. Here, technology transfer professionals are key.
- Last, but not necessarily least – licensing revenue is shared between inventors and their respective institutions. Distribution schemes and percentages vary, but typical ranges are usually anywhere between 30 and 50% of net funds received by the institution that go to the inventors as an incentive to continue innovating. While only a selected few institutions can claim royalty revenues contributing to a significant share of research funds, even small amounts are important, if only because these funds do not have many restrictions on how they need to be spent, other than that such spending should advance education and research at the university.

Here are a few numbers that illustrate the scale of technology transfer activities at US academic centers. The 2005 AUTM US Licensing Survey, the most recent data currently available [2], notes:

- More than $42 billion in research & development expenditures
- 527 new products introduced into the market in 2005 – 3,641 introduced from Fiscal Year (FY)98 through FY05
- 628 new spinoff companies created in 2005 – 5,171 since 1980

- 28,349 current, active licenses (each single license represents a one-on-one relationship between a company and a university)
- 4,932 new licenses signed in 2005

The survey also notes that "university technology transfer adds more than $33.5 billion to the economy and supports 280,000 jobs each year." It is worthwhile to point out that about 70% of successful technology transfer transactions are directly related to biotechnology and bioengineering.

That said, the utmost goals and primary objectives of academic technology transfer are to ensure that research ultimately reaches and benefits the public, improving people's lives locally, nationally, and, in the end, globally. Naturally, all of these efforts require people – bright, well-educated, and knowledgeable in various aspects related to the technology transfer process. These people can have backgrounds in many different fields, the including science, engineering, business, and economics.

What Do Technology Transfer Professionals Do?

- Facilitate new technology disclosures by working with faculty, research staff, and students to assist them in determining whether they may have come up with something that is worth protecting, and guiding them through the disclosure process. Educational aspects play a very important role in relationships between tech transfer officers and researchers, with each party patiently educating the other – on one side, involving important details of cutting-edge research being conducted in a lab, and on the other side, regarding the importance of proper protection of intellectual property and the role that protection plays in determining chances of bringing a new technology to people. Companies, in particular in biotech and pharmaceutical fields, will need to invest millions of dollars to further develop a technology coming out of a university lab, and most likely will pass on even a very promising invention if they can not recoup those development costs. Putting a new drug formula straight into the public domain may well result in that drug never being manufactured, and, therefore, never having a chance to improve people's lives. Tech transfer officers often speak to groups of faculty and students on issues related to innovation management and intellectual property, and provide necessary advice, direction, and support.
- Provide guidance in patenting and other means of protecting intellectual property by working hand-in-hand with an outside or in-house patent counsel in devising a sound protection strategy for a patent, copyright, or trademark, and taking care of all of necessary details. As mentioned above, proper protection of intellectual property is critical for improving chances of any invention actually making it to the market. Covering all potential applications and products in patent claims is paramount. Tech transfer officers also face the need to maintain a good balance between protecting everything they see fit and the financial limitations of their institution. This need also comes into play when a decision

needs to be made whether to continue maintaining (and paying a maintenance fee to the US Patent and Trademark Office) an issued patent that has not been licensed yet.

- Negotiating material transfer agreements (MTA), non-disclosure agreements (NDA, sometimes called confidentiality disclosure agreements, CDA, or proprietary information agreements, PIA), evaluation agreements, inter-institutional agreements, and other agreements to ensure proper handling and protection of information throughout the process of preparing the invention for commercialization. Researchers like to collaborate with their colleagues. They also like to be recognized by their peers for reaching certain attainments before others. These desires are very important, and they provide a potent driving force. But it is important to maintain a good *balance* (this word becomes a theme for this article!) between the need to publish and the need to keep information confidential until at least a provisional patent application has been filed. Tech transfer officers have to keep constant contact with their researchers, and often make a very quick (sometimes requiring an overnight turnaround) decision on filing a patent for something that is ready to be published or presented at a conference. Maintaining good relationships with a patent counsel, who is usually "outside" the university or company, is very important, as he or she may have to do an overnight turnaround as well. If information has not been made public yet, it has to be protected from becoming such when a need arises to start talking with a potential licensee or an industry partner about collaborating in commercializing that technology. Thus, a need for an agreement such as an NDA or MTA to keep the information protected. Failure to do so and allowing the information to become public will immediately result in a loss of all intellectual property rights outside of the United States, and start a one-year clock within the United States (the US patent system differs from the rest of the world and allows a one-year grace period for filing a US Patent application after publication or public presentation).
- Licensing and providing help with market analysis, technology assessments, active marketing of new technology, negotiating terms and conditions of licensing agreements with industrial partners, and ensuring that technology gets into the marketplace efficiently and provides all parties with a win-win deal. Going back to the issue of patenting, making a decision on whether to have something patented does not have any bearing on the academic value of the invention. Companies often patent simply to protect their marketplace. For academic institutions, on the other hand, it would only make sense to invest time and money when there is a good chance of having that invention licensed, or, even better, having a licensee for it already. A US patent application alone may easily cost $15K to $20K or more before the patent is issued, and international protection may cost ten times that.
- Helping with formation of start-up businesses based on university-developed technologies, and assisting entrepreneurs with access to various resources, including financing, various incentives, and business planning. Quite often technology transfer offices manage incubator facilities that house these start-up companies.

- Providing guidance to senior management and various administrative units throughout the institution in regards developing and implementing sound business and legal policies that encourage innovation and do not detract industry from coming to your institution for new technologies and other products of research.

Technology transfer offices can be quite large (40+ people) or very small ("an army of one"), or just about anything in between, with the vast majority having fewer than ten people. These offices often work very closely with contracting and sponsored research units, but are sometimes separate from "contracts" functions.

I would like to conclude by noting that the way technology transfer process works – from building good relationships with researchers and innovators to handling patenting decisions, from marketing inventions to negotiating licensing agreements – can vary from institution to institution. One thing remains constant though: this process needs to be carefully managed by people who are knowledgeable and *passionate* about the role that they play in making a transition between an idea and innovation that has found its way to do public good. As our economy continues its transition from manufacturing-based to becoming more knowledge-based, the importance of skillful management of intellectual property in improving quality of life and social and economic well-being, will grow dramatically.

References

1. *The Association of University Technology Managers – About Technology Transfer*, www.autm.net/aboutTT/index.cfm
2. *The Association of University Technology Managers – FY 2005 Licensing Survey*, www.autm.net/surveys/dsp.surveyDetail.cfm?pid=33

24 Politics and Legislation

Assemblyman David R. Koon

135th Assembly District, Fairport, New York, and Legislative Offices, New York State Capitol and Assembly, Albany, New York, USA

"We cannot solve our problems with the same thinking we used when we created them."

—*Albert Einstein*

November 13th, 1993 will always remain an unforgettably tragic day for my wife and I. Our 18-year-old daughter, Jennifer, who was a college sophomore, decided to stop for bagels at a nearby shopping plaza after finishing her shift at the psychology clinic where she worked. As she was leaving the shop at about 11:30 a.m., she was abducted. After being held captive for a time, Jennifer managed to dial 911[†] on her car phone, but her urgent call for help was for naught—she was raped and shot to death in bright daylight. In this moment of life and death urgency, the 911 dispatcher and her supervisor could not even determine which cell tower was transmitting Jennifer's signal; they had no idea where she was. All they could do was helplessly listen to the last twenty minutes of our daughter's life.

Jennifer's phone, as it turned out, did not contain location-enabled technology–what is now commonly referred to as the global positioning system, or GPS, a technology that was not available until 1995. Though the potential live saving benefits of this technology had been realized across the society, it was still awaiting broadband implementation. If we had had today's technology in 1993, my Jennifer might still be alive today.

Our family's tragedy prompted my involvement in public service—I became determined to help make my state, New York, a safer place. Since that life-altering episode, I carry a copy of the tape from Jennifer's 911 call in my briefcase at all times. It serves to keep me grounded, but more importantly, it strengthens my determination and motivates me not only in my day-to day activities, but in my life's quest for the safety of others.

After the trial, Jennifer's assailant was convicted and is now in a New York state penitentiary, serving 37 1/2 years-to-life. It was right after the trial that my first duty

[†] The standard emergency telephone number in the United States.

G. Madhavan et al. (eds.), *Career Development in Bioengineering and Biotechnology*, DOI: 10.1007/978-0-387-76495-5_24, © Springer Science+Business Media, LLC 2008

as a citizen-protector began, as I approached my county legislator about getting security cameras installed in the parking lot of the shopping plaza where Jennifer was abducted. I pointed out that we use security cameras to protect the *merchandise* in the stores, but we do nothing to protect the *people* buying the merchandise. The legislator never acted on my plea for public safety.

At the time, I was working as an engineer for Bausch and Lomb, an international corporation known for its ophthalmic engineering and vision health products. By training, I am an industrial engineer, with twenty-eight years of experience in manufacturing engineering, as well as process and product systems analysis. My responsibilities in technical project management have included time-series analysis, operations, cost justification, scheduling, efficiency investigations, and modeling of technical requirements. *Engineer* is an identity that *defines* me—my personality, my interests, and my abilities. In some sense, you could say engineering is my way of life.

Even though I was an engineer and not a politician, I believed that I could do a better job than the incumbent legislator—at the very least, I knew I would get back to people. So I decided to run for the legislator's seat in an election that year. After extensive campaigning while working full-time at Bausch and Lomb, I lost that election by about 600 votes; a relatively close margin, considering no one had contested that seat in the past fourteen years. Buoyed by that constructive experience, I ran for the state assembly in 1996 during a special election. This time, I campaigned for several weeks, going door-to-door in six to ten inches of snow and at harsh winter temperatures below freezing. With great determination, I explained my life's quest and the reason for it.

I won.

Suddenly, as an assemblyman, I found myself in a position to influence change. I went right to work on a public safety initiative involving enhanced 911 or E-911. Unfortunately, my fellow politicians did not understand the technology. I then extensively educated my colleagues and got legislation moved to the governor's desk, but he simply and repeatedly vetoed this vital initiative.

Undaunted, I turned to my project management skills, which were to prove as useful in politics as they had been in engineering, I worked hard to push a new bill that involved redirection of monetary funds from various sources through the state assembly into E-911 programs. But the bill rested in the senate for long periods without any action.

I am a typical engineer, and I will admit that sometimes I have been impatient, frustrated, and disappointed by the length of time involved in typical political processes. But politics has taught me the value and benefits of negotiation—this knowledge has been a benefit for me ever since. After thorough research involving each and every line of the state's operating budget (thanks to the critical skills I learned from my cost justification and efficiency analyses in engineering), I proposed a cost-cutting deal to the state governor.

Again the governor declined. And then again.

But the legislation, as it turned out, was poised for implementation—it needed only one last tragic push. Four teenagers drowned in January 2003, when their rowboat capsized off City Island in the Bronx, New York. Like my daughter

Jennifer, one of the boys managed to make a 911 call using a wireless handset. According to published reports, the boys told a dispatcher they were on Long Island Sound and their rowboat was taking on water. But the dispatcher could not determine their location, and a supervisor allegedly decided there was not enough specific information to notify rescue authorities. This caused an unconscionable fourteen-hour delay in the search for the boys—and hence, a media firestorm.

I, too, expressed my outrage, telling the press that those deaths were on the governor's shoulders for his lack of support on the E-911 initiative. That statement made headlines. My persistence was finally recognized by the state senators, who appropriated $100 million of the 2003–2004 federal government budget.

But the governor exercised yet another veto.

The recent tragedy had helped the legislature to override the governor's veto. The dollars have since then been spent on deploying E-911 technology throughout New York state. New York will be the first large state to meet the Federal Communications Commission's regulations on E-911. It had taken me nearly a decade to achieve my goal, but it was well worth the struggle. Now New York has an E-911 facility with triangulation and GPS features for anyone dialing the emergency number. What is even more satisfying is that the rest of the country has followed suit in implementing this crucial public safety system. Hopefully soon, anyone, anywhere within the United States, can have the safety net that might have saved our dear Jennifer's life.

Why Does Politics Need Engineers and Technologists?

My engineering training and experience have given me the knowledge I need to be able to help ensure public safety. However, as importantly, politics has given me the ability to *exercise* that knowledge. My day-to-day responsibilities as an assemblyman make full use of my engineering skill set—I serve as a member of the committees on economic development, job creation, commerce and industry, local governments, small business, alcoholism and drug abuse, and library and education technology. I also co-chair the legislative commission on rural resources, as well as the legislative commission on toxic substances and hazardous waste within the assembly constituency. My constituents and I meet periodically in my assembly district, and my duties also involve weekly visits to the state assembly and capitol to complete the link between the government and the public. All of these responsibilities provide a platform that capitalizes on my engineering training and skills for the greater good of everyone.

All national, regional, and local governments in the world could undoubtedly make use of efficiency enhancement, and who better to improve that efficiency than engineers? Manufacturing, service, and even healthcare industries have adopted the principles of total quality management (TQM) and Six-Sigma. I truly believe these engineering tools can help the efficiency and functionality of politics as well. Engineers are experienced at enhancing cost-effectiveness and process efficiency; they, perhaps more than any other group, have the potential to cut away the bureaucratic tape that is the bane of the political process. Politics desperately needs new blood and novel thought processes to solve problems from a systems perspective, and to simply get things done. Engineers can capitalize on this opportunity, much as they do when involved in new product development and entrepreneurship. But a word to the wise—in the world of politics, it takes days instead of hours, and months or years instead of weeks, for action. This different way of operating can be frustrating for engineers and technologists. It is prudent to learn the art of patience.

In many parts of the world, people have given up on their deeply dysfunctional governments. Instead of healing the system, they end up alienated from it. Unfortunately, it is all too easy to look at a dysfunctional process and cynically blame politicians. This alienation and public cynicism is unsafe for the future health of governments and nations. Engineers, I believe, can play an important role in the healing process.

Perhaps surprisingly, engineering and information technology have been very useful in recent years in creating awareness amongst the public about political processes. Two very successful recent technology examples include the successful deployment of open source software for monitoring legislative earmarks, as well as blogs, which improve the capability for political critiques, decentralization, and transparency. If interdisciplinary thinkers and public servants from the fields of engineering and technology join hands and become directly involved in politics and governance, the effectiveness and efficiency of these systems are bound to experience dramatic improvements. Legislative processes and actions could be completed in a more timely fashion, and decisions could be reached systematically and effectively.

Entry Strategies into Politics

My entry into politics, which happened because of Jennifer's tragedy, was unconventional; however, I would ask a single, basic question of any engineer or technologist who is thinking about the possibility of serving the public through a career involving politics and governance. That question is—*are you involved in your community*?

I invariably ask this question of those I interact with—even constituents who approach me seeking help, regardless of their party affiliation(s) or vote. By the word *involved*, I mean, do you know your neighbors? Do you try to help them? Do you volunteer on community boards? You really cannot complain to your elected official if you are not trying to make the community a better place by doing something yourself. I think this is an individual responsibility for all of us. Vaclav Havel, the first president of the Czech Republic, described the essence of this idea:

> "Genuine politics—the only politics I am willing to devote myself to—is simply a matter of serving those around us: serving the community and serving those who will come after us. Its deepest roots are moral because it is a responsibility expressed through action, to and for the whole."

My performance in politics today flows directly from my passion and involvement in the community. Even before Jennifer's death, I had always devoted myself wholeheartedly to such activities; I was a church councilman, a Mason, a member of the Optimist Club, and a Boy Scout leader. I always kept myself engaged in community service wherever our family moved, be it on an executive board for scouting and or as the president of a Parent-Teacher Association or Lions Club. In 1994, after I began my activism to prevent crime, I was appointed by the mayor of Rochester, New York, to a team to mitigate crime and violence in the community. Ultimately, I became a leader on the Task Force to Reduce Violence. The group also worked repairing homes, planting gardens, and cleaning streets, demonstrating that community members do have a desire to bring an end to violence and are absolutely willing to devote their time and energy working together to achieve peace.

So my fundamental advice to students or professionals who want to enter into politics is to engage themselves in any of the many community projects that are available. Start becoming a change agent at a local level. This will also help to improve your networking and people skills, which are valuable for engineering work as well. Always remember, politics is all about people and their networks; you will always need this support if your political career is to survive and prosper.

Many institutions, policy centers, and local political offices offer internship opportunities for college students. You can make use of those internship opportunities to gain experience with politics and policy research. This type of work can help you learn how to analyze and distribute information and become familiar with financial issues, campaign activities, and international and domestic issues. Interning at a state or federal office can also help you understand the "statics and dynamics" of bill formulation, proposal, and passage; as well as the process of holding public hearings prior to the formation of new laws. Educational and experiential service opportunities in diplomacy, international relations, foreign affairs, and global relations are also available. The pre-doctoral and post-doctoral congressional fellowships offered by many state governments are also a worthwhile way to get a feel for political processes.

Recommended Qualities for "Political Engineers"

Politics is a very difficult field. (If anyone says it is easy, then you can be sure he or she is not doing the job properly.) Based on my experience, I recommend the following five qualities for an individual from engineering and technology to succeed in political service.

- ***Honesty:*** Regardless of your political affiliation, a fundamental duty for anyone in this field is to be honest and upfront. History offers abundant examples of politicians who have done extraordinary work to help those they serve—these politicians are remembered with gratitude for decades and even centuries after their deaths. But history also holds examples of shyster-politicians who used lies and fakery to fulfill their selfish desires. Do not have your name go on that latter list.
- ***Passion:*** Only the most passionate will survive in politics. If you are genuinely committed to a cause, and work diligently in relation to that cause for public welfare, you are bound to gain people's support. In the story of an egg and bacon breakfast, the chicken was involved but the pig was committed.
- ***Persistence:*** Politics means constantly fighting battles for your cause of serving people. This means you must be persistent and resilient. Giving up is an option only for those who are not fully committed to their cause. Inspiration can be

derived from the countless high-minded political leaders who have clung persistently to their mission and changed the course of history, and the destinies of nations and societies.

- ***Public Relations:*** Politics is a profession of helping and serving people. A politician will learn about—and be influenced by—many different lives. In some cases, the politician may in turn influence the lives of others. A career in politics demands individuals who genuinely care for people and relate to them on a personal level. Good public relation skills are important, especially when it comes to educating fellow politicians; your skills in this area will play a dominant role in determining how many political friends you will make during your career, and whether your colleagues will support your bills. Effective relations with others is vitally important in gaining support and spreading your professional cause. You do not have to be born with public relations skills, but much like any other skill, public relations can be learned and developed through practice.
- ***Systems Thinking Skills:*** Finally, engineers with an ability to solve problems by considering all components of a system are truly a great asset to the political system. Unfortunately, most political issues are non-linear and emerge unexpectedly. This requires engineers to understand non-linear systems and the blind spots from which they emanate, in order to develop reasonable provisions to help prevent, or react appropriately to, these unexpected problems instead of simply managing their uncontained aftermath. In politics, systems-based thinking is key. Engineering is one of the very few professions that teaches and develops this mental framework.

Concluding Remarks

Among the very best inventions of the Greeks and Romans over two thousand years ago, was the democratic process—a way to help people more effectively govern themselves. Elected officials in this model were primarily *servant* leaders. But over time, the situation has shifted. Elected officials often come to believe that it is the people who are beholden to them. Politics is a platform of great power; individuals can become self-absorbed and begin playing games of manipulation and exploitation. Whatever one's position in the political hierarchy, be it a council leader or the president of a nation, it is his or her fundamental duty to serve the people. Engineers and technologists can effectively serve as a new breed of individuals to resurrect the true intent of the democratic process and change the way systems operate. It is my great pleasure to invite my fellow engineers and technologists to consider public service through a career in politics. It is high time we got *real* problem solvers on board to address important social issues.

> My Country! When right keep it right; when wrong, set it right!
>
> —*Carl Schurz*

Suggested Reading

Taleb, N., *The Black Swan: The Impact of the Highly Improbable*, Random House, 2007.

25 Social Entrepreneurship

Robert A. Malkin, PhD

Engineering World Health and Department of Biomedical Engineering, Pratt School of Engineering, Duke University, Durham, North Carolina, USA

Everyone told David Green that it was impossible. No one could make intraocular lenses for $6, and such devices certainly could not be manufactured in the developing world. In fact, some of the most prominent specialists in global health told David it was inappropriate to even *think* about using intraocular lenses to cure blindness from cataracts in the world's poorest nations. Nevertheless, six months after obtaining initial funding, David Green's company, Auralabs, was producing intraocular lenses for $5 a piece and selling them in India using a sustainable business model.

David Green is a social entrepreneur.

According to Bill Drayton, CEO of Ashoka, a *social entrepreneur is someone who decides that something in society must change, then figures out how to change it – no matter what.* If you think that describes you, then you could be a social entrepreneur.

What Is (and Is Not) a Social Enterprise?

A social enterprise, at least as it is practiced in bioengineering, is a business delivering, or enabling the delivery of, a medical device, pharmaceutical, or service. First and foremost, however, a social enterprise is a business. Like all businesses, a social enterprise must obtain revenues that exceed expenses, or it will fold. However, unlike a typical for-profit business, a social enterprise:

- has a primary mission to create social value, not profit;
- typically operates with exceedingly limited resources, and
- exhibits a heightened accountability to its stakeholders.

It can be easier to recognize a social enterprise than to define it. The Goodwill, the Boy Scouts, the American Heart Association, and a local soup kitchen are all easily recognizable examples of social enterprises. For bioengineers, organizations like PATH, Project Impact, Benentech, and Engineering World Health typify the social enterprise.

G. Madhavan et al. (eds.), *Career Development in Bioengineering and Biotechnology*,
DOI: 10.1007/978-0-387-76495-5_25, © Springer Science+Business Media, LLC 2008

However, precisely defining the border between the social enterprise and the typical business can be difficult. General Electric Corporation (GE) has a large program for donating medical equipment to Africa. However, few people consider GE a social enterprise. Likewise, Auralabs, described above, sells their lenses strictly for-profit. Yet, many consider Auralabs a social enterprise.

There are a few things that are commonly (some incorrectly) considered hallmarks of the social enterprise.

Myth #1: Social enterprises do not sell a product for a profit. This is definitely not true. In fact, the definition of a non-profit corporation is not related to the price at which a product is sold. You can be sure that the cookies you buy from the Girl Scouts cost less to produce than you paid for them. The restriction is that non-profits cannot distribute any portion of the profits to shareholders.

Myth #2: Social enterprises are non-profits. In fact, not all social enterprises are non-profits. Some are for-profit. Although still outside of mainstream social entrepreneurship, some for-profit startups include in their articles of incorporation clauses placing the need to be socially responsible equal to the need to return profits to shareholders.

Myth #3: Social enterprises are tax exempt. Many engineers misunderstand the meaning and implications of this aspect of US law. Also, there is no single, fixed relationship between a corporation's tax status and the tax implications for its donors. Fortunately, for bioengineering social enterprise startups, obtaining 501(c)(3) status (a US Federal designation that affects donors' taxes and may affect the corporation's state or federal taxes) is more a matter of gaining legitimacy than a significant startup roadblock. Tax exempt status does not matter at first.

Myth #4: Social enterprises use a volunteer labor force. For most social enterprises, this is far from true. Labor costs are too small a fraction of total expenses, and once infrastructure and insurance are considered, using volunteer labor is not that great a savings. However, bioengineering is an exception. The medical device industry is a still largely a low-volume industry, leading to a relatively high fraction of total expenses devoted to labor costs. And the delivery of medical care is often dominated by labor costs, especially professionals like doctors and nurses.

Who is a Social Entrepreneur?

Many bioengineers selected their profession because they were interested in helping people. So it makes sense that bioengineers would be more inclined toward social entrepreneurship. While this inclination may be there, it may not be sufficient to make a bioengineer a successful social entrepreneur.

To determine whether you could be a successful entrepreneur, consider the qualities that Gregory Dees and William Bygrave ascribe to social entrepreneurs. Does this describe you?

- *Dreamer:* Social entrepreneurs are driven by a vision for social change.
- *Decisive doers:* Social entrepreneurs do not procrastinate. Rather than looking for reasons to avoid action, they seize opportunities to act.

- *Determination and dedication:* Starting a social enterprise is extremely difficult. Total commitment and personal sacrifice are required.
- *Devotion:* Social entrepreneurs love what they do, and it shows.
- *Details:* Social entrepreneurs must be able to connect the vision to the details, because the details can devour the fledgling enterprise.
- *Dollars:* Getting rich is not a strong motivating factor. Bioengineers who become social entrepreneurs should expect a significant cut in salary.

In this list of attributes, no mention is made of a business degree. The reason is that in many business schools, the critical skills of social entrepreneurship are not taught. There is no formal education that is typical of the successful entrepreneur. However, there is one skill that is critical.

The most critical skill of an entrepreneur is *salesmanship*. More than any other single skill, entrepreneurs must be able to sell their ideas. Social entrepreneurs have an even more difficult hurdle to overcome than most salespeople, because they must sell their idea to both clients and donors, often for very different reasons. Indeed, the inability, or perhaps unwillingness, to sell is probably the most common reason engineers fail to be successful entrepreneurs.

If you are not a natural salesman, there is still hope. There are classes in selling. And there are books, such as *3 Steps to Yes* by Gene Bedell, that place selling in a framework that is comfortable for engineers.

However, many engineers are more comfortable sitting in the back of the classroom, tend to eat alone at conferences, sit quietly through clearly flawed diatribes, or have trouble (or worse yet, fear) convincing their neighbor to turn down the music. This tendency to shrink from social interaction, and particularly from social interactions that involve conflict, can undermine an engineer's ability to make the leap to social entrepreneurship.

On the other hand, engineers can be exceedingly good at mastering the details of a project. Also, engineers tend to have innate problem-solving abilities, and can be very devoted to finding and executing those solutions.

Nevertheless, a social entrepreneur must pitch their idea hundreds, perhaps even thousands, of times before convincing the first donor to contribute. If the thought of one hundred people telling you *no* – or much worse – before the first one telling you *yes* seems repugnant, then being a social entrepreneur is probably not for you.

The Story of Engineering World Health

In 1999, I dreamed of a way my students and I could use our unique talents as engineers to help those in need. I had lived in Thailand, and my colleague, Dr. Mohammad Kiani, had lived in Iran. We knew that there were places that could use our help, but we did not know how to match our skills with their needs. After two years of talking, visiting, and meeting, and many hours of frustration, Engineering World Health was born.

Since the birth of the organization, I have served as the Director. I am proud to have overseen the organization as in five years it went from budgets that were far short of needs, to sustainability and the ability to fund new initiatives. Most of that effort was devoted to non-traditional fundraising, meaning fundraising that was not obtained through grants or a donor campaign.

After several years of delivering refurbished medical equipment and training students to repair medical equipment in the developing world, I became convinced that bioengineers could do more if they were partnered with social entrepreneurs. Our engineers could design excellent solutions to hospitals' problems, but without the businesses to manufacture and distribute them, the designs' impact was minimal. This need formed the impetus for Competition for Underserved and Resource-Poor Economies (CUREs).

Duke-EWH CUREs is a non-profit business incubator. Its vision is to create a portfolio of simple, brilliant engineering designs appropriate for the developing world; to incubate these designs, creating sufficient data to facilitate rapid commercialization; and to finance and implement business models developed at Duke University for the manufacture and distribution of these innovations to the world's neediest people.

CUREs is implemented in a five-step screening process: (1) need finding – an on-the-ground market research in developing world hospitals, (2) project selection – graduate business or management students select the most promising projects, (3) prototype design – BME students design prototypical solutions, (4) winning/incubation – the best project is selected in a spring competition for incubation for approximately one year to collect data on the safety, efficacy, and commercial potential of the selected solution, and (5) deployment – a new company or partnership is launched to commercialize the product.

As an example, in 2006, the CUREs competition winner was Vijay Anand, a Duke University engineering management major with a project to treat infantile jaundice called BlueRay. BlueRay costs substantially less and lasts longer than currently available treatment devices, although it is stripped of many of the convenience factors typically seen in US devices. Vijay has continued to work on the project since winning, to the point where 200 units have been built and are scheduled for distribution. Vijay is well on his way to being a successful social entrepreneur.

Tips for the Future Social Entrepreneur

There is no single path for the future social entrepreneur. There is not a single textbook you can read or class you can take to ensure success. However, here are a few tips that may help improve your chances.

Live in your space: Once you feel you have a dream you are devoted to and ready to commit to, spend time volunteering for the organization that most closely matches the one of your dreams. Better yet, quit your job and go work for them. It will be just the first sacrifice you will need to make, and it would not be the last or the largest. In either case, you will learn about your customers and clients and the possible funding sources in your space.

Sell your idea 1.0: You should sell your idea to everyone you know, starting with those who would benefit from your idea. Make sure you understand the poorest member of the community you want to help. Meet with them and know them by name. Make sure they support your idea. Get photos with them (you will need them when you pitch). You can judge their support with their willingness to commit time (and sometimes money) to your project. Can you get the community to give up a Sunday afternoon to work on your idea? If not, then refine your idea.

Sell your idea 2.0: Almost all social enterprises require a separate sales pitch for those who are funding the idea. Start with your closest family. If you cannot look at your family and ask for money for your idea, then you would not be successful with a stranger. While money is a good measure of how convincing you are being, time is an even better measure. Can you get your friends and family to give up a Sunday afternoon to work on your idea? If not, then refine your pitch.

Do not focus on funds: Despite what the newspaper or television might have you believe, most business do not start with venture capital or angel investors. Of the one million businesses started in the US last year, only a few thousand were started with any significant influx of cash. Most successful entrepreneurs bootstrap their companies. In fact, as Gianforte and Gibson point out in *Bootstrapping Your Business* having lots of cash only delays the onset of the sales learning process.

Do not focus on accounting: As an engineer, you can understand everything you need to know about the finances of your small company with a cash flow budget (forecast). This one tool is easy to use and quite powerful. In fact, cash flow is far easier to understand than energy flow in a thermodynamics problem. Learn this accounting technique and leave all the other statements and forms to a professional accountant (and you can probably ignore them – the forms I mean, not the accountant).

Do not focus on "cool" technology: Engineers are most comfortable with technology. So, they tend to feel that their business needs a good website, perhaps MySQL support, a PBX, and perhaps an upgrade or two to the product. It does not. A new social enterprise needs donors and recipients who are sold on your idea.

Do not focus on the legal: For most social entrepreneurs, the best advice is not to sweat the legal aspects of the business. What you need to know is common sense and something you can learn from one good book. In particular, do not pay attention to intellectual property (IP). IP is an important aspect of some for-profit businesses, but social enterprises very rarely add social value by protecting intellectual property. They also very rarely can afford the large costs associated with protection.

However, bioengineers do have one regulatory concern: the Food and Drug Administration (FDA). Many social ventures involve drugs or medical devices. The FDA regulates the manufacture, sale, and use of medical devices and pharmaceuticals, even when those devices are exclusively for export. Even approved devices that are repackaged or resterilized are considered to be remanufactured and fall under the auspices of the FDA, even when exclusively for export. Unless you are strictly using previously approved medical devices in an unaltered state, your social enterprise will need to interact with the FDA at some level.

The good news is that one of the alumni of your engineering school is probably working at the FDA. Also, in many cases, a simple set of memos suffice.

However, failure to contact and consider the legal implications of your enterprise can be disastrous.

Conclusion

As a bioengineer thinking of a career path as a social entrepreneur, the thing that is most likely standing between you and success is sales. Not sales in the sense of hawking used medical equipment on the street corner, but sales in the sense of getting other people excited about your idea. If you are hesitating, then practice your sales pitch over and over again until you are sure you can convince those that will benefit from your idea of its value. Then try the pitch. The excitement and support you get from those who will benefit will carry you through the difficult times that lay ahead. Most importantly, just try it!

Suggested Reading

1. Bornstein, D., *How to Change the World*, Oxford University Press, 2005
2. Bedell, G., *3 Steps to Yes: The Gentle Art of Getting Your Way* Crown Business, 2000
3. Gianforte, G., Gibson, M., *Bootstrapping Your Business*, Adams Media, 2005
4. Robinson, A., *Selling Social Change (Without Selling Out)* Chardon Press, 2002
5. Dees, G., Emerson, J., Economy, P., *Enterprising Nonprofits* Wiley, 2001
6. Skloot, E., *The nonprofit entrepreneur* The Foundation Center, 1988
7. Drayton, B., "Everyone a Changemaker" *Innovations Journal*, 2006, 1, 80–96

26

Technology and Management Consulting

Guruprasad Madhavan, MS, MBA

Department of Bioengineering, Thomas J. Watson School of Engineering and Applied Science, State University of New York, Binghamton, New York, USA

"I keep six honest serving-men
(They taught me all I knew);
Their names are 'What' and 'Why' and 'When'
And 'How' and 'Where' and 'Who' "

—*Rudyard Kipling*

A pivotal moment came in my junior year of engineering studies at the University of Madras, India, in October 1999. By that time, I had already completed industrial training internships in the agrochemical, petrochemical, power equipment, and thermal energy sectors—this had helped me gain practical skills related to instrumentation and control engineering. Through a network referral, I was selected as a project trainee in the food processing plant of a company that was famous throughout Southeast Asia for its wide variety of confectionary products.

Six weeks into the training, I attended a meeting where the engineering team was reviewing the technical, support, and production aspects of the toffee, candy, and chocolate bar lines. We found that the efficiency of the candy line process had decreased during the previous few weeks because several hundred kilograms of candy had been scrapped—a disaster, in view of the approaching holiday season. After a great deal of exploration, experimentation, and thought to determine the "root cause" of the issue, we determined that the microfilm cooker, which was powered by steam from a boiler and blended the ingredients into a thin chocolate film, was experiencing intermittent temperature fluctuations. This sometimes resulted in the candy being over- or undercooked.

I was a silent observer as my colleagues concluded that an "intelligent" temperature controller was required. However, the engineers were hesitant to take on the job themselves because of their many other commitments. Remembering what I had learned from an earlier internship at a thermal power station, as well as my first hand

G. Madhavan et al. (eds.), *Career Development in Bioengineering and Biotechnology*,
DOI: 10.1007/978-0-387-76495-5_26, © Springer Science+Business Media, LLC 2008

experience with such a system in one of my courses, I proposed a schematic for an intelligent adaptive control system. My supervisor welcomed this idea, and the lead project engineer said that he would like to appoint mc as a "*consultant.*" I did not really understand the precise meaning of the term at the time.

In the following days, I studied the process layout and instrumentation until I had a good understanding of the system. Then, I created a proposal that detailed mechanical, electrical, biochemical, and environmental specifications related to the new control system I was envisioning. Next, I obtained a commercially available controller and customized it for the microfilm cooker. The process had to be shut down for brief periods for our testing, and sometimes I had to work the night shift when the cooker was not in use.

On the evening of December 31, 1999, the rest of the world was celebrating the arrival of the new millennium, but I was running the final tests on the controller that I had devised for the cooker. By the end of the evening, I was certain that the controller was able to predict and adapt to temperature fluctuations so that a consistent syrup texture was maintained. My project was successful! A week later, I purchased a pack of candies from the new batch process in a grocery store and took it home to give my parents and friends a taste of my "science project." Shortly thereafter, the candy process line yield increased nearly five percent. In the candy business, five percent is a big deal.

This project helped fine-tune my technical skills, but it also improved my human relations skills by requiring me to interact with a variety of people—from janitors to senior administrators. It gave me a great deal of satisfaction to see how practical and useful my undergraduate education had been.

I was then asked to address the temperature control issue of a milk storage unit in a different segment of the plant. This time, a visiting confectionary engineering team from Nairobi, Kenya happened to observe my work. After reading my reports, one of the engineers asked me to work on a similar project in one of their Nairobi plants. And so what began as a chance assignment led me to pursue a passion and career in consulting, especially in health technology innovation and non-profit strategic management.

Ultimately, my advice to any student interested in consulting is to do as many industrial trainings and internships as possible to gain a *breadth* of experience, not only in his/her field, but in as many fields as possible. One never knows where and how such experiences—and the fascinating people one will meet—may prove helpful in the future.

Consulting as a Practice

"A consultant is someone who takes the watch off your wrist and tells you the time."

—*Anonymous*

There are now consultants for almost every conceivable activity—dating, wedding, divorce, weight loss, media interactions, and even funerals. Some consultants are worth far more than their pay—others, as the above quip indicates, are not worth a dime. In that sense then, it is helpful to understand a good consultant's true role. In my experience, *consultants offer strategic innovation through methodical reason, inquiry, and the ability to pinpoint changes that need to be made within a system.* In many cases, consultants simply serve as functional assistants to help companies (clients) help themselves. The role of consultants can vary from being an independent expert (say, catching a fish for the client), to an instructor, or to an interventionist (teaching the client how to fish). They can be counselors, analysts, commentators, exemplars, proactive advisors, or collaborators [1]. Companies tend to hire a consulting firm that has consultants with varied experience, training, and interests, or an independent consultant to provide general or specific contributions. I have found that disciplined consultants are preferable to those who are smart but less organized.

An alumnus from McKinsey and Company, a premier international consulting firm, compared a consulting process to that of a diagnostic process in a doctor's office [2]. For instance, a patient (client) thinks he has the flu, and describes his symptoms, such as a runny nose, headache, and scratchy throat. The doctor (consultant) takes note of all the symptoms and history (business or process variables) by asking probing questions, and incorporates an organic, diagnostic approach (strategic analysis). In some cases, the patient may have something more serious than the flu, so it is the doctor's duty to avoid jumping to the same conclusion as the patient. Some situations can be challenging, as when patients show up saying "I don't feel well." This can be compared to the situation when clients approach consultants with an equally vague objective such as "We would like to increase our profitability." A consultant must therefore be proactive and holistic in listening to, assimilating, diagnosing, analyzing, and integrating information.

Consulting can be pursued in different ways. For example, a consultant can work for a firm that specializes in a range of industries. This type of work generally provides the ability to progress from an associate (entry-level analysis and research position), to consultant (client communication and project management), to senior consultant (industry-level practice), to manager (team leading and client relationship management), and finally, to partner (new business and firm governance). Big corporations may also have a system of financial compensation and benefits based on the size of client projects. As one climbs the ladder, salary and responsibilities grow commensurately.

An alternate path is that of an independent consultant. This is most often accomplished through a sole proprietorship or a small partnership. Projects are secured as a result of technical expertise and networks related to people one knows. Independent consultants have more freedom in the type of work they do and in how they accomplish their consulting duties. The drawback, of course, is that the revenue stream may not be stable.

Contracts between clients and consultants are crucial and can take different forms, but they are usually preceded by signing a non-disclosure agreement that lays down the legal ground rules relating to information protection and non-dissemination. Three types of consulting arrangements are common in the market place—full-time, part-time, and retainer (on-call/per-need) basis—I have had experience consulting with all three arrangements. Experiences and remuneration tend to be different in each based on the company stage, industry, laws, and regulations. People who are working full-time in another job, or students interested in consulting for technical or financial reasons, typically work part-time or on a retainer-basis.

The structures of consulting projects are variable and tailored to the clients' needs, so there is little point in giving specifics. However, it might be useful to provide a sample outline of a recent project I undertook for a small business that wanted to develop a marketing plan to launch its healthcare product across different medical conditions and customer segments:

Sample Marketing Plan Outline for a Start-Up Company

1. Executive Summary
2. Business and Technology Review
3. Strategic Focus and Plan
   - Mission and Vision
   - Goals (Non-Financial and Financial)
   - Competencies and Competitive Product Advantage
4. Situation Analysis
   - Internal Analysis (Strengths and Weaknesses)
   - External Analysis (Opportunities and Threats)
   - Industry Trends and Competition Analysis
   - Customer Analysis
5. Product-Market Focus and Objectives
6. Target Market Strategic Analysis
   - Pricing
   - Promotion
   - Positioning
   - Distribution
7. Financial Projections
8. Strategic Recommendations for Immediate, Short, Medium, and Long Range
9. Implementation Plan for Recommendations
10. Evaluation and Measures

The idea of the plan, in the lines of Rudyard Kipling, was to simply bring in structure and discipline to help the company address the following questions:

- Where are we going?
- Why are we going there?
- How are we going to get there?
- Who are we fighting against (or partnering with) to get there?
- What needs to be done to get there?
- When do we have to leave to get there?

Technical consulting projects vary considerably, and generalization is not possible the way it is for management consulting projects, as described in the sample marketing plan.

Consulting as an Industry

Consulting is a lucrative industry estimated to amount to $400 million in the United States by the year 2010 [3], with experienced technical consultants (those with an engineering background) reported to have a median hourly billing rate of $110 (as of 2004) [4]. In contrast, management consultants earn about $143 an hour at an entry level to almost $300 at an advanced professional level [5]. Given the myriad of challenges that healthcare systems face, consulting in this arena is burgeoning [6], particularly for clinical trials, information technology, human resources, regulatory affairs, and operations management. Other standard areas of consulting practice in this sector can be found in:

Life sciences, including pharmaceutical, biotechnology, medical device, and other supplier organizations;

Provider side including hospitals, health systems, physician practices, and out-patient centers;

Insurance/Payer side including healthcare insurers, payers, health maintenance organizations, and related organizations; and

Government marketplace, including local, state, and federal health-related agencies, as well as associations working in various aspects of healthcare, and other areas, such as facilities planning, health policy, and non-profit consulting entities.

Project Management Skills for Consulting

Some skills that are very useful for a consultant include:

- ***Time Management:*** Arguably the most crucial skill for a consultant is time management. Time nowadays is more difficult to find than energy, ideas, and inspiration. It is critical for individuals in consulting—or those choosing consulting as a career—to be good at time management, be it through task prioritization methods such as flowcharts, "to-do" lists, project logs, and task management software. As one of my mentors said, "Consulting is not for procrastinators." The words of William Matthews aptly summarize this:

  > "Nothing inspires confidence in a business man sooner than punctuality, nor is there any habit which sooner saps his reputation than that of being always behind time."

- ***Client Relationship:*** Do not enter this field for the wrong reasons. Consulting relies heavily on building a sound relationship with clients, through care and attention. Consulting is only for service-oriented individuals. It is important not to jeopardize any relationships through a major conflict or even a simple miscommunication or misinterpretation of information.

"I never tell one client that I cannot attend his sales convention because I have a previous engagement with another client; successful polygamy depends upon pretending to each spouse that she is the only pebble on your beach."

—*David Ogilvy*

- *Team Work:* One of the best aspects of consulting is that it provides a great platform to work in diversely skilled teams or groups. Individuals should be comfortable working with people from varied backgrounds. Maintaining a healthy positive attitude is necessary, be it to learn from others or to defuse any conflicts that may arise.

"Coming together is a beginning. Keeping together is progress. Working together is success."

—*Henry Ford*

- *Data Acquisition and Analysis:* Consulting involves arriving at or making useful decisions for clients. Data drives decision-making and knowledge-formation. Consultants need to be comfortable with experimental study design and statistical and data presentation methods that will provide the client with the most information for effective decision-making.

- *Communication Skills:* A main objective of professional communication that people often fail to recognize is not to deliver information, but rather to stimulate the right action. Both verbal and non-verbal faculties play an important role in this and have been receiving enormous attention in recent times—in the context of globalization, where cultures converge or collide. Hundreds of books and articles have been written about this, and universities and schools are working equally hard to train students in good communication skills. A good consultant needs to know the right tone, timbre, timing, and presentation techniques for delivering what needs to be said. Non-verbal communication via facial expression, posture, and gesture can be critical, in a multi-cultural environment.

- *Bookkeeping:* Finally, consulting involves maintaining a systematic time log and invoice system to support on-time payment by the client. This system is also useful when filing taxes, especially for independent contractors or self-employed individuals.

Creating the Consulting Mindset

"A pessimist sees the difficulty in every opportunity; an optimist sees the opportunity in every difficulty."

—*Winston Churchill*

As the 21st century and globalization continue to grow and impact our businesses and lifestyles, in the lines of Henry Kaiser, I have come to believe that there are no problems, but only opportunities appearing in work clothes.

Innovation is not about thinking "outside the box," because there is no box! In the words of the "father of modern management," Peter Drucker, innovation is "*purposeful and organized search for changes, and in the systematic analysis of the opportunities such changes might offer for economic or social innovation* [7]." To exploit opportunities, we must be able to identify the changes that will lead to those opportunities. In *Innovation and Entrepreneurship*, Drucker identifies at least seven sources of innovative opportunities that are relevant to consulting. In reverse order of importance, opportunities stem from:

7. new technology and scientific findings
6. changes in public perception
5. demographic shifts
4. industry or market structure change
3. changes in process need
2. changes based on incongruities
1. the unexpected

Consultants should have a positive attitude toward change and be receptive, reactive, and adaptive to opportunities that can lead to innovative outcomes for their clients.

The traditional methods of problem solving do not apply to businesses and trade anymore. A holistic view is required. In this framework, the principles of *complex systems* offer a revolutionary way of thought for studying non-linear dynamic patterns and depart significantly from conventional mechanistic and reductionist points of view. Complex systems do not mean complicated systems, and do not always involve undecipherable mathematical formulae. They involve rather a new style of viewing and thinking about systems and how those systems behave. In short, complex systems are adaptive and typically composed of a large number of independent agents/components that behave and are organized according to self-determined principles. This leads to leading to local interactions that create unpredictable global patterns. These emergent patterns cannot be characterized by simply adding the effects of individual agents/components or by creating a "blueprint," but require thinking in terms of interconnections and effects over time. This type of thinking is crucial when consultants are dealing with political, social, psychological, cultural, human, ethical, technical, and regulatory aspects in any dynamic business system or social process that seeks to grow its portfolio and multiply its value offerings to its customers.

> "And if history has for its object the study of the movement of the nations and of humanity and not the narration of episodes in the lives of individuals, it too, should seek the laws common to all the inseparably interconnected infinitesimal elements of free will."
>
> —*Leo Tolstoy*

> "The task is not so much to see what no one else has seen, but to think what nobody yet has thought about that which everybody sees."
>
> —*Arthur Schopenhauer*

How and Where to Begin?

Consulting, like many disciplines, is mainly about marketing and branding oneself in the service of others. Here are some pointers on how to begin working toward being a consultant:

- You should have a high degree of self-awareness, as consulting is an intensely collaborative state of putting yourself in other people's shoes and helping them reach their goals. Numerous cognitive inventories, such as the *Myers-Briggs Type Indicator* and the *Keirsey Temperament Sorter*, may help you to arrive at an approximate idea of your own nature and personality traits.
- To avoid missing out on unexpected opportunities, you may want to brand yourself *broadly* in the area of your interests and capabilities (for example, "a specialist or generalist in medical devices"), yet with a marketable depth (for example, "cardiovascular implantable technologies," but not too specific such as "a specialist in patent foramen ovale catheters," unless it is absolutely necessary).
- Practice identifying changes and resulting opportunities. Innovation is a skill that can be learned like any other and can be strengthened by practice. Some standard ways to find changes include perusing periodicals and journals from varied fields, attending multi-disciplinary conferences, and constantly interacting with other people.
- Build on your core competencies, list the *value-added services* that you can provide to clients, and be sure to have an up-to-date résumé or vita at all times.
- Work on your timing and time management skills, as these are the fuel for a consulting business.
- Constantly practice communicating, and update or refine your communication skills and tactics as you work with different people.

For those interested in corporate (for-profit) consulting, I would suggest making or having friends in business schools. Discuss world issues with them to expand your horizons and thought processes, and learn to appreciate their viewpoints. Further, for engineers and technologists, do not confine yourself to technical job fairs, but also attend or "hang out" at business school career fairs to get a feel for the environment, or to just meet and socialize with new people. McKinsey and Company, the leading international consulting firm recruits accomplished graduate and doctoral degree holders in science and engineering especially for traits including critical thinking skills, integrative analysis of issues, and fact-based decision making. An internship at any of the numerous consulting firms may assist to you in deciding if consulting is for you. So get out there and socialize! An additional option would be to take a course or two in management to help you acquire a broad understanding of business dynamics. A graduate degree in business administration will be useful, but is not required for consulting.

Another interesting way to get a feel for consulting is to work with a non-profit or social organization while you are still in school. Many of these organizations struggle with understaffing, financial, and strategic management issues, and if you volunteer (as I did), the likelihood of your work for them being appreciated is very high. In addition to providing contacts for the future, this may also serve as a non-threatening learning environment and can help you develop some critical thinking and human relations skills.

The Beauty and the Beast Scenario

Here are some great reasons to become a consultant [8]:

- Independence means that you are your own boss!
- The freedom to choose an area of interest.
- A platform for reputation and professional growth.
- Financial rewards proportional to effort (especially in large consulting firms).
- Acquisition of multi-disciplinary skills.
- Frequent world travel– at the client's expense!
- A professional network for future development and success.

And some caveats:

- An occasional difficult client.
- Financial uncertainty and insecurity.
- Long, hard hours of work and killer deadlines.
- Competition where one has to adapt or lose.
- Possible feast or famine situations, in terms of contracts.
- Situations where one never discovers what the client did with findings or proposed recommendations.
- Additional tax work and extensive documentation, if operating independently.
- No company benefits.

'A FIRST STEP' to Consulting

In this concluding section, I would like to summarize key attributes and qualities to help you determine whether consulting might be a suitable career for you, and, if so, how you can achieve success in this field, using a helpful framework called "A FIRST STEP."

Adaptiveness is an absolute necessity for a professional who wishes to have a lasting presence in the domain of consulting. Change is inevitable, and resulting opportunities can be systematically explored and exploited only if the consultant has the ability to adapt to changes. Large-scale corporations lose the ability to adapt to changes and innovate. As a consequence they frequently bring in an external set of eyes to look at issues and possibilities from a fresh perspective. Adaptiveness entails, not a compromise of the consultant's own personal values and work ethics, but an *open-minded* approach with a *positive attitude* to satisfying the wants and needs of clients.

> "Adapt or perish, now as ever, is nature's inexorable imperative."
>
> —*H. G. Wells*

Focusing on the client with dedication and a clear mental framework is what differentiates an outstanding consultant from rest of the pack. A consultant who realizes that each client is unique, and that no cookie-cutter approach should be used, is well on his or her way to success in this competitive profession. Focus is also vital when making recommendations to clients, be it immediate, short, medium, or long range. Recommendations that lack focus, substance, residual message, and utility are bound to be discarded, or worse, forgotten immediately.

> "Concentrate all your thoughts upon the work at hand. The sun's rays do not burn until brought to a focus."
>
> —*Alexander Graham Bell*

Information is key. A consultant must constantly seek information from trade periodicals, newspapers, and journals across a range of disciplines. Ideas and innovation emerge out of information that enables you to exploit opportunities.

> "Information is the seed for an idea, and only grows when it's watered."
>
> —*Heinz Bergen*

Resilience is essential in consulting, and, needless to say, in any profession. The sooner you start developing resilience through persistence, the better off you are in terms of career and character development. As Napoleon Hill wrote, "*Edison failed ten thousand times before he made the electric light. Do not be discouraged if you fail a few times*," to which Thomas Edison, responded, "*I have not failed, I have just found ten thousand ways that do not work*"—a true illustration of indomitable resilience.

Strategy is systematic planning and thinking that can lead to action. As Dwight Eisenhower once opined, "*Plans are nothing, planning is everything*." Planning can be mastered through practice. The next time you play chess, keep putting yourself in your opponent's position *as well as* your own. The tactics you may develop from this strategy will be very different than when you focus on just your side of the game. One such historical war hero who excelled in this was Napoleon Bonaparte.

> "Strategy without tactics is the slowest route to victory. Tactics without strategy is the noise before defeat."
>
> —*Sun Tzu*

Trust is the hardest thing to gain, and the easiest to lose. At some point in time, you may have heard the maxim that says "*It takes years to build up trust, and only seconds to destroy it*." The trust between consultant and client is the backbone of long-term consulting relationships, especially for retainer arrangements.

> "Trust is like a vase, once it is broken, though you can fix it, the vase will never be same again."
>
> —*Anonymous*

Structure and (self-) discipline are necessary to analyze and address issues systematically. From my experience, most issues in modern businesses arise from a lack of vision and structure. Consultants are frequently hired just to bring discipline and a new perspective to the process through carefully structured thinking.

> "Nothing is more harmful to the service, than the neglect of discipline; for that discipline, more than numbers, gives one army superiority over another."
>
> —*George Washington*

Timeliness and punctuality show that you care for what you do and that you have what it takes to achieve and set a high standard. When I was young, my father distilled his wisdom for me by saying, "*Never be on time. Always be before time and you will be ahead of the world.*"

> "I owe all my success in life to having been always a quarter of an hour before my time."
>
> —*Lord Nelson*

Enthusiasm, a word of Greek origin, means a whole-hearted, energy-filled, and undeterred devotion to a cause. If you are not enthusiastic about your work, no one else is going to be. I am not the first person to say that enthusiasm is like fire. It spreads and inspires contagiously.

> "Enthusiasm is the leaping lightning, not to be measured by the horse-power of the understanding. Every great and commanding movement in the annals of the world is the triumph of enthusiasm."
>
> —*Ralph Waldo Emerson*

Professionalism, last but not least, is one's commitment to life-long learning and excellence in service and scholarship. Be it soft or hard skills, the ability to commit to professionalism and professional development is what determines the growth of discipline. Bioengineering and biotechnology include classic examples of great accomplishments in the history of mankind, truly as a result of passion and professionalism. With any career path we choose, life-long learning is what is going to determine our ultimate development as human beings.

> "Live as if you were to die tomorrow. Learn as if you were to live forever."
>
> —*Mohandas K. Gandhi*

Selected Resources and Readings

Professional Organizations and Consulting Networks

- Association of Professional Consultants, USA, www.consultapc.org
- Institute of Management Consultants, USA, www.imcusa.org

- IEEE-USA Consultant Network, USA, www.ieeeusa.org/business/consultants
- Small Business Administration, USA, www.sba.gov
- LinkedIn Business, Professional, and Social Network, www.linkedin.com
- FaceBook Social and Emerging Professional Network, www.facebook.com

Consulting Magazines and Trade Journals

- *Fast Company*, published monthly.
- *Inc.*, published monthly.
- *McKinsey Quarterly*, published quarterly by McKinsey & Company
- *Consulting to Management*, published quarterly by Journal of Management Consulting, Inc.
- *Consultants News*, published monthly by Kennedy Information Group.

Useful Books Related to Consulting and Project Management

- L. Stroh, H.Johnson, *The Basic Principles of Effective Consulting*, Lawrence Erlbaum Associates, 2005
- P. Block, *Flawless Consulting and Fieldbook*, Pfeiffer, 2000
- S. Biswas, D. Twitchell, *Management Consulting: A Complete Guide to the Industry*, Wiley, 2001
- E. Rasiel, P. Friga, *The McKinsey Mind: Understanding and Implementing the Problem-Solving Tools and Management Techniques of the World's Top Strategic Consulting Firm*, McGraw-Hill, 2001

Useful Books Related to Management, Strategy, and Complexity

- P. F. Drucker, *Innovation and Entrepreneurship*, Collins, Reprint 2006
- P. F. Drucker, *Managing the Non-Profit Organization: Principles and Practice*, Collins, 1992
- J. C. Collins, *Good to Great: Why Some Companies Make the Leap and Others Don't* Collins, 2001
- R. Stacey, *Complex Responsive Processes in Organizations: Learning and Knowledge Creation*, Routledge, 2001
- P. M. Senge, *The Fifth Discipline: The Art and Practice of the Learning Organization* Currency, 2006
- H. Eisner, *Managing Complex Systems: Thinking Outside the Box*, Wiley-Interscience, 2005
- I. Prigogine, *Order Out of Chaos*, Bantam, 1984
- S. Tzu, *The Art of War*, Dover Publications, 2002
- R. Bruner, M. Eaker, R.E. Freeman, R. Spekman, E. Teisberg, S. Venkataraman, *The Portable MBA*, 4th Edition, Wiley, 2002

Useful Books Related to Networking and Negotiations

- K. Ferrazzi, T. Raz, *Never Eat Alone: And Other Secrets to Success, One Relationship at a Time*, Currency, 2005
- R. Fisher, W. Ury, and B. Patton, *Getting to Yes*, Random House Business Books, 2003
- Barabasi, A., *Linked: How Everything Is Connected to Everything Else and What It Means*, Plume, 2003

References

1. A. Weiss, *Million Dollar Consulting*, McGraw Hill, 1997
2. Adapted from E. Rasiel, *The McKinsey Way*, McGraw-Hill, 1999
3. *Global Consulting Marketplace 2007–2010*, Kennedy Information Group
4. *Profile of IEEE Consultants*, IEEE-USA Report, 2004
5. *Fees, Utilization and Other Key Metrics 2006: Managing Profitability in the Consulting Profession*, Kennedy Information Group
6. *Healthcare Consulting Marketplace 2007*, Kennedy Information Group
7. P. Drucker, *Innovation and Entrepreneurship*, Collins Reprint, 2006
8. Partly adapted from W. Swartz, "Why you might want to consider consulting," Annual Meeting of the American Chemical Society, Philadelphia, 2004

Expert Witness and Litigation Consulting

John G. Webster, PhD

Department of Biomedical Engineering, University of Wisconsin-Madison Madison, Wisconsin, USA

In trials, "fact" witnesses may testify about facts, but may not give opinions. A special type of witness is the expert witness, who has credentials that permit him or her to offer opinions. Bioengineers are frequently called upon to serve as expert witnesses in product liability cases where a subject is injured by a medical device, or in patent litigation where a first company sues a second company for infringement of a patent owned by the first company. A patent on an invention is granted by the government for twenty years, which grants the holder an exclusive right to practice the patent in exchange for explaining the invention to the world in a patent published by the US government patent office.

The Cold Call from an Attorney

I had been working about twenty years teaching and doing research in medical devices and instrumentation when an attorney phoned and asked if I would be interested in serving as an expert witness in patent litigation. He asked if I would read some patents and offer a written opinion on their claims. It sounded pretty good to me; they would pay me to read patents and do some writing. It was a real eye opener. I soon learned that in order to get a patent, the patent holder must reveal everything and not hold anything back. I really liked that, as I learned so many details about how medical devices and instrumentation worked. Afterwards, I could quickly turn around and teach my students or write articles or books to inform others. In fact, I am always looking to develop books written from a bioengineer's point of view. At one time, I wanted to develop a book on cardiac pacemakers, which I consider one of the most important medical devices. But when I approached designers in companies they all clammed up because their company attorneys told them not to give away any company secrets in this competitive market. So I gave up. Later, I was asked to serve as an expert witness in cardiac pacemaker patent litigation. As I read all of the patents, I realized, "Here is all the information I need!" So, I took the detailed information from the patents and was able to develop the book.

G. Madhavan et al. (eds.), *Career Development in Bioengineering and Biotechnology*,
DOI: 10.1007/978-0-387-76495-5_27, © Springer Science+Business Media, LLC 2008

How Do You Become An Expert Witness?

You need a skill and you need to be unbiased. You may work for a medical device company and be skilled in design, but if so, you cannot serve as an expert witness because you would be biased. So attorneys look for skilled, unbiased expert witnesses. Frequently, they ask university faculty who have established reputations through writing books or journal articles. Private consultants are also chosen. It is most helpful if you have published extensively on the subject matter. It is also helpful if you are a teacher and can explain complicated devices to a jury with limited technical background. Holding a license as a professional engineer may be useful as well. I am a biomedical engineer with an electrical engineering background. If an attorney phones me about a case involving biomaterials, I tell them, "I am not qualified in biomaterials." It is important to accept cases only in areas where you can confidently say, "I am an expert." Otherwise the opposing side can challenge and get you disqualified.

Sometimes a broker will call you, describe a case, and offer to arrange a phone meeting with attorneys. Whether you work through a broker or directly with attorneys, the results are the same to you, except the broker will mark up your rate to the attorneys. The broker earns his markup by working for attorneys to search for and find the best expert witness. So in the beginning it is like a job interview. You should realize that the attorneys may be considering other expert witnesses and are looking for the best one. If you get the job, the work begins. If the attorney asks you to give an opinion that you do not agree with, you should decline and resign from the case.

So What Does an Expert Witness Do?

You may be asked to sign a confidentiality agreement that states that you will not disclose any confidential information supplied by the opposing side. Your first question to the attorneys should be, "Do you have an agreement with the opposing side that expert witness report drafts are not discoverable?" If yes, you can send drafts back and forth to your attorney at will. If no, be sparing in writing anything down because the opposing side may demand everything you have ever written on the case, and ask you questions on why you changed this or that. Your attorney will send you the complaint, a pile of patents, patent file histories, prior art publications, depositions, and devices. You read and examine these critically to understand them, and then must write a report by the date set by the judge. In your report, you describe your qualifications, fee, depositions and trials in the past four years, and publications for the past ten years. You attach your resume and other information as exhibits.

If you are on the plaintiff side, you analyze the language of the claims as interpreted by the specification (the main part of the patent). The judge may define the meaning of the claims. You examine and test the accused device to determine whether it is covered by the claims. If so, you write an opinion that the accused device infringes the patent claims. Infringement may be literal: the words in the claim may describe the device. The infringement may be under the doctrine of equivalents if the accused device or process performs substantially the same function, and operates in substantially the same way to achieve substantially the same result as the claimed invention. You develop as an exhibit a two-column chart with each phrase of the claim in the left column and each matching part of the accused device in the right column.

If you are on the defendant side, you list reasons why the patent is invalid, for example, it was obvious, not novel, not useful, invented earlier as described in prior art (anticipated), the specification does not have an adequate description of how to implement the invention (lack of enablement), the patent applicant held back important information that he should have provided to the patent examiner and did not tell all (inequitable conduct), or the file history describes how the applicant restricted the meaning of the claims during patent prosecution. Not only the listed reasons above, but in addition that you have actually tested the medical device, and it does not infringe the patent because it does not contain every element of the claim or under the doctrine of equivalents.

It is best to err on the side of putting too much in your report, because if it is in your report, you can use the information at trial. If it is not in the report, you cannot bring it up later and use it. There should be no surprises.

It is not common, but you may receive a subpoena from the opposing side requiring you to deliver to them copies of all documents you have used in preparation of your report.

What Is a Deposition?

About a month after opposing sides exchange expert reports, you will be deposed. The deposition is usually for a day, but sometimes two days, with videotaping and a court reporter transcribing. With you under oath, the opposing attorney will ask you questions that you must answer. You cannot win anything in a deposition, you can only lose something. If you are asked if you know what time it is, you do not answer, "two o'clock," you answer "Yes." This is not the time to demonstrate your great learning. Just listen very carefully to the question, wait a few seconds to think about it and give your attorney time to object, then answer briefly. If the attorney begins a question with, "Is it not true that . . .?" make sure you preface your answer with, "Under the following conditions, it is true."

At the deposition you should be ready to answer many questions: all your personal information regarding education, employment, memberships, which publications are pertinent to this suit, who engaged you and when, who wrote or assisted writing your report, what bases and reasons helped form your opinion, when you formed your opinion, and when and how you prepared for this deposition.

What Happens at the Trial?

The vast majority of cases settle just before trial because litigants prefer to avoid the large expense of a trial. But a few cases do go on to trial. Before the trial, you should help prepare demonstrative exhibits. They may include large charts that help you explain complicated concepts. Large scale models are also helpful, particularly when they include movement, sound, and/or light. One time I had to explain what derivative (slope) meant. I drew a rising curve that showed the response of an under-the-tongue temperature sensor to a step increase. When the slope decreased to a certain threshold, the thermometer calculated the final temperature. I took a home mercury thermostat in a small box and moved it up the curve parallel to the line and showed how when the slope was steep, nothing happened, but when the slope of the box decreased to a certain threshold, the mercury made contact and an internal buzzer sounded. This is the kind of demonstration that will wake up the jury and make them pay attention and remember something during a four week trial where mostly what they hear are words.

At the trial, your attorney will ask you friendly questions that elicit the testimony that he or she wants to get into the record. In one trial, I testified for five hours on direct. Then the opposing attorney cross examined me for ten hours, challenging my testimony and trying to make me appear a fool. You do not have to memorize everything, but can ask for documents to help refresh your memory. If the opposing attorney asks for a yes or no answer to an ambiguous question, you can answer, "If by your question you mean this, then the answer is yes." When expert witnesses for the opposing side are on the stand, you can pass a slip of paper to your attorney to suggest questions he might ask.

What about Personal Injury?

When patients undergo surgery, they expect good results. When bad results occur, they may sue the physician, the hospital, and the medical device manufacturer. Modern surgery involves medical devices such as electrosurgery, anesthesia machines, and large imaging devices. Problems associated with such devices have resulted in serious burns, loss of limbs, and death. Because the bioengineer may be an expert on one of these medical devices, he or she may become an expert witness in one of these cases and offer an expert opinion on the cause of the problem.

What are Expert Witness Fees?

As of 2007, if you have a master's degree and several years of experience, $200 per hour is reasonable. If you have a PhD and ten or more peer-reviewed publications, $300 per hour is reasonable. If you have a PhD, have many publications and books, and are one of very few experts in your specialty area, $400 per hour is reasonable. These are typical fees, and must be stated in the expert witness report. When an attorney calls you, I suggest that you send your resume, your previous expert witness work, and your fee schedule similar to the sample below.

Sample Fee Schedule Letter

John G. Webster, PhD
Expert witness and consultant in medical devices and instrumentation
Fee schedule
For work performed in "Location," $$$/hour.

Includes research and investigation, testing of devices, reading depositions, deposition and trial preparation, telephone counseling, oral and written reports, literature searches in extensive personal library, online literature searches of all books in all campus libraries, journal literature and patent searches from computer in my office, searches in University of Wisconsin Engineering Library (which contains all patents).

Work outside "Location" will include regular hourly charge for travel time. I will use this time as efficiently as possible to review literature on your project and to prepare documents for your project.

Includes research and investigation, consultations, oral reports, travel, depositions, and court testimony. To be followed with a written report, as required.

Actual expenses reasonably and necessarily incurred, such as travel, subsistence and lodging, long distance telephone charges, professional support requirements, etc., are additional to the consulting fee and will be billed to the client at cost.

Terms: Letter of agreement that states who guarantees payment of the fees and expenses.

Billed monthly and payable net 30 days from date of invoice.

Summary

If you have expert qualifications, career or sideline work as an expert witness can be interesting, educational, remunerative, and enjoyable.

Suggested Reading

1. Poynter D. *The Expert Witness Handbook: Tips and Techniques for the Litigation Consultant*, 3rd ed., Para Publishing; 2004.

28 Public Relations

Cynthia Isaac, PhD

Healthcare Practice, Ogilvy Public Relations Worldwide, New York, New York, USA

My interest in the business of science started when a friend who was in the finance field asked me one day what "genomics" was, and I did not know the answer. He wanted to invest in a company called Millennium Pharmaceuticals. That was in the late nineties, when biotechnology companies were coming to the attention of investors on a larger scale than before. Companies like Millennium and Amgen were headlining the news, and chief executive officers (CEOs) like William Haseltine of Human Genome Sciences were being interviewed by CNN. Reading about these companies opened a whole new world. I had come to the USA from India with an interest in cell biology, and was doing my doctoral research at the Albert Einstein College of Medicine, New York. I was surrounded by doctors and scientists, but this was my first introduction to the nexus of science and business. I was hooked.

From Lab Rat to PR Flack?

I enjoyed the intellectual rigor of my life in the laboratory, but wanted a career with more human interaction and with a more immediate impact on people around me. The question was, what could I do that would involve my background in science, and allow me to use other "soft skills"—writing and public speaking—that I possessed but had little use in the laboratory? Public relations (PR) came to mind when I read about someone with my background who was working at an agency that specialized in biotechnology. With a vague idea of getting some work experience, I sent my resume to various PR firms specializing in biotechnology, playing up my science background as well as my communications skills, and asked for an internship. I got a phone call from a firm that I had heard of, but never approached. Apparently, my resume was sent from one HR director to another, and after a short interview with the CEO the week after, I was offered an entry-level position. With almost no idea of what I was getting into, I was stepping onto the first rung of my PR career.

G. Madhavan et al. (eds.), *Career Development in Bioengineering and Biotechnology*, DOI: 10.1007/978-0-387-76495-5_28, © Springer Science+Business Media, LLC 2008

PR 101

Public relations tells an organization's story, and assists the organization in developing effective relationships with its customers, employees, shareholders, other institutions, and with society at large. The week I first started working, I was in charge of the mind-numbing but important task of uploading press releases to the wire. This involved sending the press release to the vendor that disseminated them, and checking the final version word by word to make sure that it contained no errors when it was made public. I soon graduated to actually writing the press releases, and I still write them to this day. Press releases are one very important method of communicating with reporters and editors from around the world and, through them, to either a targeted audience—like doctors or investors—or to the general community. In this Internet age, minutes after a release is disseminated, it is online and available to the public.

Many options in the PR field are open to people with scientific and technical backgrounds. These include investor relations, which is the interface between the company's management and the capital markets, (i.e., analysts and investors); media relations, which specializes in building relationships with editors, reporters, and producers at newspapers, television, magazines, and blogs; product marketing, which influences consumer behaviors and helps build loyalty to a particular product; and crisis management, which involves forecasting and planning for potential crises and taking steps to minimize damage to the company's reputation in the face of a crisis. There are many other specializations such as public affairs and social marketing that you can look into if you are interested. I have recommended some resources below where you can find more information on this subject.

There is also diversity in the type of organization you can work for: many PR practitioners work in-house as a public relations manager or press officer at a single organization. Others work in PR agencies where they juggle the needs of one to several clients. Once you have more experience and develop contacts in your chosen specialty, you can also work as a freelancer. It would make sense with a science or engineering background to consider working with a company or agency whose clients have some connection to your field of study. For instance, with my background in the biological sciences, I worked in several agencies with clients in the biotechnology sector and then switched to an agency whose clients included the top ten pharmaceutical companies worldwide.

So how difficult was it to make the career transition, and what hurdles did I have to overcome to succeed in this field? The first thing I discovered was, as with any field, PR abounds with jargon I had to master. Second, communications, especially healthcare communications, demands specialized knowledge. While I knew a lot about medicine from a scientific point of view, I also had to learn much about drug discovery, development and marketing, and the role of the Food and Drug Administration (FDA) in overseeing the testing and marketing of these drugs. I also started in an entry-level position with corresponding pay (ironically about equal to what a post-doctoral fellow in the biological sciences earns in the USA!). You too will probably have to start low on the totem pole, in spite of your advanced science degree(s), working alongside much younger colleagues, just out of college, many with journalism or English degrees. While I was promoted quickly over the years, I had to show that I learned quickly on the job, and also had to pay my dues in terms of doing the boring, sometimes repetitive work that starting at the bottom sometimes entails. Uploading press releases are a case in point!

So, Are There Any Advantages to My Degree?

There certainly are many advantages to a background in science and research when working in PR, especially in a specialized field of PR similar to the kind that I do. You will be able to quickly grasp the technicalities, which is a great first step to translating those concepts for easier understanding by a lay audience. For instance, I am often called upon to explain complex clinical trial data to consumer-focused press though various means such as background documents, websites, presentations, and of course, the ubiquitous press releases. You will also be able to easily analyze and synthesize a great deal of information. Occasionally, when I have to comb through thousands of pages of background on a prospective client, I am happy I practiced my research skills in graduate school. You will relate easily to many of your client contacts. I found that the CEOs and chief scientific officers (CSOs), as well the medical directors that I work with share the same background, and that helps build rapport as well as mutual trust and respect.

In contrast, there are some disadvantages to your background which, with some focus, you will be able to overcome. Most science students do not have a deep understanding of business or PR. Many have never worked in an office environment before, and may not have mastered business etiquette, although I believe etiquette is more a matter of common sense and good manners than anything else. Moving from a somewhat "flat" academic organizational structure to a more hierarchical corporate structure might be difficult. In addition, you are trained to be a specialist, not a generalist, and you will have to balance your inclination to learn everything about a subject with the need to learn just enough to finish a project successfully within a deadline.

Deadlines are hugely important in the agency environment. If you worked in a laboratory and took years to finish your research, publish papers, and finalize your thesis, you have to be prepared for the fact that almost no project takes that long in the PR world. Corporations are machines with an internal rhythm: quarterly calls, monthly internal reports, annual reports, and annual conferences—interrupted by small and large crises—a reporter calling with a difficult question, or unexpectedly negative data. Keeping up with your company's rhythm is important. If you work in an agency, you must keep up with the deadlines of multiple clients, each one convinced that their business is more important than others'. Working in the corporate world, especially in a service-oriented industry like PR, is not a 9 to 5 job. You will need to juggle many projects at the same time and not become overwhelmed.

When interviewing a scientist who wants to make the career transition into the field of communications, I look for certain skills sets and personality traits. These include writing ability, people skills, computer proficiency, planning and organizational ability, presentation skills, and attention to detail. I will be looking to see if you will be able to interact effectively and tactfully with other staff, clients, and vendors; that you can be a team player with a positive attitude; that you can work independently, but will know when to ask for help, and that you have a professional demeanor.

When I interviewed at various firms, some of the prejudices that I had to overcome as a former scientist included: "She will write like a scientist," "She will not be a team player," and "She is probably not a people person." Take those biases into consideration when writing your resume and interviewing for your first job.

A Day in the Life of a PR Executive

7:30 AM Read email on my Blackberry to see if my clients in Europe have any urgent needs.

9:00 AM Arrive at work, catch up on other emails and voicemails from colleagues and clients.

9:30 AM Join three account team members on first conference call of the day with client team. Since I work on a global account, the clients dial in from the USA as well as different cities in Europe. We discuss an important piece of clinical trial data that is soon going to be made public. The client medical team provides us with a summary of the data, and we discuss broad strategies on how best to publicize it.

10:30 AM One of my team members wants me to review some background materials that we are developing for a PR campaign geared toward consumers and patients. I go through it in detail with him and offer suggestions.

11:30 AM Management meeting at the agency to discuss whether we will meet the financial numbers for the month, and to project a budget for the next month. Since everyone in the agency records the time they spent doing different tasks using special software, we are able to bill their work accurately to various clients. By reviewing these timesheets, we are also able to see if someone has too much or too little to do. As team leaders, our job is to make sure everyone is productive, but no one is overwhelmed.

12:15 Lunch at the company cafeteria with my colleague.

1:15 Write a communications plan for the launch of a new indication for the drug we are publicizing. A new indication means the drug will be available to a larger group of patients. How do we communicate this to the physician and patient community? How soon after the decision by the FDA can we release the information? What if the FDA does not approve the indication—what do we say then? I will discuss the plan with a few of my colleagues before it is submitted.

2:45	New business brainstorm with a team of ten. They are preparing for a presentation to a prospective client and want ideas from a range of people. Hopefully some of my ideas are useful.
3:30	Second client call of the day. I send the agenda to the client just before call. The whole account team updates the client on all the projects we are working on, and I speak to the team after the call to make sure everyone knows what the next steps are.
4:30	Take phone call from client. A reporter has called with some questions regarding drug reimbursement. He asks if I can help him write a statement in response. I draft a note that he sends to be reviewed internally before it is sent to the reporter. I offer to follow-up with the reporter tomorrow to see if she has everything she needs for her story.
5:45	A client calls—it's only June, but we need to prepare a PR plan for next year. In two weeks, we will have to present the plan to the client business team. They will then decide what our budget allocation will be. I work on the plan until the end of the day, and I will likely spend most of tomorrow on it as well.
6:30	I leave work after completing my timesheets.

Getting Your Foot in the PR Door

If this sounds like something you might be interested in pursuing, what should you do to gain entry into the field? Having no previous communications experience can be a major barrier. First, hone your writing skills and practice communicating without jargon. For instance, can you explain what you do in the laboratory to your friend who knows nothing about HPLC or DNA? Then look around for activities you can do while still in the lab that will show prospective employers that you are genuinely interested in communications. Take advantage of the opportunities offered by scientific magazines and your university newspaper—offer to write articles. Gain insight into how your university press office promotes its science research. Try to get some work experience via an internship—perhaps with a company that was spun out of your university. Organize various activities, perhaps at the student council level, so you can show that you can plan and execute an event successfully.

Learn more about the field that you want to work in. If you are interested in working within the biotech sector for example, read trade publications and follow biotechnology stories in the newspaper or on the internet. Periodically review the websites of biotechnology companies, especially the section geared toward the press. A good way to learn more about the field is to talk to people—does your university have a career office that can put you in touch with people in the field? Are there career conferences held by local organizations where you can meet people? Can your mentor or anyone else in your department or organization put you in touch with anyone who is a good contact for you?

Work on your resume. Your resume for a job in PR will be very different from the curriculum vitae you use when applying to a post-doctoral position or an academic position. Highlight your communications skills, organizational experience, and writing ability. Keep on hand examples of newsletter articles you have written, and take them along to the interview. Make sure you brief anyone you indicate as your reference so that they understand the type of job you have applied for. If you are applying for a job at an agency, take the time to read their website and learn about their clients so you can have a good conversation during the interview.

Oh, the Places You Will Go

So now you have a foot in the door—what next? Well, it all depends on you. Are you interested in investor relations? Would you like to work in-house as the PR Director and public spokesperson of a company? Do you want to be in charge of PR for a product on the market? Would you like to specialize as a science writer, and perhaps work from home? Do you want to be in charge of a PR department at an agency, and manage a lot of people? Would you like to eventually start your own PR agency? You will find that a lot of options are now open to you.

Will you miss academia? I certainly do not. Being in scientific communications gives me the opportunity to learn about a variety of science and healthcare subjects that I would otherwise never have been exposed to. I know that I contribute something positive by educating physicians and the public about diseases and their treatment options. I never know what each day might bring. It might be hectic at times, but never boring.

Resources

1. Public Relations Society of America, www.prsa.org
2. Next Wave, www.nextwave.sciencemag.org
3. Health and Science Communicators Association, www.hesca.org

29 Sales and Marketing

Jason M. Alter, PhD

Aureon Laboratories, Yonkers, New York, USA

When I entered a doctoral program in molecular biology, I never intended to spend my entire career at a lab bench, nor was it my goal to pursue a marketing career. A decade ago, it was not as common to find scientists who had made the transition to sales and marketing. Although I knew that I would use a science degree in my future career, it was not initially clear to me where my path would lead.

After a satisfying post-doctoral position with a large pharmaceutical company, I became involved with marketing through a fortuitous accident. An unusual advertisement caught my eye for a *scientist-lecturer* doctoral-level position. The person did not need specific topic expertise, but would become an acknowledged scientific expert in specific areas for which the company sold products. In addition to analytic ability, a scientist-lecturer needed to be an excellent public speaker. The ability to clearly communicate complicated science to interested audiences was very appealing. Critically important, the role was considered a stepping stone into a sales or marketing position within the company. I knew that that this was the bridge I had been looking for!

If you are reading this book, you may be looking for a bridge or transition mechanism from science and technology into an alternative career. Are you currently pursuing an advanced degree in the life sciences or engineering? Perhaps you are toiling away in an academic post-doctoral position? Why not consider branching from science and taking an alternative path into the business world? Even though scientists often view sales and marketing roles with skepticism and just a hint of elitism, marketing and sales can be a very rewarding career in a fast-moving, exciting field.

The good news is that a strong scientific or engineering background is a tremendous asset to a marketing professional. You possess knowledge and a perspective that is not easily matched by people with a traditional business background.

Why Sales and Marketing?

There are many alternative career paths for scientists and engineers. Why go into sales and marketing? Two immediate answers: a sales and marketing career can be tremendously rewarding and fun. Second, there is a tremendous need for qualified

G. Madhavan et al. (eds.), *Career Development in Bioengineering and Biotechnology*,
DOI: 10.1007/978-0-387-76495-5_29, © Springer Science+Business Media, LLC 2008

scientists and engineers to work in sales and marketing roles. In my opinion, it is easier to teach a scientific professional the basics of business than it is to educate a business professional on genomics, chemistry, or engineering. However, this does not mean that everyone is suited for a sales and marketing role. Not everyone has or wants to develop the skills necessary to sell products or concepts to customers. This chapter describes a career in marketing using the anecdotes and experiences of one scientist who made the transition.

Non-marketing people, including many scientists, think of marketing as advertising, or the production of pens and pads for trade shows. This is an extremely myopic and unfortunate view of the marketing professional. The classic textbook definition of marketing is "the process of planning and executing the conception, pricing, promotion, and distribution of ideas, goods, services, organizations, and events, to create and maintain relationships that will satisfy individual and organizational objectives." This is an accurate, if somewhat dry, definition. I like to shorten the definition of marketing to "the strategic planning and tactical execution necessary to ensure a product/company's success in the marketplace." Success is ultimately measured by sales revenue and profitability, but success can also be evaluated by harder-to-measure metrics such as brand awareness or customer preference.

In the life sciences marketplace, the type of marketing you do depends on the segment you are in, the resources you have, and your specific corporate environment. Job opportunities for marketing professionals range from pharmaceutical and biotech companies to medical device and diagnostic firms, as well as computer software and hardware businesses. Different market segments have different needs, focuses, and audiences. As an example, pharmaceutical marketing differs from the marketing necessary to sell reagents to laboratory researchers. In addition, pharmaceutical marketing requires knowledge and experience in dealing with a strict regulatory environment. Finally, the marketing budget and resources of a pharmaceutical company cannot be compared to a device company that is also selling to researchers.

Company size and marketing philosophy often dictate how marketing is conducted. Many large companies operate marketing organizations by segregating marketing functions into discrete groups: marketing intelligence, marketing management, and marketing communications are examples of some of these specific niches. In these types of organizations, marketing professionals are expected to have specific expertise in their niche. Employees are expected to work only on marketing that falls under the umbrella of their specific job description; for example, if you join the marketing communications group, you will not be conducting research on market size, pricing, or competition. In contrast, small firms do not have the resources or the need to discriminate between marketing functions. However, in my experience, many small companies believe that marketing is primarily focused on producing literature, advertising, and managing tradeshows: all important activities, but not a true depiction of marketing value.

Transitioning to a Sales or Marketing Role

First, decide whether sales or marketing is the right path for you. You have invested considerable time and expense in your scientific training. It is important to conduct your own research. Read as much as you can. Talk to people who have made the transition. Ask them what they do, and how they like their positions. What was unexpected? What skills do they believe are important?

Determine whether you need additional education or training. I am often asked: do I need a Masters of Business Administration (MBA) degree? There is no clear-cut answer, but having an MBA combined with an advanced science degree will clearly open doors for you. Early in my career, I interviewed with a manager obsessed with the need for an MBA. Although I had the pre-requisite work experience and industry knowledge, an MBA was required. Unfortunately, this approach requires that you stay in school for another degree, or take courses while working. This may or may not be feasible based upon your circumstances. Although it is possible to work in a sales and marketing environment without an MBA, it is more difficult to break in. Lack of a business background necessitates a cross-over role: a position that allows you to transition from a scientific background by gaining practical experience. My cross-over role was the scientist-lecturer position. It allowed me to prove my value to the organization and network with key sales and marketing professionals. In addition to my job responsibilities, I volunteered to help with a variety of marketing projects, such as literature and article development and sales training. Over time, these activities, as well as my track record, smoothed my transition to the marketing team.

The Scientific/Engineering Advantage

A strong scientific background is a tremendous advantage for a marketing professional. In my opinion, although a bachelor's degree is useful, a master's or doctorate is more useful. The more complicated the arena (e.g., genomics, proteomics), the more useful your technical background and training becomes. I believe one of the primary reasons selling in the scientific marketplace is different from many other industries is because of the education, training, and skeptical nature of scientists. A scientist is trained to question and be skeptical. This cynical nature becomes stronger (if possible) when interacting with sales and marketing people. A scientist or engineer will easily dismiss a sales person who can not understand their problems or speak their language. But a sales person who can speak with scientists about their work and provide possible solutions to problems will find the process much easier. Credibility must be earned, but the knowledgeable, academically trained individual has an advantage.

Previously, I worked for a large information technology company that started a group focused exclusively on selling information technology solutions into the life sciences. The organization hired many knowledgeable sales and marketing professionals who had worked extensively in the area. However, many of the senior managers were MBAs, moved in from other business units. These people had not been exposed to science since college, or even since high school. It was an interesting experience to observe senior level managers attempting to learn basic biology to gain a better grasp of the marketplace. Scientists/marketing personnel intuitively understood concepts and subjects that were alien to managers attempting to gain a grasp of basic biology terms and concepts.

All marketing professionals understand that successfully selling to customers requires knowledge about their needs and problems, work environment, and thought processes. A science background is not a substitute for speaking directly with your customers or closely working with your sales teams on an account. But it does provide you with insight as well as credibility when you do interact with customers. As an example, I worked with an ad agency to develop a print advertisement which would be my company's initial introduction to a biology-focused audience. The agency took our input and came back with an ad that they thought was perfect for speaking to laboratory researchers. It featured scientists in "spacesuits" working in a P4 containment laboratory. It was obvious that this ad would communicate to customers that our company did not understand their business! This is something that a scientifically-trained advertising or marketing professional would have understood instinctively; the specific ad would never have been developed. In another instance, a colleague sent a note to biologists inviting them to a seminar to learn about the "value proposition" of a new "best-of-breed" product. It was classic marketing speak and an alien language to the audience. It was also a major turn-off: no one came to the seminar.

Typical Marketing Activities

As mentioned, marketing activities are varied, and what you do and the extent you are involved in any particular activity depends upon the company or business unit. There are wide-ranging strategic marketing activities and more focused tactical activities. In larger companies, you may be involved with one specific marketing discipline or participate in planning strategy for an entire division. In smaller companies, marketing professionals tend to wear many hats. Activities range from strategic to tactical. Here are the highlights of a broad range of typical marketing tasks:

Strategic Level

- *Market intelligence:* Who is your audience? What are their problems and needs? How do they receive information? How many potential customers could buy your product? Who makes decisions? Successful marketing and sales programs require a thorough understanding of your customers and the marketplace.
- *Value propositions and messaging:* What value does your product or service provide? Value is determined by understanding your audience (see market intelligence). You must understand the value of your product and how it fits a

customer's need. After defining value propositions, you must build effective messages to articulate that value.

- *Branding:* How do you want to be perceived by your customers? Branding defines how your business and your products are perceived.
- *Competitive analysis* (also filed under market intelligence): Who is your competition? What are the existing and pending threats to your business?
- *Pricing:* How much should your product cost? What is the appropriate price point?
- *Sales channels:* Should you sell direct or via distributors? How your products/services reach the market is based upon your business model and the nature of your product.
- *Alliances/co-marketing:* With which companies should you partner? Resources are limited – what are you looking for in effective partnerships? Which co-marketing relationships will significantly increase sales effectiveness?

Tactical Level

- Integrated marketing communication: All external communications should be part of an integrated campaign where the individual tactics work together synergistically to provide a result greater than the sum of the individual approaches. Marketing communications has several goals, including but not limited to:
    - *Awareness:* Awareness campaigns are designed to make your company and product more visible. Revenue generation is not expected as a direct outcome of awareness activities. However, any marketer worth his or her salt understands that awareness campaigns must lay a strong foundation for future revenue growth.
    - *Demand generation:* Demand generation campaigns are designed to drive prospects to leads and convert to revenue. Typically, demand generation programs are constructed once awareness has been built in the marketplace. These campaigns have closely-watched metrics and are all about driving dollars.
- Sales training and enablement: The sales force must be superbly prepared for success. Marketing personnel often train the sales team on marketplace issue and specific product features, functions, and benefits, overcoming customer objections, providing sales tools, etc. In addition, technically competent marketers train the sales team on relevant journal articles and technical issues. Sales enablement consists of providing field representatives the necessary tools to succeed (literature, demos, case studies, etc).

Behaviors Critical to Success

A corporate environment requires a number of specific behaviors that are not emphasized in an academic setting. Many of these attributes are accentuated in the marketing profession. Here are important personality traits to develop if you want to succeed as a marketing professional:

- *Flexibility:* The ability to start and stop projects as business requirements change. In research and academic settings, years are spent on a specific project with fixed goals. Marketing needs as well as projects can change very quickly. Marketing professionals need to be able to swiftly adapt to change.
- *Teamwork:* Arguably the most critical character trait. Research is often a solitary, competitive profession. Although collaboration does occur, scientists are trained to conduct independent research and competitively publish their research. Marketing specialists need to work together as part of a larger team. The size and nature of the team depends upon the corporate structure and available resources. In small companies, I have coordinated with a handful of people in the development group, as well as field sales representatives. In much larger corporations, I have coordinated with many people working in product development, sales, different marketing sub-disciplines, ad agencies, and corporate management, as well as separate divisions focused on other vertical industries. Regardless of company size, intra and inter-divisional cooperation is essential to success.
- *Communication:* Sometimes thought of as part of teamwork, but I consider communication to be its own behavior. It is not enough to speak – it is more important to listen. Often, people focus on their inner voice and do not apply critical thought to someone else's ideas. In marketing, this is especially relevant because marketing communicates both internally and externally. Internally, we speak to "stakeholders" across disciplines within the company. Marketing needs to align internal considerations and development plans with what the market wants. This can be difficult, as internal processes and needs take on a life of their own. Marketing's external communications with customers and close cooperation with field sales people are critical to success. Sales people and customers provide that important window into "market reality."
- *Thick skin:* Thin-skinned, sensitive people will have a difficult time in marketing. Everyone has an idea on what is needed, what works, and what does not. Everyone will make sure you have their input and look to you to incorporate their approach in your activities. Always be open to new ideas and new directions: no one has a lock on absolute truth. At the same time, a marketing professional uses his or her knowledge, background, and experience to select particular approaches most likely to have a successful result.
- *Education/training:* Marketing experts who want to stay on top of their game and continue to excel stay current about the marketplace as well as the science. Tom Peters talks about each person as a brand. Branding is not just critical for a business, it is also vital to build and refine your personal brand. Breaking into a marketing role will require you to market your technical background as well as your ability to learn, adapt, and bring something extremely valuable to the position. Personally, I believe it is vital to be constantly learning. Marketing to science and engineering professionals demands that you stay current with the science, certainly in the field in which you market.

Closing Thoughts

This is an exciting time for scientists and engineers. Alternative careers abound. The number of science and engineering doctoral recipients working in sales and marketing has doubled from 1993–2003, outpacing total doctoral growth. [1] As mentioned, scientists who choose sales and marketing as a career possess an extra dimension that is invaluable.

There are many books and articles that can provide you with background to help expose you to marketing. Although not directly about marketing, I refer you to Steve Jobs' commencement address delivered at Stanford University. [2] Mr. Jobs, CEO of Apple Corporation, is arguably one of the best marketers around. His commencement address provides a perspective on business success that you will not find in any traditional marketing text. On a more prosaic level, you can enjoy a bird's eye view of marketing strategy and tactics in the *Harvard Business Review*. [3] It is a very complete look at the elements that comprise marketing.

If you believe that you have the necessary skills, drive, and desire to purse a sales and marketing career, I encourage you to so. It is an exciting time to be a sales and marketing professional in the sciences!

Bibliography

1. Scientist and Engineers Statistical Data System [SESTAT on the Internet]. Arlington (Virginia): National Science Foundation. c1993–2003. www.nsf.gov/statistics/ doctoratework/
2. Jobs, S. "Stay hungry. Stay foolish," *Fortune*. 2005 Sept 5: 31–32.
3. Dolan, RJ. "Note on marketing strategy." Boston (Massachusetts): Harvard Business School, *Harvard Business Review*; 1997 Nov. Report No.: 9-598-061.

30

Sports Engineering

Celeste Baine, MAEd

Engineering Education Service Center, Springfield, Oregon, USA

I have always been athletic. When I was young, my brother was considered the prodigy child, as he excelled in art and reading. Not knowing how to compete, I chose to go my own way and excel in running, jumping, and playing. In sixth grade, I could run faster than all but one boy, and by my junior year in high school, I was captain of the volleyball team, had a bowling average of 240, was swimming open water competitions, playing softball, and riding a unicycle all over town. Having always loved technology, gadgets, and mechanics, my gravitation towards bioengineering and sports equipment was obvious. Unfortunately, bioengineering was a relatively new field when I went to college, and finding a way to link sports to engineering was challenging.

The sports industry provides entertainment, physical fitness, and health awareness for millions of people around the world. This field is wide open and growing so rapidly that opportunities are plentiful and imaginative. Sports engineering is an excellent way to impact athletes, sports, and businesses around the world, and sports engineers are some of the most dynamic, innovative, and creative engineers on the planet. Not only is this industry full of diversity, fun, and intriguing opportunities, most of the engineers working in the sporting goods industry became engineers because they love sports and want to either increase their performance or enhance the sport overall. A company that wants to design a new golf club for Tiger Woods would prefer to hire an engineer who has some experience with golf. A company that is designing new high-performance mountain bikes would prefer to hire an engineer who has a keen interest in bike design or racing bicycles. The industry offers outstanding careers for athletically inclined engineers!

Sports equipment traces its roots back 4,500 years to a ski found in Hoting, Sweden. However, only recently has the sports engineering industry boomed as athletes sought improved equipment and technology to better their performances. Sports engineering helps to break down each sport into its most basic elements (stride length, foot placement, swing speed, etc.) so that engineers can make tiny adjustments that result in bigger improvements in performance.

G. Madhavan et al. (eds.), *Career Development in Bioengineering and Biotechnology*,
DOI: 10.1007/978-0-387-76495-5_30, © Springer Science+Business Media, LLC 2008

Sports engineering is the bridge between classical engineering and sports science. Sports engineers use technology to analyze the equipment, materials, and mechanics of sports, athletes, and movement. They strive to prevent injuries and to increase the achievement of athletes around the world. As a sports engineer, you can work to improve performance for athletes, increase the fun and recreation of dedicated fans, and improve the image of engineers.

Sport science is the study of how the body works, through biomechanics, human anatomy, human physiology, movement anatomy, and exercise physiology. To succeed in sports science requires understanding biology, chemistry, physics, psychology, sociology, coaching, education, motor skill acquisition, nutrition, resistance training, exercise programming, sports injury, and exercise rehabilitation.

Engineers in the sports industry must have a solid and well-rounded knowledge of materials, equipment, and technology, as well as familiarity with sports and athletics. Interpersonal skills are also a must because the sporting goods industry is a people-oriented business. You must be able to work with people and athletes to ensure that your products meet the needs and requirements of the sport.

Many people do not realize that engineering plays such a big role in the sporting goods world. Millions of dollars are spent each year on research and development for new equipment and technology. Thousands of professional engineers contributed to the previous Olympic Games, and many more contribute daily to make professional sports more enjoyable and fun. For example, civil engineers spent two years building the bobsled, skeleton, and luge tracks for the 2002 Olympic Games in Salt Lake City. A mechanical engineer developed the machine to make the snow that appears on the ski jumps at the Utah Olympic Park. Electrical engineers designed the state-of-the-art timing systems to ensure accuracy to 1/1000 of a second. Other engineers designed the bicycle track for the 1996 Olympics in Atlanta, and many more work on devices or technology to make all of the other events in those 17 days of fierce competition more exciting and dramatic. This field is driven by innovation and the creativity that comes from applying knowledge already achieved into new situations to find new solutions.

Suppose for a moment that you are on the college basketball team. You are not sure if you have the skill to play professionally, but enjoy the sport of basketball more than anything else in your life. By using your engineering and problem-solving capabilities, you can contribute to the sport in many ways. For example, as a materials or chemical engineer, you may find new materials for court floors or shoes that can lessen knee and ankle injuries. As a mechanical engineer, you may work hard to find new materials for bats that will enable a batter to hit the ball much farther and reduce the unwanted sting of the bat associated with hitting the ball off-center.

Better training techniques, nutrition, and coaching have helped to improve the performance of athletes. Improvements in training and physiology, in conjunction with technological advancements, have helped to take the athletes much farther. Very often, the introduction of new technology to aid a sport can increase performance in an unexpected way. For example, pole-vaulters have increased their vaults from just over 117 inches in the late 1800's to 241.75 inches in a 1994 competition. In 1964, glass fiber composite poles, although expected to give the jumper a higher jump, worked to help jumpers change their style to go over the bar feet first and thereby increase the vaulted height. Other engineers were needed to support the increased vaults by designing improved landing areas.

Another important aspect of sports engineering concerns how the evolution of new technology helps to prevent injuries. The advances of technology that make sports more enjoyable and more dynamic are wonderful for athletes and spectators but sometimes come at a cost. In the 1950s, football players were at high risk of head injuries because they did not wear helmets. The engineers that designed the plastic helmets never imagined that they would create a much more dangerous sport because the players perceived that the helmets made them safer, so they could take more risks. As explained by J.N. Gelberg, author of *The Lethal Weapon: How the Plastic Helmet Transformed the Game of Football* "Now that players can also tackle each other below the waist, football neck injuries have more than tripled and cervical spine injuries have doubled." The engineering design must always evolve to keep up with the changes in the sport and attitudes of the participants.

Sports Engineering: What it Takes to Succeed

Creativity and innovation are the keys to bringing designs to the marketplace. To be a successful sports engineer, you must understand the sport, the performance criteria, and the perception of the fans. You must appreciate the mental state of both the athlete and the audience. Equipment must be designed to keep the athlete safe and to lessen injury. The social appeal of color, function, and trend must act as a unit. The materials should be the best for the conditions or developed to advance the sport. Lastly, the equipment must be thoroughly tested and evaluated. If problems are detected, you must work on redesign until your designs pass the important test and evaluation process.

Before every Olympic Summer Game, the scientific community from around the world gathers to exchange ideas, developments, and findings in the Pre-Olympic Congress. The Congress is designed to bring all sport fields together to share research and developments in the field. It is one of the largest gatherings of cutting edge research and performance improvements in the sports industry.

Sophie Woodruff, an undergraduate student in Sports Product Design at South Bank University in London said, "In my brief two years as a design student I have learned that sports design does not just entail the use of typical engineering principles, as is the case in most product or engineering design situations. To design a great sports product, the designer must understand why the product is of such great importance to the athlete, and how to maximize the success of such products whilst knowing how they work in conjunction with the human body."

Bioengineering Career Opportunities in Sports and Sports Equipment Design

- *Swimming* – Engineers may design new suits or model a swimmer's performance using computer systems that analyze stroke capabilities. To help swimmers swim faster, scientists and engineers decided to study fish. Engineers at the Massachusetts Institute of Technology in Cambridge, Massachusetts, built a robotic fish to study its movements. Researches at other institutions are creating other models to study strokes to better understand where the power comes from. Engineers and scientists at Adidas found that the following variables enhanced a swimmer's performance:

  1. *Body temperature* – Studies indicate that athletes have more power when they run a slight fever. The higher temperature induces chemical reactions that make muscles contract faster.
  2. *Drag* – When swimming, water sticks to your body and forms a boundary layer that causes surface drag. However, researches found that this did not happen to the Mako shark. The shark was able to move quickly through the water because of a V-pattern in its skin. The bodysuits that were developed with this pattern had less resistance than human skin.
  3. *Muscle oscillation* – This is a problem in many sports that causes fatigue. Muscle oscillation is when an athletes muscles in motion look like Jell-O. Bodysuits help reduce oscillation because they aid in compressing the muscle.
  4. *Tight suits* – Wearing a tight suit enhances a swimmers performance, and is also called proprioception. Proprioception refers to the nervous system's interpretation of its body position based on how the suit presses against the skin. According to Adidas, "the suit creates compression that allows the nervous system to provide better feedback on where the body is positioned." Athletes who feel stronger and more in tune with their muscles perform better.

Aside from suits, there is tremendous engineering that goes into designing a swimming pool for Olympic competitions. The governing body for Olympic swimming pool design is the Federation Internationale de Natation (FINA). FINA specifies the depth, width, length, space between the lanes, and temperature of the pool. Engineers who designed the solar heating system at the aquatic center that was used during the 1996 Summer Olympic Games worked tirelessly to determine how to maintain one million gallons of water at 78 degrees. Other specific areas of research in sports include the following:

- *Baseball and football* – Researching impact to determine padding requirements; designing systems to analyze the human body wearing the helmet to better understand how to prevent neck and spine injury; researching the motion of many sports to determine the requirements for helmet design.
- *Skiing and snowboarding* – Designing systems to analyze the human body while skiing or snowboarding to aid in injury prevention; researching the

motion of many sports to determine the requirements for product design and to help keep people safe. Engineers who love to ski and snowboard naturally gravitate toward work in the snow sports industry. Traditionally, when an idea for a new ski or snowboard design came along, engineers would build a prototype, perform laboratory tests for stiffness, and test it on the slopes. Based on the test experience, engineers would make design changes and retest the equipment. This method of design was a slow and tedious process.

The perfectly crafted ski or snowboard is not perfect for everyone. The needs of a 5'2" female snowboarder are much different from the needs of a 6'0" male snowboarder. The snowboarder's height, weight, and skill level, as well as the snow conditions and the angle of the slope, all need to be taken into consideration when trying to fit the perfect board to the enthusiast.

Structural strength of a snowboard is very important. Engineers determine the strength by figuring out the acceleration of the rider. Acceleration is determined by finding the coefficient of friction, the incline plane, and the combined mass of the rider and board. On fresh fluffy snow, a heavier boarder on a short board will cut deeper into the snow and reduce the speed. On a long board, the edges may not stay in contact with the snow at all for a light person. If the edges do not stay in contact, this results in a loss of control and reduced speeds. In general, a narrow snowboard allows for turning control and a wide snowboard goes faster.

- *Tennis* – If tennis is your passion, as an engineer, you can design racquets, strings, balls, and other new technologies to advance the game and reduce injuries. Every year, manufacturers come out with multiple racquet designs to make the game better for professionals and more enjoyable for recreational players. There is a wealth of opportunity in the tennis industry, but for bioengineers specifically, the demand is for injury prevention.
- *Inline skating* – The beauty of skating is that whether you are use skates for transportation, for playing hockey or basketball, for dancing, for fitness, or for sailing with the wind across empty lots, you have the ability to thoroughly express and enjoy yourself. This enjoyment is expanded and increased by having the proper equipment, such as knee and wrist pads, a helmet, and well-maintained skates. Engineers are at the forefront of making the sport safer and more fun by applying engineering principles to creating the wheels, bearings, frames, boots, and safety gear. The creation of an inline skate is no small feat of engineering. For example, just to create the wheels of a skate, engineers go through meticulous design and testing phases that depend on the type of skating an individual wants. An aggressive skater wants speed and maneuverability. Recreational skaters need comfort and stability. Fitness enthusiasts desire a good low-impact and cardiovascular workout. Inline racers want precision engineering for maximum speed, and some want to transverse mountains in the quest for the ultimate trail or activity. There are unlimited possibilities for the engineer with an interest in skating and a creative streak to take skate design to the next level.

Getting Started

Sport engineering is not an accredited degree at any university in the United States. However, there are several programs in the UK. If you want to work in the sporting goods industry, contact manufacturers or design companies that look interesting to inquire about temporary, summer work or co-op opportunities. For example, if you love to bowl, contact Columbia Industries, Brunswick Bowling and Billiards Corporation, Ebonite, Storm, AMF Products, AZO Bowling, Lane#1, Morich Motion, Track Bowling, Visionary Bowling Products, or one of the other many companies that support bowling, and ask about opportunities. If the company is local, you can even volunteer your time to get your foot in the door.

Tour facilities and begin talking to other sports engineers for tips on getting in the door. Look for online forums about sports engineering, and get on the sports engineering listserv of the International Sport Engineering Association (ISEA). ISEA also holds conferences on sport engineering.

In college, the most important thing that you can do to get yourself prepared for a job in the sports industry is to get a co-op position. For example, at Purdue, the top 1/3 of the class is offered the opportunity to be in a co-op program. In this program, a student alternates between working for a few months at a company and going to school. Basically, it turns a 4-year program into a 5-year program, but you graduate with work experience. The top 1/3 of the class gets a flyer about what companies are coming, and you select your top ten co-op choices.

Consider sports engineering as a gateway. Bioengineers are at the forefront of sports equipment design, and opportunities are plentiful for those with an engineering degree and an interest in sports. From passionately designing new and improved products for athletes and spectators, to the design and manufacture of skateboarding, golf, swimming, and skiing equipment, to shoe design and much more, there is something for everyone who has a love of sports and an interest in engineering, human physiology, kinesiology, chemistry, medicine, rehabilitation, injury prevention, and engineering.

Bibliography/Recommended Reading

1. Baine, Celeste. *Is There an Engineer Inside You? A Comprehensive Guide to Career Decisions in Engineering*. Belmont, California: Professional Publications, 2004.
2. Baine, Celeste. *High Tech Hot Shots: Careers in Sports Engineering*. Alexandrai, Virginia: National Society of Professional Engineers, 2004.
3. Bolles, Richard Nelson. *What Color is your Parachute? A Practical Manual for Job Hunters and Career Changers*. Berkeley: Ten Speed Press, 2001.
4. Brody, Howard, Rod Cross, and Crawford Lindsey. *The Physics and Technology of Tennis*. New York: Racquet Tech Publishing, 2004.
5. Byars, Mel. *50 Sports Wares: Innovations in Design and Materials*. Switzerland: Rotovision, 1999.
6. "Careers in Science and Engineering: A Student Guide to Grad School and Beyond." National Academy Press, 1996.

7. "Congressional Commission on the Advancement of Women and Minorities in Science Engineering and Technology Development," *Land of Plenty*, Arlington, Virginia, Sept. 2000.
8. Davis, Susan, Sally Stephens, and The Exploratorium. *The Sporting Life: Discover the Unexpected Science Behind Your Favorite Sports and Games*. New York: Harry Holt and Company, 1997.
9. Ferguson, Eugene S. *Engineering and the Mind's Eye*. Cambridge: MIT Press, 1997.
10. Ferrell, Tom. *Peterson's Job Opportunities for Engineering and Computer Science Majors*. United States: Thomson Learning, 1999.
11. Field, Sally. *Career Opportunities in the Sports Industry: A Comprehensive Guide to Exciting Careers Open to You in Sports*. New York: Checkmark Books, 1999.
12. Gabelman, Irving. *The New Engineer's Guide to Career Growth and Professional Awareness*. New York: IEEE Press, 1996.
13. Haake, Steve. "Sports Engineering Journal." International University of Sheffield on behalf of the International Sports Engineering Association. 2002.
14. "The Green Report: Engineering Education for a Changing World." American Society for Engineering Education, 1998.
15. Landis, Raymond B. *Studying Engineering: A Roadmap to a Rewarding Career*. Burbank, California: Discovery Press, 1995.
16. LeBold, William K. and Dona J. LeBold. *Women Engineers: A Historical Perspective*. American Society for Engineering Education, Washington, District of Columbia, 1998.
17. Love, Sydney F. *Planning and Creating Successful Engineered Designs: Managing the Design Process*. Los Angeles: Advanced Professional Development Incorporated, 1986.
18. Peters, Robert L. *Getting What You Came For: The Smart Student's Guide to Earning a Master's or Ph.D.* New York: Farrar, Straus and Giroux, 1997.
19. Peterson, George D. *Engineering Criteria 2000: A Bold New Change Agent*. American Society for Engineering Education, Washington, District of Columbia, 1998.
20. Petroski, Henry. *Invention by Design: How Engineers Get from Thought to Thing*. Cambridge: Harvard University Press, 1998.
21. Petroski, Henry. *To Engineer is Human: The Role of Failure in Successful Design*. New York: Vintage Books, 1992.
22. Petroski, Henry. *The Evolution of Useful Things: How Everyday Artifacts-From Forks and Pins to Paper Clips and Zippers-Came to be as They Are*. New York: Vintage Books, 1994.
23. Subic, A.J., and S.J. Haake. *The Engineering of Sport: Research, Development and Innovation*. Oxford: Blackwell Science, 2000.
24. Tietsen, Jill S. and Kristy A Schloss with Carter, Bishop, and Kravits. *Keys to Engineering Success*. New Jersey: Prentice Hall, 2001.
25. Zolli, Andrew. *Catalog of Tomorrow: Trends Shaping Your Future*. Indianapolis: Que and Tech TV, 2002.

31 Writing Non-Fiction Books

Barbara A. Oakley, PhD, PE

Department of Industrial and Systems Engineering, School of Engineering and Computer Science, Oakland University, Rochester, Michigan, USA

Introduction

Writing non-fiction books related to your area of expertise can be an extraordinarily satisfying experience. However, such writing is best thought of as a supplement to your career, rather than a career in and of itself, since the financial rewards are typically small. In that sense, then, this chapter describes an activity that compliments the many careers discussed in this book.

Writing books is not for the faint-hearted. It can often take years of meticulous attention and feedback from others to create a useful, readable volume. On the other hand, those who are fascinated by their work often find that writing books allows them to obtain a better overview of their area of interest or expertise. It also provides a mechanism for meeting many talented individuals who share the same interest. This can supply a valuable boost for any career.

Types of Non-Fiction Books

Non-fiction books are generally written for three different audiences: specialists, general readers, or students. Here are some examples and additional information related to each of these categories:

- ***Non-fiction for specialists.*** Such books provide an understanding of the subject area at hand, often from an in-depth perspective—they can be either *authored* or *edited.* An authored book has one author or a small team of authors who do all of the writing. An edited book has many contributors whose efforts are coordinated by an editor or small team of editors.
  - *Ion Channels and Disease: Channelopathies*, by Frances Ashcroft, (authored book).
  - *Career Development in Bioengineering and Biotechnology*, by Guruprasad Madhavan, Barbara Oakley, and Luis Kun, (edited book).

G. Madhavan et al. (eds.), *Career Development in Bioengineering and Biotechnology*,
DOI: 10.1007/978-0-387-76495-5_31, © Springer Science+Business Media, LLC 2008

- ***Non-fiction for general readers.*** This type of non-fiction provides a readable and entertaining explanation of general science or engineering-related topics that can be understood by non-specialists.
  - *The Singularity is Near*, by Ray Kurzweil.
  - *Evil Genes: Why Rome Fell, Hitler Rose, Enron Failed, and My Sister Stole My Mother's Boyfriend*, by Barbara Oakley.
  - *Nanotechnology: A Gentle Introduction to the Next Big Idea*, by Mark Ratner and Daniel Ratner.
- ***Textbooks.*** Textbooks are one of the few areas of non-fiction book writing that can be financially remunerative. However, the more specialized and higher the level of the textbook, the smaller the sales. Very high level textbooks can sometimes also be classed as non-fiction for specialists.
  - *Introduction to Biomedical Engineering*, by John Enderle, Susan Blanchard, and Joseph Bronzino.
  - *Fundamentals of Biomechanics*, by Duane Knudson.

What Publishers Are Looking for

Non-fiction books are often sold to publishers based only on a book proposal and a sample chapter or two. Before you begin writing, you will want to read more detailed explanations about how to write a book proposal. Several excellent sources include *Handbook for Academic Authors*, by Beth Luey, a good reference source for any type of non-fiction; and *How to Write a Book Proposal*, by Michael Larsen, which is geared toward general non-fiction book proposals [1, 2].

Publishers usually have their own descriptions of what information they would like to see in a book proposal, particularly in relation to specialized non-fiction. The following are commonly required:

a. Book title and subtitle
b. Reason for the book

This section explains, in one or two paragraphs, why the book is needed and who it is being written for.

c. Topic

This section provides an overview of the book's subject matter, as well as a more specific explanation of coverage. Little or no technical jargon should be used. This is where you make the case for how useful your book will be to readers.

d. The book's audience

The most important aspect of finding a publisher for a book is simply the number of people who may potentially want to buy it. Publishers want excellent content, but if there is no one willing to shell out cash to buy a book, there is little sense in publishing it. There should generally be at least 10,000 potential buyers for a book to be deemed marketable. Deducing the number of potential buyers can sometimes involve a little detective work. You may find yourself investigating the membership numbers of various societies, the number of subscribers to specific technical journals and number of attendees of related conferences, or results from the Bureau of Labor Statistics. For this section, it can be helpful to make a table of your findings.

e. Competition

Here you would list the title, author, and publisher of books that are similar to the one you are proposing. These competitive books are used as a gauge for the marketability of your book. You will want to indicate how your book differs from and is superior to the other books.

f. Estimated final size and related statistics

This is where you indicate final word count for the manuscript, and how much of the manuscript you have written. (One double-spaced manuscript page is about 250 words.) You will also provide a rough indication of the number of illustrations, including computer generated drawings, line drawings, and photos. (A rough rule of thumb is that the illustrations should be 300 dpi when sized at least 4 by 6 inches, and that the illustrations should be submitted in jpg, tiff, or psd formats.) If you will be needing assistance with your illustrations, you would indicate that need in this section.

Everybody realizes at this stage that it's impossible for the author to know exactly what the final book will be like. So if you have indicated that the book will have 40 illustrations, for example, and the final manuscript has only 35, it is not typically a problem. (However, it is always a good idea, as far as illustrations go, to provide an maximum estimate that won't be exceeded.)

g. Outline and Sample Chapter

A table of contents or outline gives the editors a good feel for the coverage of your proposed book. You will want to include a brief paragraph that describes each chapter's contents. This sample chapter provides an indication of your writing ability, so it is important that it be sharp.

h. Reviewers

Editors often need help determining the value of specialized non-fiction. Therefore, proposals for this type of non-fiction are often farmed out to respected external experts who might be willing to provide an independent opinion. You are often expected to provide names of five to ten of these external reviewers. It is not a bad idea to contact these external reviewers before you even submit the proposal and ask them if they'd be willing to have their name listed as a potential reviewer. If your credentials look strong and the book looks interesting, these individuals may even be willing to "pre-review" your proposal before you submit it to the publisher. After all, even these experts have a need to keep abreast of their developing field. Such pre-reviews can markedly improve the quality of your proposal.

i. Biographical information

Publishers generally want the full curriculum vitae of authors and co-authors. Naturally, your experience plays a critical role in what kind of expertise you might have to write the book you propose.

Other Considerations Related to Non-Fiction Book Publishing

Matters of Style

Good non-fiction books explain a subject in simple, clear terms. This is harder to do this than it seems. Beth Luey discusses this problem in her *Handbook for Academic Authors*:

> "Good academic writing is clear and succinct. (To use myself as an example, I first wrote that sentence: 'For the purposes of academic writing, writing well is writing clearly and succinctly.' I read it, saw that it was neither clear nor succinct, and rewrote it. Reading and revising are essential to good writing. Had I not spotted the problem, the manuscript editor would have fixed it, after a good laugh at my expense.) If you can move beyond clarity to grace and elegance, you are to be congratulated. Editors will happily settle for clarity, however [1]."

It is important to have other people review your writing. One reviewer should have good technical expertise in the book's subject matter, the other should have only general familiarity. You may be surprised to find that your crystal clear prose is not quite as clear as you had thought. *The Elements of Style*, by Strunk and White, can be an invaluable aid for improving your writing [3]. *The Chicago Manual of Style* is another important writing tool you should have on hand [4].

Impact of Authoring Books on a Career

Surprisingly, people in industry, as opposed to those in engineering academia, have very different perceptions of the value of publishing books. For industry, publication of a non-fiction book is an asset to any resume—it provides instant credibility for the author. For academia, on the other hand, publishing a book is sometimes seen as rehashing, instead of creating, knowledge. Thus, there can be a bit of snobbery from other academicians surrounding what is seen as the "routine" authoring effort. Because of these perceptions, many engineering academics wait until later in their careers, after they've been tenured, to devote time to writing a book. Textbooks are particularly harmful in this regard—although difficult and time-consuming to write, they are generally regarded as lightweight, non-research caliber work.[1]

[1] Occasionally, this attitude can backfire. One textbook author of a popular textbook was apparently denied promotion to full professor because his research was not deemed strong enough. Since the author had become independently wealthy as a result of his textbook's strong sales, the author simply quit the university. Later, the university realized the value of this individual's name and requested that he be listed as still affiliated with the university. The former professor refused the request.

In summary then, writing books that people will buy, read, and, perhaps even enjoy, isn't easy. But, as long as you keep your "day job," it can be as rewarding as it is demanding.

References

1. B. Luey, *Handbook for Academic Authors*, 4th ed. New York, New York: Cambridge University Press, 2002.
2. M. Larsen, *How to Write a Book Proposal*, 3rd ed. Cincinnati, Ohio: Writer's Digest Books, 2004.
3. W. S. Strunk and E. B. White, *The Elements of Style*, 4th ed.: Macmillan, 1999.
4. *The Chicago Manual of Style*, 15th ed.: University of Chicago Press, 2003.

32 Emerging Innovative Careers

Guruprasad Madhavan, MS, MBA

Department of Bioengineering, Thomas J. Watson School of Engineering and Applied Science, State University of New York, Binghamton, New York, USA

Jennifer A. Flexman, PhD

Department of Bioengineering, College of Engineering, University of Washington, Seattle, Washington, USA

Aimee L. Betker, MSc

Department of Electrical and Computer Engineering, University of Manitoba, Winnipeg, Manitoba, Canada

"Keep away from people who try to belittle your ambitions. Small people always do that, but the really great make you feel that you, too, can become great."

—*Mark Twain*

Bioengineers and biotechnologists are uniquely poised in terms of career flexibility, diversifiable skill sets, and a broad perspective that compliments a multitude of other disciplines. All of this combines to provide for an extraordinary number of career opportunities, as this volume has already shown. However, the possibilities are actually even wider than you might imagine; this chapter briefly surveys ten additional innovative career options.

Science Writing and Journalism

Imagine spending a week learning about the latest research in stem cells, climate change, and newly discovered galaxies. Then imagine the next week going on to an entirely new set of topics. Science writing and journalism provide just that kind of intellectually challenging and stimulating lifestyle.

"The bottom line goal for science and technology journalism is to take a large issue that is outside the scope of daily understanding and translate it

G. Madhavan et al. (eds.), *Career Development in Bioengineering and Biotechnology*, DOI: 10.1007/978-0-387-76495-5_32, © Springer Science+Business Media, LLC 2008

for others in a way that they can understand not only the subject matter, but its impact on their daily lives. There are two challenges for effective writing of general articles in science and technology. First, you have to understand the science that you are writing about. That is not always easy—for example, you may understand health very well, but the underlying physics could be challenging. Second, you have to cut through and translate the scientific language that is being used and make sure that the average person can understand it."

—Mark Senak, JD, Fleishman-Hillard, Washington, District of Columbia, USA

Science writers can work as public information specialists at organizations related to science, such as a university, museum, or biotechnology firm. Public information specialists relay information about their organization through materials such as press releases. A second role for the science writer is as a journalist. Science journalists typically work for a media outlet, such as a newspaper, writing for the public. But they may also write for the scientific community or as a freelancer. The science journalist must be able to stay abreast of major scientific discoveries and write the story in a way that engages the reader. Some science writers produce advertising materials or manuals for scientific equipment.

Science writing and journalism require many skills beyond simply understanding the science. Unfortunately, the education of a bioengineer or biotechnologist does not typically involve the development of strong writing skills. Bioengineers and biotechnologists can overcome this hurdle through natural ability, practice, or complementary education in journalism or writing.

"To prepare for becoming a science writer, you should constantly be reading popular science. One helpful assignment is to search for science news articles written by different authors on the same topic. Examine how different science writers approach the same topic and what information each feels is the most valuable. Think about how you might write the story given the facts in front of you.

—Abby Vogel, PhD, Office of Research News & Publications, Georgia Institute of Technology, Georgia, USA

University newspapers and news offices may offer opportunities to gain experience while you are a student. Formal university programs offer a comprehensive education in communications and journalism, and often provide internships. For those interested in science journalism, working as a non-specialized reporter provides practice interviewing, putting a story in context, and writing under strict deadlines.

"If you find that you have a lot in common with other science writers or journalists and enjoy the kind of work that they do, then go forward with the transition by going to a science writing meeting and then a scientific meeting. Ask to shadow some of the writers or journalists and try writing up the sessions and workshops that you attend together. Ask for feedback."

—Robert Frederick, MSc, Science Magazine, Washington, District of Columbia, USA

Just as the technologies of communication have changed our everyday lives, they have also changed the mediums of science writing and journalism. With the "dead tree" newspapers and magazines eroding in favor of electronic mediums such as blogs, podcasts, and websites, modern science writers and journalists must be ready to multi-task in their careers and embrace diverse mediums and formats for their writing. They may also consider acquiring skills in many types of multimedia, including website design, animation, and videography.

Selected Resources: Science Writing and Journalism

National Association of Science Writers, www.nasw.org/
Council for the Advancement of Science Writing, www.casw.org/
American Association for the Advancement of Science, www.aaas.org/
Society for Environmental Journalists, www.sej.org
American Medical Writers Association, www. amwa.org
Blum, D. and Knudson, M. eds., *A Field Guide for Science Writers: The Official Guide of the National Association of Science Writers*, Oxford University Press, 1997.
Communicating Science News: A Guide for Public Information Officers, Scientists and Physicians. Available from the National Association of Science Writers.
Directory of Science Communications Courses and Programs in the United States. Compiled by Dunwoody S, Crane E., and Brown, B., www.journalism.wisc.edu/dsc/allEntries.php.

"Read broadly and constantly practice writing. While writing, also focus on the slants and angles, and develop a captivating writing style. Be creative, accurate and thoughtful. Keep sending out ideas and never give up. Once you break into publishing, either in print or on the Web, momentum builds and eventually, editors may be calling you."

—*Lynda McDaniel, Freelance Creative Writer and Coach, Washington, USA*

Overall, the future of science writing and journalism provides many dynamic opportunities for bioengineers or biotechnologists with creative drive, strong writing skills, a commitment to learning, and a dedication to science.

Informatics and Information Technology

Picture yourself visiting your doctor and presenting a credit card-sized chip that contained your entire medical history, or a surgeon in the US operating on a patient in Russia. Suppose you were responsible for managing the electronic health records at a large hospital with hundreds of thousands of patients. These scenarios represent current and future possibilities for health information technology.

The volume and complexity of information captured in biomedical research and healthcare has increased exponentially in recent years. As a result, new tools are needed to capture this information and extract meaning. Bioengineers and biotechnologists are well positioned to fill this niche, since they already straddle the line between technology and biomedical science. The US Department of Labor considers the information technology industry to be one of the fastest growing sectors in the economy. The rising importance of information technology and informatics has created new and exciting opportunities in a wide range of areas, including electronic patient health records, telemedicine, biosecurity, genetics, and high throughput analysis for the development of new drugs for disease.

Certain major advancements have spurred new industries within information technology and informatics. For example, the completion of the human genome project has produced the code for approximately *three billion pairs* of nucleotide bases constituting the complete human deoxyribonucleic acid (DNA). This immense data set is being decoded to identify the secrets of human identity and disease through the simultaneous development of sophisticated informatics tools. In the future, extracting patient-specific information about the likelihood of disease and optimal treatment strategies will be possible through further advances in information technology and informatics. Bioengineers and biotechnologists will play a critical role in genomics by combining skills in computer science and information management with genetics and basic biology.

Bioengineers and biotechnologists can prepare for careers in health information technology by acquiring a sound understanding of computer science, strategic technology planning and assessment, information security, and clinical workflow. These skills can be acquired through mechanisms such as formal schooling, industrial experience, and interactions with diverse health professionals.

"Students and young professionals in bioengineering and biotechnology who are interested in the area of information technology and informatics may benefit from taking a course (or minor) in computer science or management information systems. I would suggest paying particular attention to topics related to computer methodologies and information management. Those topics will require an understanding of a variety of subjects, which is key for success in this area. Furthermore, information technology and informatics will increasingly involve diverse cross-cultural teams, many times across continents, so developing a solid leadership vision and collaborative skills in an electronic environment will become inevitable."

—*Luis Kun, PhD, Information Resources College, National Defense University, Washington, District of Columbia, USA*

Information technology and informatics hold many opportunities for individuals trained in bioengineering and biotechnology. As a rapidly growing and interdisciplinary field, success will require a dedication to continuous learning, a team-oriented perspective, and a strong desire to improve the delivery and efficiency of healthcare.

Selected Resources: Information Technology and Informatics

American Medical Informatics Association, www.amia.org/

International Medical Informatics Association, www.imia.org/

Health Information Management Systems Society, www.himss.org/

H. Bidgoli (Ed) *The Handbook of Information Security* (3 Volumes), Wiley; 1 edition, 2005.

A. Salazar, S. Swayer., (Ed) *Handbook of Information Technology in Organizations and Electronic Markets*, World Scientific, 2007.

J. H. Van Bemmel. The young person's guide to biomedical informatics. *Methods of Information in Medicine.* 45(6):671–80, 2006.

J. Raymond, and C.A. Quinsey. Career opportunities in information technology. *Journal of the American Health Information Management Association (J AHIMA).* 75(8):56–62, 2004.

J. Sims. Career focus: medical informatics. *British Medical Journal.* 317(7173): 2, 1998.

M. Ruffin. Where does informatics fit in health care organizations? *Physician Executive.* 23(5):62–4, 1997.

Biohealthmatics Career Center, www.biohealthmatics.com/careers/

Regulatory Law

While career development opportunities in intellectual property law are now well recognized, a highly specialized and emerging pathway within law relates to the regulation of food, pharmaceutical drugs, and medical devices. With training and experience in law, bioengineers and biotechnologists can work in positions of high responsibility, including monitoring regulatory and legislative initiatives, notably in genomics and biologics; coordinating industry policy positions; managing development of industry consensus; interacting with regulatory agencies; counseling industries on the interpretation of complex areas of law and requirements, and their impact on product development; and licensing and compliance activities related to product quality.

Selected Resources: Regulatory Law

Food and Drug Law Institute, www.fdli.org/
Piña, KR., Pines, WL., *A Practical Guide to Food and Drug Law and Regulation,* 2nd Ed, Food and Drug Law Institute; 2002
Hutt, PB., Merrill, RA., Grossman, LA., *Food and Drug Law: Cases and Materials,* 3rd Ed, Foundation Press; 2007
Pisano, DJ., Mantus, D., *FDA Regulatory Affairs: A Guide for Prescription Drugs, Medical Devices, and Biologics,* Informa Healthcare; 2003
Pisano, DJ., *Essentials of Pharmacy Law,* CRC Press, 2002
Academic Programs Offering Food and Drug Law Courses, www.fdli.org/academic/lawschls.html

> "The advice I have for anyone wanting to practice regulatory law is to work as a consultant in the field, at the agency, or in industry on the regulatory end for at least a year prior to beginning in any legal practice. The most successful new attorneys appear to have one or the other of these types of employment experiences. Additionally, deep familiarity with government laws and guidance documents will become essential very quickly after entering practice. It would be very helpful to arrive with this knowledge. An entry-level associate with a training in engineering or science who is not 100% clear about what kind of practice they want should either try for recruitment to a smaller practice where the boundaries are not entirely calcified, or should try for recruitment to a firm that allows them to rotate among food, drug, or device law area for the first two or three years."
>
> —*A leading food and drug lawyer, Washington, District of Columbia, USA*

Many law firms that have multiple practice areas are recruiting bioengineers and biotechnologists to work on cross-cutting areas to help clients protect their inventions and products through patents; as well as to obtain the US Food and Drug Administration's (FDA's) approval for products through an extensive clinical trial process; to comply with post-market statutory requirements of the FDA; and to defend themselves during litigation. This trailblazing area is certain to keep bioengineers and biotechnologists intellectually challenged and financially rewarded, while creating an area of expertise within law and regulation.

Defense Sector

For engineers and scientists who prefer a career pathway in public service, working in the defense sector may be an exciting career choice. Three different types of service opportunities currently exist for career development in the defense sector: civil, intelligence, and military. The civil service may involve working across a broad range of disciplines, including traditional scientific research, engineering management, and government advising. Service opportunities in the intelligence area quite often allow engineers to function as investigators (agents) in wide ranging areas, such as counter-terrorism, cyber-crime, civil rights, corruption, violence, drug trafficking, and threat monitoring. On the other hand, armed military forces, such as navy, air force, marine, and border protection, may also offer insightful field experience while enhancing national security.

Selected Resources: Defense Sector

Federal Bureau of Investigation, www.fbi.gov/intelligence/di_career.htm
Central Intelligence Agencies, www.cia.gov/
National Security Agency/Central Security Service, www.nsa.gov/
Defense Intelligence Agency, www.dia.mil/
Respective government agencies across different countries

> "When it comes to hiring in the defense sector, I particularly look for new recruits to have an insatiable intellectual curiosity and appetite for life-long learning. They must be able to work effectively on interdisciplinary teams – teamwork is the coin-of-the-realm in defense. They must be creative. They must have a passion for what they are doing. Further, they must do every job to the best of their ability, and not be afraid to fail. Rather, they should be willing to use any failure as a learning experience."
>
> —*Barry L. Shoop, PhD, Colonel, U.S. Army, U.S. Department of Defense, and the U.S. Military Academy, New York, USA*

Seeking out the counsel of a former or present military professional can be useful when deciding on a career in the defense sector. Note that nearly all positions within the defense sector require individuals with a security clearance. Further, many agencies are financially supportive of higher learning during service in defense. This sector offers fine opportunities for integrating current knowledge, higher learning, and national service.

Venture Capitalism

The word "venture" can be defined as *undertaking risks.* This is an activity characteristic of the vast majority of entrepreneurs. In addition to their risk-taking personas, entrepreneurs typically rely on individuals known as venture capitalists (VCs). VCs are professionals who, through skillful judgments, invest money to assist in the growth of start-up or early stage companies launched by entrepreneurs. In return, VCs look for high growth and return on investments, with interest rates in the range of 40% to 75% in return for the uncertainty of the venture's success. VCs generally perform three main functions: (1) evaluate investment opportunities for the current fund, termed "due diligence," including negotiation and business agreement development; (2) monitor the investment portfolio via holding seats on the board of directors, making additional investments in the company, and reviewing milestones and financial performance; and (3) raise capital for future investments. The venture industry is increasingly recruiting engineers and scientists for their quantitative skills, specialized technical knowledge and expertise in innovative industries, and structured thinking processes. After all, these are precisely the skill sets that most investment projects demand.

> "Engineers have a unique advantage in the venture industry, because their technical skills can be easily expanded into business management through learned experiences. Additionally, engineering and venture capitalism have one fundamental aspect in common: both are engines of innovation and entrepreneurship. Without them, economic development is hard to imagine."
> —*Melissa Carrier, MBA, Venture Investments and Social Entrepreneurship, Dingman Center for Entrepreneurship, University of Maryland, Maryland, USA*

With training or exposure to such areas as business development, management strategy, or corporate finance, bioengineers and biotechnologists can launch into the exciting career area of venture capitalism. Such a career involves staying current on technology and innovation trends, especially in fast growing sectors such as the life sciences. Aspiring VCs should try to meet and network with a wide range of investors involved with endowments, foundations, philanthropic activities, hedge funds, and mutual funds. Success in venture capital depends on the ability to screen potential entrepreneurs and firms for investment, and an ability to work across various markets and cultures. Typically, aspirants for the venture industry do not jump in directly from college, but first gain experience across a broad range of industries or careers.

> "Venture capital is something to do at the end of your career, not the beginning. It should be your last job, not your first. You do not plan for it as you would a career in law or medicine. My theory is that when you are young, you should work eighty hours a week to create a product or service that changes the world. You should not sit in board meetings listening to entrepreneurs explaining why they missed their numbers while you read emails on a Blackberry. People should enjoy a rich and varied career of operational roles and then consider venture capital at the end."
> —*Guy Kawasaki, MBA, Garage Technology Ventures, California, USA*

With the mix of high risks and high rewards, the venture industry is certainly not an easy field to break into. However, a competent, highly experienced bioengineer or biotechnologist could find this area to be the ultimate destination in a satisfying career. With luck, over time, one could even serve on the board of directors of companies that succeed as start-ups. This is a rewarding and lucrative way of helping others achieve success.

Selected Resources: Venture Capitalism

Careers in Venture Capital: Wetfeet Insider Guide, WetFeet, Inc; 2006
Bartlett, J., *Fundamentals of Venture Capital*, Madison Books; 1999
Cardis, J., *Venture Capital: The Definitive Guide for Entrepreneurs, Investors, and Practitioners*, Wiley; 2001
Fast Company Magazine, www.fastcompany.com
Inc., Magazine, www.inc.com
The Wall Street Journal, www.wsj.com
National Venture Capital Association, www.nvca.org/
Young Venture Capital Society, www.yvcs.org/

Environmental Sector

With the growing public consciousness of the importance of environmental conservation, careers in environmental sciences and engineering are burgeoning. Research in this sector typically cuts across disciplines and requires scientists and engineers to have a good foundation in biology, chemistry, physics, and ecology. Work in the environmental arena will be exciting and satisfying, as you will help to develop sustainable practices for use in industry and our everyday lives.

So where do bioengineers fit into the equation?

> "Sustainability work can roughly be divided between (a) environmental/conservation work aimed at protecting the natural world, and (b) economic/social development work aimed at providing jobs, health, and justice for local people in a global economy. Although the two worlds are growing together all the time, it is still likely that individual people will approach the work professionally from one side (e.g. forestry, hydrology, conservation biology, soil science), or the other (e.g. entrepreneurship, finance, economic development, job training, ecotourism). The world of environmental/public health straddles the two approaches, and requires both scientific/engineering professionals and social/behavioral expertise."
>
> —*Kevin Doyle,*[1] *Environmental Careers Organization, Massachusetts, USA*

There are many notable areas of research in the environmental sector. A fast growing topic is bioremediation, in which living organisms and their byproducts are used to remove contaminants from the environment. This includes managing different water sources, hazardous waste, air quality, and pollution. Climate change is another hot topic in both research and in the news. Pattern recognition and signal analysis methods can be used to aid in climate prediction and outcomes. Other areas where bioengineering skills are useful are environmental protection, aquatic biology, oceanography, energy resource management, and geological hazard management. From being a private consultant or working for the government,

Selected Resources: The Environmental Sector

The Canadian Society for Engineering in Agricultural, Food, Environmental, and Biological systems: www.bioeng.ca/
Environmental Engineering Society (Australia), www.ees.ieaust.org.au/
The Society of Environmental Engineers (UK), www.environmental.org.uk/
The Association of Environmental Engineering and Science Professors, www.aeesp.org/
American Academy of Environmental Engineers, www.aaee.net/
Environmental Careers Organization, www.eco.org
Environmental Careers Organization, *The Complete Guide to Environmental Careers in the 21st Century*, Island Press; 1998.
Wang, LK, Tay, J-H, Tay, STL, Hung, Y-T., *Environmental Bioengineering*, Humana Press; 2008.

[1] Quote extracted from K. Doyle, "*Ask ECO*" Career tips at www.eco.org/site/c.dnJLKPNnFkG/b.942793/k.CFFE/Ask_ECO.htm

career development opportunities in the environmental sector appear to be robust for quite a foreseeable future.

Selected Resources: The Agricultural Sector

- American Society for Agricultural and Biological Engineers, www.asabe.org
- The Canadian Society for Engineering in Agricultural, Food, Environmental, and Biological systems, www.bioeng.ca
- European Society of Agricultural Engineers, www.eurageng.net
- Nagata, T, Lörz, H, Widholm, JM., Biotechnology in Agriculture and Forestry (Series), Springer, 1988–2007.
- Food Engineering Magazine, www.foodengineeringmag.com/
- Murphy, D., Plant Breeding and Biotechnology: Societal Context and the Future of Agriculture, Cambridge University Press, 2007.
- Applied Engineering in Agriculture, Journal of the American Society for Agricultural and Biological Engineers

Agricultural Sector

Agricultural science and engineering is a permanently important discipline that has begun to attract bioengineers and biotechnologists. Research in this area offers possibilities for the development of new food products, as well as new mechanisms for controlling pathogenic agents in foods for the benefit of public health. There are many exciting career opportunities in both the private and public sectors. Working conditions in this field can be quite variable, ranging from a sophisticated and pristine research laboratory to wallowing through mud ourdoors looking for bugs and mold.

The agricultural application of science and engineering applies the pragmatic approach of engineering to a fundamental staple of our livelihood – the production of agricultural equipment and food supplies. Careers in this sector include applied research, development, quality control activities in gene discovery, seed development, plant growth agents, crop yield analysis, water conservation, and mineralogical studies. Biological systems knowledge can also aid in recycling through engineering new methods to convert waste and byproducts into useable products. Software development and analysis skills can be useful for developing agricultural informatics and information management.

How do you know if the agricultural sector is right for you?

> "For the student who enjoys science and mathematics, biological and agricultural engineering offers a unique opportunity to combine those scholarly interests with the challenge of providing food and other goods for a growing world population while protecting our natural resources.
>
> Regardless of the specialty, [biological and agricultural engineering] students enjoy a distinct advantage when it comes time to enter the workforce. Their well-rounded engineering experiences enable them to function exceptionally well on the multidisciplinary teams in today's workforce. And only biological and agricultural engineers have the training and experience to understand the interrelationships between technology and living systems – talents needed to succeed in engineering positions today and in the future."
>
> *—American Society of Agricultural and Biological Engineers,[2] Michigan, USA*

[2] "Biological and Agricultural Engineering – Finding Solutions for Life on a Small Planet," at www.asabe.org/membership/beengin.html

Social Sciences Sector

Extraordinary opportunities now exist for cross-fertilization of the social sciences with cutting edge results from the sciences and technology. This sector offers a broad array of research areas, making a background in bioengineering and biotechnology useful in many of the social science fields. For example, one might ask what the significance of biotechnology is when applied to geography:

Selected Resources: The Social Sciences Sector

The National Geographic Society, www.nationalgeographic.com
The American Geographical Society, www.amergeog.org
Beyond Academe, www.beyondacademe.com
American Anthropological Association, www.aaanet.org careersbroch.htm
American Sociological Association, www.asanet.org
Non-Academic Careers for Scientific Psychologists, www.apa.org/science/nonacad_careers.html
American Economics Association, www.aeaweb.org/joe

> "Biotechnology has a significant role to play in making resource use more efficient than it is at present. This is a priority for sustainable development, another theme which is achieving prominence in geography and elsewhere. On this basis alone biotechnology should have a place in geographical studies. There can be few geography syllabuses that do not provide a niche for the teaching of aspects of biotechnology. For example, as an agent of environmental change biotechnology has a place in courses on environmental issues; as a tool for manipulating energy flows and biogeochemical cycles it has a role in courses in biogeography and/or agriculture; as a mechanism for harnessing biotic and mineral assets it has a place in courses concerned with resources; as a means of curbing some forms of pollution it could be included in courses examining environmental quality."
> —*Antoinette Mannion, PhD,[3] Department of Geography, University of Reading, UK*

Bioengineering and biotechnology are now playing roles in many other social science fields. In anthropology, particularly in the areas of biological and cultural anthropology, we see biotechnology linked with genomic research and a changing perception of ourselves. Knowledge of medical device timelines, including their cultural impact and how they are developed in different ethnic regions, can be applied to the historical analysis of medicine. Biotechnology has a large presence in the many different branches of psychology, including industrial, organizational, developmental, social, and neuropsychology. The economics sector, including international, developmental, welfare, business, environmental, agricultural, and applied economics in such areas as banking and labor relations, benefits from an engineering background at the intersection of business and technology. A bioengineer's ability to analyze benefits, success rates, and business plans relating to health and biotechnologies will help to provide the top economical benefit.

Governments, private foundations, and industry fund science and engineering at sums in the billions of dollars each year. For example, the ability of science

[3] Quote extracted from *A. M. Mannion*, "Biotechnology: Its Place in Geography," GeoJournal, vol. 31, no. 4, pp. 347–354, 1993.

and technology to drive economic growth through commercialization plays a prominent role in funding strategies. There is a growing need to understand the process of scientific research and the development of technology through social science. Bioengineers and biotechnologists can work on interdisciplinary teams to study the process of science, or at funding organizations to promote specific scientific and technological objectives related to their disciplines.

The vast arena of this sector amplifies the inter-disciplinary aspect of bioengineering. A resourceful bioengineer can come up with many new fields to apply their knowledge and create new and exciting careers.

Imagineering and Entertainment

A recent brochure from the Walt Disney Corporation states that nearly 100 disciplines come together in their "Imagineering" group. Yes, Imagineering has emerged as a breakthrough and attractive career pathway in the entertainment industry for the most creative, thrill seeking, and artistically perseverant individuals. A common understanding of Imagineering is to create and realize what one can imagine, and to offer a thrilling and memorable experience to the audience. The field includes such areas as visual journalism, animation, screenplay, character formation, audio-visual graphics, and thrill ride design. Following are insights from Imagineers themselves.[4]

> "Never pass up the opportunity to see new things, draw things, build things, talk to experts and learn new skills. I learned how to invent machines of all kinds over the years. I have worked as an auto mechanic, machinist, carpenter, factory worker, artist, concept engineer and many other trades. Some were for money and some were just for fun, but I learned from every one of them."
>
> —*Bruce Johnson, Walt Disney Imagineering–Research and Development, USA*

> "[Executive designer and longtime Imagineer] Rolly Crump told me of some advice Walt Disney had given him: Become a student of life, be interested in everything. Be a life sponge, soaking up, observing and recording anything and everything of interest. Develop an attitude where you never stop learning."
>
> —*Joseph Lanzisero, Walt Disney Imagineering–Concept Design, USA*

Selected Resources: Imagineering

The Walt Disney Company—Disney Imagineering, www.corporate.disney.go.com/careers/who_imagineering.html
Pixar Animation Studios, www.pixar.com/
The Imagineering Way, Disney Editions; 2003
The Imagineering Field Guide to Disney's Animal Kingdom at Walt Disney World, Disney Editions; 2007
Walt Disney Imagineering: A Behind-The-Dreams Look at Making the Magic Real, Disney Editions; 1998
Imagineering, www.imagineering.themedattractions. com/
Nintendo-Wii, www.wii.nintendo.com/

[4] "An Unofficial Look at Imagineering," www.imagineering.themedattractions.com/become_an_imagineer. The material at this link was originally adapted from *The Disney Magazine*, Winter 1995, p. 49.

"If you really want to be part of Imagineering, you will naturally keep growing while practicing and expanding your knowledge. Wander far and wide in your quest for experience. Do not just limit it to what you perceive as the world of Disney. Imagineering is always growing, too. It is always looking for new realms, styles and possibilities."

—*Larry Nikolai, Walt Disney Imagineering–Show Design, USA*

Though Imagineering is primarily associated with careers only at the Walt Disney Corporation, there are many similar career opportunities in the entertainment industry, especially in the conceptualization and development of interactive video or computer games. Engineers and technologists who like the idea of linking their imagination and engineering skills should explore this career route.

Fashion Design

Recollecting the geek-engineer stereotype, you might think that fashion is antithetical to engineering. It may come as a surprise to realize that fashion design offers a fast-paced, extremely competitive and demanding environment for engineers and technologists. But, in fact, the two disciplines have a great deal of overlap. Both fields prize creativity, innovation, imagination, and drive to develop better products. Beyond employment with an established fashion house or company, bioengineers and biotechnologists can also succeed as freelance or entrepreneurial designers. Opportunities also exist in cosmetology, aesthetics, and merchandising industry.

Have you ever wondered what went into the production of your new ski jacket or running shoes? In addition to creativity, the fashion industry depends on fabric, process engineering, and modern manufacturing and materials technology to create innovative products. Engineers skilled in areas such as chemical processing, materials science and manufacturing can introduce new concepts to an industry that is always seeking the next big thing.

"Look for a store in your neighborhood that makes and sells its own clothes and see if you can meet the owner and a designer (often one and the same). Try to talk your way into a tour of their studio or factory or ask for the opportunity to see what a typical day is like. Ask as many questions as you can from as many people as possible: this will help you gain a real sense of the industry. Ask about the hours they work, how long it took them to get their own store, if they went to school, what were their greatest hurdles, what is the toughest part of the job, you get the picture. Some may be too busy preparing for a show or new clothing line, but do not be discouraged. You really should make every effort to get a feel for the business – I assure you it is not all

Selected Resources: Fashion Design

Hartsog, D., *Creative Careers in Fashion*, Allworth Press; 2006

Han, H., *Career Guide to the Fashion Industry*, Vault, Inc; 2003

Goworek, H., Careers in Fashion and Textiles, Wiley-Blackwell; 2006

Fashion Schools, www.fashion-schools.org/fashion-careers.htm

24|Seven, www.fashion-careers.biz/

Fashion Careers, www.fashioncareers.com/

Style Careers, www.stylecareers.com/

Chartered Society of Designers, www.csd. org.uk/

catwalks and photoshoots. If you still want to be a fashion designer, then dig in some more. Go to your local fabric store and see if they offer any lessons, or try buying some fabric and making something for yourself. Contact your local community college to see what courses or programs they offer in fashion. Check to see if they have any visiting professionals that are willing to give a workshop or talk at your school. Check out books on fashion design at your local library and the Internet – it is a wealth of fashion information waiting to be tapped into."

—*James Fowler,[5] The International Academy of Design, Ontario, Canada*

For those who have a flair for creative stand-out design and a knack for understanding market trends and customer preferences, fashion design can be a career track with endless opportunities.

[5] J. Fowler, "Is Fashion Design For You?" *FAZE Magazine*, Summer 2000 Issue at www.fazeteen.com/summer2000/fashion.htm

Part IV

Career Development and Success Strategies

33

Holistic Engineering: The Dawn of a New Era for the Profession

Domenico Grasso, PhD, PE, DEE

College of Engineering and Mathematical Sciences, University of Vermont, Burlington, Vermont, USA

David Martinelli, PhD

Department of Civil and Environmental Engineering, West Virginia University, Morgantown, West Virginia, USA

The Golden Gate Bridge, in San Francisco, California, was the longest suspension span in the world when it was completed in 1937. At the time, it was widely recognized as an engineering marvel and a symbol of technology in harmony with its surroundings. When the bridge opened to a ceremonial trickle of cars, it would have been hard to imagine that an estimated 100 million tons would eventually cross annually between San Francisco and Marin County. Even less foreseeable, however, were the nearly two suicides per month, facilitated by this testament to the power of engineering thought. As we say in the profession, the bridge has exceeded its "design specs."

The Golden Gate Bridge is a useful metaphor in considering the scope of the challenge faced by every engineer beginning a design project: how to design for a specific objective without creating unintended consequences. Avoiding unintended consequences has never been more difficult or important than it is today, as population soars and technology, ever more complex, becomes increasingly embedded in human experience.

In this evolving world, a new kind of engineer is needed, one who can think broadly across disciplines and consider the human dimensions that are at the heart of every design challenge. In the new order, narrow engineering thinking will not be enough. American higher education is in an unusual position to create the 21st century engineer. Engineering and technical education are very much in the public eye now. For more than a year, Congress has debated how to best respond to the National Academy of Engineering's report *Rising Above the*

G. Madhavan et al. (eds.), *Career Development in Bioengineering and Biotechnology*,
DOI: 10.1007/978-0-387-76495-5_33, © Springer Science+Business Media, LLC 2008

Gathering Storm: Energizing and Employing America for a Brighter Economic Future. The report is powerful in its statement that the "scientific and technical building blocks of our economic global leadership are eroding at a time when other nations are gathering strength." Among the recommendations it proposes are increased investment in research and education in technical disciplines. In response to the report, President Bush announced, in his 2006 State of the Union address, the American Competitiveness Initiative, which United States Congress has been considering in various versions ever since. But investing resources in simply encouraging a technical-education paradigm developed in, and best suited to, the 20th century, would be shortsighted and ineffectual. Congress might be well advised to use the opportunity to encourage the major transformation that the new century demands—and that American engineering schools are distinctly positioned to supply.

In the global marketplace, engineers are proliferating at an astounding rate. The past decade and a half has seen the rapid economic development of half a dozen countries in Asia and Eastern Europe, which were once mired in poverty or slow-growing controlled economies. Now millions more people are embracing capitalism, and the technological engine that drives it. A previously untapped global human resource is being extracted like oil from new wells, yielding first a manufacturing capability, and now a staggering number of new engineers and scientists. According to some estimates, Asia alone graduates more than 10 times as many engineers annually as the United States does, many of them as qualified as our top graduates.

The emergence of a new global engineering work force and its threat to the US economy have been the topic *du jour* in engineering and business circles, but responses tend to focus on increasing the number of traditional engineering graduates so that we can go head-to-head with other countries in the technological marketplace. Such a goal alone, however, would do little more than drive down the price and value of engineering services, leaving the United States no better equipped than other nations to solve the increasingly complex problems facing society.

The answers lie in improving the quality of the product rather than increasing the quantity of output. The crucial question facing academia is whether we are adequately preparing our future engineers and designers to practice in an era that requires integrated and holistic thinking. Or are we instead needlessly limiting their solution spaces to those that contain only technological answers, with scant or passing consideration of many other influencing and dependent factors.

Where should current and future educators turn in preparing high-quality engineers who are better equipped to serve in the changing global marketplace? As engineers are often taught, solutions to new problems are found in returning to first principles. In that context, "first principles" means examining the definition and role of the engineer in their purest forms.

For centuries, society's problems have been assumed to be sufficiently linear, mechanistic, and discrete to be addressed by engineers responsible for "solving problems through the application of math and science," the classic definition of engineering that has served us well—until now. By many accounts, 80% of our economy is now information based. Yet if one were to pursue an undergraduate engineering degree from a typical state university, the result would be courses and curricula not significantly different from those offered during the middle of the past century, when we were largely a manufacturing-based economy.

Pursuing the holistic concept of the "unity of knowledge" will yield a definition of engineering more fitting for the times ahead. The unity of knowledge—first proposed by James Marsh, president of the University of Vermont in the early 1800s, and resurrected by the Harvard sociobiologist E.O. Wilson in his book *Consilience: The Unity of Knowledge* (Knopf, 1998)—is fundamentally about integrating knowledge across disciplines to deal with complex problems and better serve humanity. Many thoughtfully constructed versions of core curricula, sometimes referred to as general-education requirements, attempt to teach multiple modes of reasoning or ways of knowing. However, colleges rarely take the next step and encourage students to understand the connections among their courses, and to integrate, or "unify," their learning.

In engineering, a discipline that purports to design for humanity and improve the quality of life, the unity of knowledge should be a *sine qua non* that asks engineers to look outward, beyond the fields of math and science, in search of solutions to entire problems. To better serve humanity, engineers must at least attempt to understand the human condition in all its complexity—which requires the study of literature, history, philosophy, psychology, religion, and economics, among other fields.

Such a perspective on engineering education need not be restricted to the undergraduate curriculum. The educational philosophy embodied in the unity-of-knowledge approach also has a research analog in one of the most promising areas of investigation today: complex-systems analysis (recently identified in the National Science Foundation draft strategic plan, "Investing in America's Futures," as an area of focus and investment). Typically, complex systems are those that change with time, do not vary in linear pattern, and demonstrate "emergence"—that is, behavior that cannot be predicted in advance from constitutive parts. Complex systems are different from merely complicated ones, such as jumbo jets or fine Swiss watches, whose behavior, though characterized by the intricate interrelationship of many parts, is determined and reproducible. While advanced mathematics is a necessary tool for working in the field of complex systems, so too is an understanding of human nature.

Complexity is especially evident when human decisions play a role in the system—for example, in the dynamic functioning of the electric-power grid. Educating engineers more broadly will not only make them better designers, but will also give them the tools to work productively alongside the other problem solvers that they will be increasingly required to collaborate with: lawyers who resolve conflicts, economists who find the incentives and disincentives that promote positive change, historians who elucidate the present through knowledge of the past, artists who have an appreciation for form and function, and politicians who reach compromise. The ability to model and incorporate elements of economics, sociology, psychology, and business to identify possible solutions to pressing problems will be a major part of the future of engineering.

Consider a rather simple example: acid rain, which results in large part from burning coal. Environmental engineers and scientists worked hard on technologies to curb the pollution, but it was economists who developed the "cap and trade" permit program—which, through tradable pollution permits, has allowed market forces to create incentives for companies to cut pollution and reduce acid rain in the Northeast. Were the mathematics in that economics program beyond the capabilities of engineers? Or did their preparation not allow them to consider all of the possible solutions, which is to say, the ones that did not depend exclusively on technology?

When Stockholm was considering ways to transport more people into and out of the city, the concept of adding one more bridge to the 57 that already connect the 14 main islands that constitute the city would have been the natural engineering extension of past practices.

Stockholm retained IBM—a company with a not-insignificant number of engineers. However, prompted by an economic realignment in the United States from manufacturing and industry to services and innovation management, IBM has already moved beyond traditional engineering thinking. Specifically, the company has embarked on a research-and-business model that applies technological and manufacturing models to the holistic delivery of services. To solve Stockholm's traffic problem, IBM designed a "tax and drive" system, in which autos are fitted with transponders and drivers are charged a fee based on the time of day their cars are in the city. In the first month of operation, the system yielded a 25% reduction in traffic, removing 100,000 vehicles from the roads during peak business hours and increasing the use of mass transit by 40,000 riders a day. Stockholm needed no new bridge and gained the concomitant benefits of reducing pollution and conserving energy. London has taken similar steps.

In a world where applied science and technology are available to practically anyone for a few rupees or yuan on the dollar, we have to ask ourselves: What will the US engineer have to offer that is not available in the global market for a fraction of the cost? If we decide to compete with other countries using the traditional definition of engineering, we will certainly succeed in converting engineers into a commodity.

A better response lies in changing the scope and significance of what engineering is, and, perhaps more importantly, who engineers are—namely, technically adept people who serve humanity through the application not simply of math and science, but of a wide array of disciplines. This new breed of engineer will be not only a truly comprehensive problem *solver*, but a problem *definer*, leading multidisciplinary teams of professionals in setting agendas and fostering innovation.

If, as many glossy college brochures say, engineers are problem solvers, we must open their eyes and minds to the range of problem solving approaches that go beyond math and science. That is not to say that engineers must stay in school for 20 years to learn multiple disciplines in depth, but that they should experience the richness of a broad undergraduate education. It is not uncommon for only about 15 percent of the typical engineering curriculum in the United States to consist of electives. There is no question that our engineering graduates are well versed in the technical aspects of their profession. But it is equally clear that many of them graduate without the breadth they will need to think through the solutions we need.

Given that many rote engineering tasks can be easily outsourced, and that engineering organizations, including the National Academy of Engineering, are calling for the master's degree to be the first professional degree in engineering, it is time to consider a major overhaul of the undergraduate engineering curriculum. At the end of the 19th century, law schools concluded that they could no longer teach all of the vast number of laws that had accumulated over time, and decided instead to teach students how to think like lawyers. So, too, at the beginning of the 21st century, should undergraduate engineering schools focus on teaching students how to think like engineers. Building quantitative-reasoning skills should still be a top priority for American engineering education, but that rigor should be complemented with developing students' ability to think powerfully and critically in many other disciplines. To be sure, it will be a challenge, but a challenge with tremendous benefits.

Recently the Golden Gate Bridge Highway and Transportation District selected an engineering firm to develop a plan to create barriers (physical or otherwise) to suicide attempts. With any luck, a well-considered and holistic solution to the human dimensions of the challenge will present itself, not one that creates unintended new problems born of the myopia of a purely technical approach.

On Searching for New Genes: A 21st Century DNA for Higher Education and Lifelong Learning

Rick L. Smyre

Center for the Communities of Future, Gastonia, North Carolina; Design Nine, Inc., Blacksburg, Virginia, USA

"It is possible to store the mind with a million facts and still be uneducated."
—*Alec Bourne*

"The aim of the college, for the individual student, is to eliminate the need in his life for the college; the task is to help him become a self-educating man."
—*George Horace Lorimer*

As a futurist, I am a collector of quotes that, when connected, may help provide a glimpse into our future because of new questions that have arisen. Two of the most important questions for our age are "*What is the future of higher education?*" and "*What future capacities will be required for an educated person in the 21st century?*"

The only thing that will be certain is that our society will be constantly transforming. It will not be the traditional society of the past on which so much of today's education is base. With this in mind, the following article suggests a 21st century "higher education DNA," so that the context of education will be aligned as best as possible with a constantly evolving society.

- The level of cognitive complexity will be increasing as technologies provide real time information and connections.
- A creative knowledge economy and society will need individuals who are able to innovate continuously, and are capable of building relationships with others at a deeper level.
- The culture will be transforming constantly as new technologies and ideas impact all aspects of the context of our society and world.

As a result, it is my opinion that we need to shift our emphasis from a past of certainty and tangible outcomes, to a future requiring a comfort with uncertainty

G. Madhavan et al. (eds.), *Career Development in Bioengineering and Biotechnology*,
DOI: 10.1007/978-0-387-76495-5_34, © Springer Science+Business Media, LLC 2008

and ambiguity. Thinking in nonlinear ways and linking multiple factors will need to complement linear thinking (the small part of reality where one answer is best).

Preparing Millennials

I recently saw an associate, Neil Howe, interviewed on a television program. He was talking about the Millennials, those students from 8–28, about whom he wrote in *Millennials Rising*. I was struck by the challenge higher education has to prepare this generation for the future. As children of the Baby Boomers, the Millennials have been conditioned to want structure, low risk, teams focused on singular objectives, defined outcomes, and instant gratification – the very opposite of what will be needed for a constantly changing society.

My observation is that much work needs to be done by higher education to adapt its focus from traditional learning methods to create a culture that will allow the environment to be more stimulating, more flexible, and more creative . . . where students have their own responsibilities to develop varied learning experiences connecting interesting people and new ideas throughout the U.S. and world. This has been evolving in certain institutions of higher education. It needs to become the norm.

In my opinion, it will be very important for higher education to realize that it needs to help its students identify trends and weak signals of the future, see newly emerging patterns and connections, and develop emotional skills to help them risk being truly creative.

Twentyfirst Century HE DNA (Higher Education DNA)

As a result of the ongoing transformation of our society, I offer the following to be considered as a DNA of ideas that will be important to the future of the quality of higher education. These suggestions need to be considered as a system of interacting genetic elements, which, when seeded effectively, will help grow and align the dynamics of higher education with a constantly shifting and increasingly complex society (HE DNA stands for "Higher Education" DNA). As you read these twelve ideas, always remember that at the age of 60+, I no longer have truth, but only opinions.

21st Century HE DNA #1: An educated person in the 21st century will need diverse knowledge, to include history, future trends, and technical skills. This person will also need the ability to ask appropriate questions in order to make connections between disparate ideas to identify new knowledge. An educated person in the 21st century will need to be introduced to new principles that will underpin a dynamic and constantly changing society. These new principles will be based upon the metaphor of a living system because of the dynamic nature of transformation. Therefore, key ideas of biology, ecology, and chaos/complexity theory will be central to the needs of an educated person in the 21st century.

21st Century HE DNA #2: Three components of learning will become the focus of building a base of knowledge and skills to ensure the capacity for continuous innovation in all college and university graduates. These components include (a) a redefined core curricula based on knowledge of 21st century science, technology, and liberal arts; (b) the skills of identifying and asking appropriate questions; and (c) the capacity to see new patterns and connections among apparently disparate ideas and factors.

21st Century HE DNA #3: Redesigning the learning process to shift emphasis from the professor to the learner as a part of individualized, reciprocal learning. Here, a learning guide concept challenges the traditional methodology of one-way information flow and standardized testing as 21st century learners develop the capacity to become "self-educating persons." The need to shift emphasis from a traditional context of education to a futures context for self-learning will require a culture which supports such a transformation.

21st Century HE DNA #4: Student services will need to create new leadership development and personal adaptive support methodologies to help the children of baby boomers learn how to (a) be comfortable with ambiguity and uncertainty; (b) identify future trends and move beyond the need for instant gratification; (c) develop the ability to take appropriate risks; (d) create the capacity to develop relationships at a deeper level; (e) understand the difference between a short-term project team and a team networked for longer-term innovation; and (f) build the ability to be non-linear thinkers comfortable with processes that are less structured – thinkers who are able to adapt to constantly transforming ideas and situations.

> "The future no longer belongs to people who can reason with computer-like logic, speed and precision. It belongs to a different kind of person with a different kind of mind."
>
> —*Daniel Pink, "Revenge of the Right Brain," Wired Magazine*

21st Century HE DNA #5: All colleges and universities will need to develop a culture in support of diversity, interconnections, and true transformation. As the economy and society move toward less standardization and more choices and pluralities, there will be a need to prepare students for a world of differences.

> "Immelt is clearly pushing for a cultural revolution. In his General Electric, the new imperatives are risk-taking, sophisticated marketing, and, above all, innovation."
>
> —*Diane Brady, "The Immelt Revolution," Business Week*

21st Century HE DNA #6: All colleges and universities will need to build "and/both" capacities in their faculties and students to ensure "connective listening, effective generative dialogue within a futures context, and parallel processes that seed transformation and continuous innovation."

21st Century HE DNA #7: Certain sacred traditional principles will need to be rethought. As an example, traditional creativity and critical thinking used to solve existing problems will need to be complemented by a new type of creativity that searches for connections among disparate ideas, to create concepts, products, methods, and services that are not a part of traditional experience and knowledge.

The focus on developing new principles and methods for 21st century creativity will be especially important for a creative knowledge economy. Any individual in the US workforce in the future who is not technically literate, with the capacity to think systemically within a futures context, will be relegated to low paying, basic service jobs that often will have little meaning.

21st Century HE DNA #8: Colleges and universities will need to coach students in how to network people, ideas and processes. We are moving from a world of hierarchies and standard approaches only, to one that will be dominated by constantly connecting and disconnecting networks and webs of diverse individuals and organizations. In the future, standards will have their place for best practices and short-term defined outcomes. However, the ability to innovate continuously will be demanded, requiring nonlinear, nonstandard thinking and behavior that allow quick adaptation while balancing increased risk.

> "We are witnessing a revolution in the making as scientists from all different disciplines discover that complexity has a strict architecture. We have come to grasp the importance of networks."
>
> —*Albert-Laszlo Barabasi, Linked*

21st Century HE DNA #9: Standardized testing at colleges and universities will continue to be done to test the understanding of knowledge content. However, if new methods of evaluation are not developed which analyze the ability of graduates to ask appropriate questions and see new patterns and connections among disparate ideas, these graduates will not be prepared for a society and economy requiring continuous innovation. In the 21st century, new knowledge will need to be identified and created. True-false, multiple choice testing "only" will actually prevent students from understanding how to identify and connect new ideas.

21st Century HE DNA #10: Integrate weak signals and future trends into all curricula. Integrate virtual technology wherever appropriate to shift the design of learning processes to the learner. Model web sites with the connections that become a natural part of the thought process of any learner.

21st Century HE DNA #11: Build webs of diverse people from different backgrounds to help build 21st century learning networks, especially for those not traditionally identified as having college and university potential. Use the concepts and methods of "transformation" to bring new candidates into the mainstream of intellectual life by building bonds among students and external support systems. It will be the success of these new types of efforts to raise the level of motivation and passion for learning among previously non-qualified candidates that will determine the successes and limitations of our economy and society in the 21st century.

21st Century HE DNA #12: Use indirect and oblique methods to create positive tension that allows students to become intellectually and emotionally equipped to deal with fuzziness, uncertainty, ambiguity, and increased risk. Help them to understand that the unit of leadership and innovation in the future will be teams of self-motivated, self-learning, deeply collaborative people of great diversity who develop new skills for building "capacities for transformation" in themselves and others.

A Counterintuitive Need

It is going to be risky to become an educated person in the future. Why? Because no one will get to "know" enough content before the context changes. We will see a counterintuitive idea emerge. In the past, the educated person had the answer to specific questions that were based on experience and traditional knowledge. But in the future, an educated person will require a new set of skills beyond just knowing content of knowledge which will remain important. It will require new skills to identify and deduce knowledge, concepts, principles, and methods that do not presently exist. It will also require taking risks and thinking differently, often at the risk of appearing foolish. And it will require humility, without which one can not be open to new ideas.

Twentyfirst Century Creativity

As a result of this transformation in society, creativity will become a core principle related to what is considered a quality education. The skills of creative thinking will be of two kinds. The traditional approach of improving existing ideas, products, services, and existing methods will continue to be important for short-term needs. However, continuous innovation will require an innovation in the concept of innovation. The ability to see connections among disparate ideas to create totally new knowledge, concepts, and methods will become the foundation for all life, whether economic, political or social. The art of 21st century creativity will integrate with the updated skills of technology as a core focus for higher education. Evaluative methods will be designed which reflect the ability of students to get outside the box of traditional thinking. These students will still need the context of past knowledge, but now as a part of the scaffolding of understanding how the past connects to the future.

In the past, standardized skills were the cornerstone for preparing graduates for an industrial society. In the future, facts and content still will be important, but only as these facts become building blocks for creating new ideas. Sal Palmisano, the current president of IBM, recently focused on the need for higher education to help students learn how to take risk and be able to develop new ideas and create totally new concepts.

In my opinion, unless higher education rethinks how to create a learning environment owned and designed by each student in individual exploration, graduates will not be prepared for a 21st century creative knowledge economy and society.

A New Approach to Diversity

Our society is moving away from the idea of a melting pot where all immigrant children wanted to learn English and were assimilated into a common culture. Even so, the knowledge of English will still be imperative. Knowing other languages will provide a competitive edge, although the use of translation technology will make even this skill less crucial. In addition, it will be increasingly important for colleges and universities to learn how to bring diversity together, not just to share ethnic backgrounds, but to realize that in thinking in common about preparing for the future, everyone will have some important idea to share. In nature, evolutionary growth comes when diversity is present.

The phrase "we are a nation of immigrants" will have even more meaning for colleges and universities as they succeed in bringing attention to the diverse talents of people of color. A great challenge will be to develop a core "futures" culture to which all students can connect, while ensuring continuance of pride of diversity of background. It will not be enough in the future to find pride in looking backward. Confidence and personal motivation will come from how well all people integrate their ideas and personalities in common cause to rethink what a new type of society will require, and then care enough about each other, our families, our communities, and our society to collaborate at a deeper level of human interaction.

Conclusion

We are entering a time of history where we will need more than intelligence; we will need the increased application of diverse intellect. We will need a workforce and citizenry filled with local, state, and national leaders who see the concept of dialoguing about new ideas as an important use of time. In the past, we bypassed the importance of the intellect, often ridiculing those interested in ideas.

How often traditional phrases continue to ring in our ears. "Those who cannot do, teach." "But are you making any money?" "Enough talking, let's do something." It is slowly becoming apparent that tomorrow's successes will be linked to those who prepare themselves intellectually in a different way, and who develop the ability to connect with others at a deeper level.

The potential of higher education has never been more promising. I also believe that there is an equal danger that many institutions of higher education will not recognize the limits of their traditional approaches, and will become increasingly irrelevant in a time of rapid change. If one assumes, as I do, that we are in a time of historical transformation for society, a key challenge will be to rethink many aspects of how we create learning experiences, so that graduates will be prepared for a different kind of society: one that is constantly shifting, interdependent, and increasingly complex. What a wonderful time to be involved in higher education and lifelong learning!

> The definition of illiteracy in the future will not be the inability to read or write, it will be the inability to learn, unlearn and relearn. . . .
>
> —*Alvin Toffler*

> I am always doing what I cannot do yet, in order to learn how to do it.
>
> —*Vincent Van Gogh*

References

1. "How the US Can Keep Its Innovation Edge," Sal Palmisano, IBM, *Business Week*, Nov 17, 2003.
2. "It Is More Than Either/Or", Rick Smyre, *Futures Research Quarterly*, 2004.
3. *Millennials Rising*, Neil Howe and Bill Strauss, Vintage, 2000.
4. *Linked*, Albert-Laszlo Barabasi, Plume, 2003.
5. "Revenge of the Right Brain," Daniel H. Pink, *Wired*, February 2005, p 70.
6. "The Immelt Revolution," Diane Brady, *Business Week*, March 28, 2005 p 64.
7. *The Third Wave*, Alvin Toffler, Bantam Books, 1980.

35

Protean Professionalism and Career Development

Steven Kerno, Jr.

Deere & Company, Moline, Illinois and St. Ambrose University, Davenport, Iowa, USA

You might not remember Proteus. He was a sea god in ancient Greek mythology. To most engineers, mythology of any kind is likely to be a distant memory—something we learned about in high school, with little perceived relevance or applicability to today's challenges and demands. Or so we thought.

Proteus was capable of altering his shape in order to fit the demands of his environment. His name is the basis of the adjective "protean," which has the general meaning of versatile, flexible, adaptable, or capable of assuming many forms—much like the career demands confronting the modern engineer [1].

"Protean," especially when applied to "career," is not simply an abstract term involving a colorful superstition, but a modern reality that for many is simultaneously exhilarating in the freedom it provides, and terrifying in the security it erodes. The protean career [2, 3], with the growing need for individual motivation and continuous, career-related learning and development [4, 5], is indeed a contemporary reality for many engineers.

Engineers have undoubtedly been the primary drivers of the industrial progress that has occurred during much of the previous century. They possess many of the skills necessary to link perceived social needs to the commercial applications that satisfy those needs. There is scarcely a product or service available that didn't require the services of an engineer before everyone decided that they couldn't live without it.

During the past 50 years, the essential ingredients of an engineering education have not seen any radical changes [6]. Until recently, newly graduated engineers received the bulk of their training on the job. Senior engineers would act as mentors, providing task-related direction and guidance, with the organization assuming responsibility for career progression and development. Such an organizationally determined structure, coupled with a more static and predictable economic environment, were the ingredients thought necessary to administer most of

G. Madhavan et al. (eds.), *Career Development in Bioengineering and Biotechnology*,
DOI: 10.1007/978-0-387-76495-5_35, © Springer Science+Business Media, LLC 2008

the career-related learning any young protégé would ever need, and to nurture the qualities deemed necessary for success [5].

But despite the indispensable nature of their work, engineers have not been exempt from today's increasingly turbulent and uncertain employment environment. Starting with the 1970s, as a result of persistent economic malaise, high energy prices, stagflation, and the resulting turmoil within many sectors of US manufacturing, a covenant that had previously existed began to change. Before that time, in what amounted to a social contract, an employer generally would provide benefits such as lifetime (or at least long-term) employment, generous pension plans, and fully paid health care to loyal employees. The arrangement assumed that both parties, through economic peaks and valleys alike, would stick together.

The following decades, and the tumult that accompanied them, have transformed the relationship of engineer and employer into a transactional contract, based upon an exchange of benefits between the two, but having a much shorter life expectancy [7]. The net outcome for engineers is that job security and its trappings will fade in importance and be replaced with marketability of skills, and the need to remain adaptable, versatile, and flexible—in short, the "protean career." [2, 3]

Engineers, despite their unique knowledge and abilities, can be affected particularly harshly by the transformed career. Why? Because the very qualities that for decades have served engineers well in the maintenance and advancement of their careers may now be a liability.

Career Concepts

A great many US organizations have traditionally defined a career as a steady progression of positions, each resulting in increasing levels of authority and responsibility. Against such a backdrop, a successful career was measured in terms of rise in position within a formally structured organizational hierarchy. While this scenario is not obsolete, it is tending to become less and less common.

It is useful to consider a career model that identifies four fundamentally different patterns of experience, each having differing trajectories, motivations, and needs within an organization [8] (Table 35.1). An engineer may tend to associate more strongly with or one; or instead equally between two, in which case he or she may possess a hybrid career concept.

The linear career typically involves a progression of steps, or promotions, within the formal hierarchy of an organization. "Climbing the corporate ladder" would describe the ideal linear career. An expert career, on the other hand, usually involves a long-term commitment to an engineer's chosen field or specialty, with the work often becoming an important component of self-identity. The department or division guru is a status (even if informal) that experts often aspire to attain.

A spiral career frequently takes the form of moving periodically, every few years, to a related or similar area of employment. Someone may learn the ropes in

Table 35.1 Four Career Concepts

LINEAR:

- Progressive series of upward steps within an organizational hierarchy.
- Rooted in the cultural emphasis that American society places on upward mobility.
- Key motivations are individual power, achievement, and the opportunity to make things happen.
- Individuals tend to be competitive, oriented toward leadership, profits, and financial success.

EXPERT:

- Lifelong or long-term commitment to a chosen occupational field or specialty.
- Focus is on development and refinement of knowledge, skills, and abilities within career.
- Nature of work performed tends to be an integral component of self-identity.
- Key motivations are expertise or technical competence, security, and stability.
- Individuals tend to be quality-conscious, and oriented toward commitment and reliability.

SPIRAL:

- Periodic (every 7–10 years) major moves across related occupational specialties or disciplines.
- Ideal career move is from one functional area (engineering, manufacturing) into an adjacent or similar one (R&D, quality).
- Previous field forms knowledge base for movement into the new one, while allowing a person to develop closely related, yet different sets of skills and abilities.
- Key motivations are a need for personal development and increased knowledge.
- Individuals tend to be creative, possess diverse skills, and are able to coordinate lateral organizational activities.

TRANSITORY:

- Frequent (every 3–5 years) major moves across unrelated occupational specialties or disciplines.
- Those pursuing transitory careers often do not perceive themselves as actually having careers.
- Key motivations are a desire for very diverse work experiences, variety, and independence.
- Individuals tend to be fast learners, adaptive to changing circumstances, and project-oriented.

Source: Compiled from Brousseau, Driver, Eneroth, & Larson (1996)

Table 35.2 Toolkit for the Protean Professional

Points to ponder

- Understand how to provide and broker your knowledge, skills, and abilities in such a way that unmet organizational needs can be satisfied by the services, engineering and otherwise, you are capable of providing.
- An engineer must be able to define and to clearly articulate the nature and scope of current and future project assignments, as opposed to individual jobs. As work becomes more project-oriented, engineering career success will increasingly depend on the ability to move from project to project, and to absorb the learning and best practices from each assignment, as opposed to retaining a relatively static job title and working environment.
- Future engineering jobs will involve more challenge, more skilled expertise, and the ability to network with others who possess similarly valuable knowledge, skills, and abilities. However, do not limit yourself. Make a conscious, sustained effort to network with non-engineering professionals in adjacent and even unrelated fields (human resources, marketing, finance, accounting, logistics, etc.). These occupations face similar threats, but in different ways. Also, these professionals may have unique and fresh career insights for the engineer, and may be able to provide appropriate reciprocal assistance on projects.
- Seek out job assignments with managers who think in terms of continuous learning and development. Increasingly, engineering managers will need not only to maximize the present performance of the engineers they oversee, but also to promote a culture conducive to career-related continuous learning. This will help better confront likely challenges in the future.
- Engineering talent is not as easy to find or develop as you might think. The engineer who adopts a proactive attitude is more likely to find the right challenges "in house" than one who simply tries to blend in with the crowd.
- Trusted counsel and guidance from someone with a better depth or breadth of experience can provide invaluable career insights. This person (or even persons) need not be an engineer, or even work for your employer. However, what distinguishes the input provided by this individual from others you may network with is the length of the relationship, the intensity (frequency and/or length) of interaction, and the level of confidence relative to the type of information exchanged. Learning to capitalize upon the knowledge provided may seem awkward at first, but with a little practice, you may be capable of being a mentor.

Questions to consider

- Do you enjoy your work, and do you see yourself *continuing* to enjoy it (at least for the foreseeable future)? If not, what steps could you take now or in the

(continued)

Table 35.2 Continued

immediate future to more closely align your talents and interests with something that will not force you to hit the snooze button every morning?

- Your resume is often your only opportunity to clearly and concisely communicate your value to others, particularly hiring managers. If you have not updated your resume or curriculum vitae recently (either due to procrastination or lack of professional achievement), figure out why, and take the time necessary to improve this all-important document. Can you fully explain any gaps in time? If your career has followed a non-traditional trajectory, do you understand how to communicate this as an asset to a prospective employer?
- Does your position provide objective, clear, and consistent feedback regarding your actions and performance? Additionally, does successful performance translate or map to superior career opportunities? If not, or if the "line of sight" is obscured by factors that are not truly objective, you may want to consider whether or not your employer values your contributions. It can be demoralizing and demotivating to have irrelevant, unclear, or inconsistent metrics gauging your performance.
- We cannot be outstanding at every job function or endeavor, and there are likely to be activities that we absolutely dread. By acknowledging this, we can take the initial steps necessary to moderate the hurdle this may create for our career. More importantly, what *actions* have you taken? Have you enlisted the help of someone more knowledgeable in the area of deficiency, such as a peer? Have you taken a class? It is not always feasible to devote significant resources in an attempt to correct a deficiency, but having a "right hand person" to provide necessary assistance is often very useful, and provides an opportunity for reciprocity on your part as well.
- Are you taking advantage of the learning opportunities available within your organization that *you* are most interested in (not necessarily those required or mandated)? Is tuition reimbursement available for outside learning activities? If so, and you are not capitalizing on these benefits, don't complain when professional opportunities fail to materialize.
- Do you remain up to date with the literature, including magazines, ongoing studies, textbooks, or online forums? Do you attend useful seminars? Do you seek out opportunities to contribute as a professional to your field of knowledge or expertise? You don't necessarily have to attain organizational guru status, but being recognized as a knowledgeable individual as it pertains to technical matters is an important bellwether for career success.
- Meetings are often a necessary evil of organizational life, but they should also be regarded as an opportunity to showcase your knowledge, skills, abilities, and talents to peers and superiors. Since meetings provide face-to-face contact with others, how you conduct yourself is critical regarding access to future career

(continued)

Table 35.2 Continued

opportunities. If you tend to avoid or, even worse, forget meetings, take the time to be honest with yourself – why? A person can learn to be effective during the time they are present in meetings, even if they rank in popularity with a root canal.
– Even if you do not particularly like your boss as a person, do you seek input for how to improve as it relates to your professional duties? Do you try to keep the lines of communication open, or do you tend to hole up in a cubicle? Let's face it, your relationship with your immediate boss often has greater influence on whether you like or dislike your position than does the actual work you do. Learn to make time to talk to your boss, as he or she will be more likely to appreciate your efforts to ultimately make them look better to *their* boss.
– No one is an island. Do you regularly exchange ideas, debate issues, or enjoy the company of peers with whom you share a common work-related interest? Yes, bowling and golfing score points for this question! Networking builds social capital that potentially can be used to build bridges to better career opportunities.
– Are you involved with charitable or religious organizations? The personal satisfaction available, coupled with the likelihood of coming into contact with someone who can provide career assistance, should be enough of an incentive for you to consider the potential upside. Besides, we receive benefits not easily quantified or articulated from helping others, and may be pleasantly surprised at the results.

Source: Compiled in collaboration with Dr. Kevin Kuznia (personal communication)

one discipline, then use the acquired knowledge and skills to gain entrance to a closely related field. The knowledge base from each discipline is used to open doors to others.

A transitory career is just that—transitioning from one experience to another so frequently that it may not seem as if a person settles down long enough to actually have a career – at least in the traditional sense. "Jack of all trades" would aptly describe the person with such a career [8].

US organizations in the past faced relatively stable economic and social environments. That is, demand for products increased year after year in a predictable manner, fueling economic growth. The dual engines of innovation and efficiency ensured that US manufacturing remained virtually unrivaled, with both workers and management sharing in the fruits of success. Regulatory agencies such as the Environmental Protection Agency and Occupational Safety and Health Administration did not exist. In such an environment, employees possessing expert or linear careers had an advantage. Experts benefited from organizational stability, allowing for focus and commitment to the knowledge and skills necessary for achieving expertise in the chosen occupational specialty, while linears could concentrate on ascending the corporate ladder. But times have changed, as have the beneficiaries of the shifting needs of organizations.

With ever-increasing globalization, a more hostile external environment with stakeholders whose agendas are frequently at cross-purposes with the organization's, and with the countless downsizings, layoffs, and off-shorings that such an environment engenders, the opportunities and threats posed to both organizations and the engineers they employ are numerous. Whereas employees with spiral or transitory careers in the past may have been overlooked by employers, they are welcome now, as they are more flexible, adaptable, versatile—more protean—than their expert and linear counterparts. They are able to change with the needs of their organizations.

There is good news and bad news in all of this. First, the bad news. The career concepts that many engineers traditionally have followed, often linear or expert in nature, are dwindling in terms of the number of job opportunities available. Engineers, by virtue of the educational demands required (longer and more intense than many other college majors), their more common and traditional association with the core competencies of their employers, and the relatively insulated working environment many have encountered [5], have enjoyed favored status in many US companies.

Such employment situations still exist, and indeed those organizations that have the luxury of longer product life cycles, significant barriers to industry entry by potential competitors, and well-established and respected brand names, can and do employ engineers in accord with these career concepts. However, engineers must also be aware that even these organizations are not immune to both internal and external pressures to redefine themselves. This may very well leave engineers to fend for themselves in a work environment that no longer cares who their previous employer was, only what their current knowledge, skills, and abilities are.

Now, the good news. Engineers can help themselves, and other engineers, by developing a higher level of self-awareness and personal responsibility regarding their careers [4]. Like many other types of learning, this one involves a learning curve as an engineer adapts to increased autonomy and decreased organizational support.

The protean career under such circumstances demands a continuous learning process—one requiring an engineer to be self-correcting to new demands from the work environment. This career will involve less organizationally-sponsored formal training, less planning and development on the part of the organization for an engineer's career, and fewer opportunities to remain in a job where mastery is the ultimate goal. However, replacing these more traditional job-related functions are several resources that will likely enrich their work experiences in the short term and job prospects in the long term (Table 35.2).

An understanding of where unmet needs exist within an organization, and how your knowledge, skills, and abilities can be leveraged to fulfill these needs, is critical. Finding a niche where you are indispensable may not guarantee long-term employment, but it will likely allow you an opportunity to showcase your abilities and may be used as leverage, if you decide to pursue other interests.

Get used to networking with other professionals, even if they are not engineers. Staying connected increases your knowledge of the challenges others face in various professions. As work becomes more project-oriented, it is these individuals who may be able to provide you with reciprocal assistance when necessary.

Also, be certain to find those managers who share a similar attitude toward career growth and development. In the future, engineering managers will need to balance the present work activities of their subordinates as well as to make appropriate investments in creating a learning environment that prepares their staff for the challenges ahead.

The occupation of engineering is not going the way of the horse and buggy, the typewriter, the slide rule, or the 8-track player. Engineers are simply too valuable, in terms of the knowledge, skills, and abilities they bring to our society. As such, engineers are likely to increase in importance as the challenges confronting organizations grow more complex.

The successful 21st century engineer will more and more frequently be the one capable of using the tools at his or her disposal to effectively orient or map their knowledge and abilities to the current and future needs of their organization, remaining flexible, adaptable, versatile—protean. After all, it's *your* career – make the most of it!

References

1. Kerno, S. (2007). "Continual career change." *Mechanical Engineering, 129* (7), 30–33.
2. Hall, D.T. (1996). "Protean careers of the 21st century." *Academy of Management Executive, 10* (4), 8–16.
3. Hall, D.T., & Moss, J.E. (1998). "The new protean career contract: Helping organizations and employees adapt." *Organizational Dynamics, 26* (3), 22–37.
4. Kuznia, K.D. (2006). *The correlates and influences of career related continuous learning: Implications for management professionals.* Unpublished doctoral dissertation, St. Ambrose University, Davenport, Iowa.
5. London, M., & Smither, J. (1999). "Career related continuous learning: Defining the construct and mapping the process." *Research in Personnel and Human Resources Management, 17,* 81–121.
6. Jones, M.E. (2003). "The renaissance engineer: A reality for the 21st century?" *European Journal of Engineering Education, 28* (2), 169–178.
7. Rousseau, D.M. (1995). *Psychological contracts in organizations: Understanding written and unwritten agreements.* Thousand Oaks, California: Sage.
8. Brousseau, K., Driver, M., Eneroth, K., & Larson, R. (1996). "Career pandemonium: Realigning organizations and individuals." *Academy of Management Executive, 10* (4), 52–66.

Suggested Readings

Mainiero, L.A., & Sullivan S.E. (2006). *The opt-out revolt: Why people are leaving companies to create kaleidoscope careers.* Mountain View, California: Davies-Black.

Buckingham, M., & Coffman, C. (1999). *First, break all the rules: What the world's greatest managers do differently.* New York: Simon & Schuster.

Leadership Through Social Artistry

Jean Houston, PhD

Human Capacities Institute and International Institute for Social Artistry, Ashland, Oregon, USA

The Leadership Dilemma

More than thirty years of working throughout the world with leaders in the fields of industry and government, education, and health have convinced me that too many of the challenges in societies today stem from leadership that is ill-prepared to deal with complexity. This is not just a matter of inadequate training in the realities of global change but, even more tragically stems from a lack of human resourcefulness: too many leaders have been educated for a different time and world. What is worse is that leaders all too often avoid working co-creatively with their constituents, thus continuing models of dependency and social apathy.

Worldwide, societies are crying for assistance in the transformation of their citizens, organizations, and institutions. New ways of looking at leadership are required, as well as new methods of developing human beings eager to serve humanity. In this chapter, I share some of my perspectives on what can be done to improve our leadership abilities through social artistry.

The Need for Social Artists

We must begin to help people, citizens and leaders alike, to bring new ideas to bear upon social change. In this way, it is hoped that we can rise to the challenge of our times and ferry ourselves across the dangerous abyss that separates a dying era from a 'borning' one.

The work of social artistry is evolving and open-ended, striving to provide a dynamic balance between inner understanding and outward expression. The social artist is one who brings the focus, perspective, skills, tireless dedication, and fresh vision of the artist to the social arena. Thus, the social artist's medium is the human community. She or he seeks innovative solutions to troubling conditions, is a lifelong learner ever hungry for insights, skills, imaginative ideas, and deeper understanding of present-day issues. Above all, the social

G. Madhavan et al. (eds.), *Career Development in Bioengineering and Biotechnology*, DOI: 10.1007/978-0-387-76495-5_36, © Springer Science+Business Media, LLC 2008

artist is one who is always extending his or her human capacities in the light of social complexity.

What follows are other qualities and capacities that characterize the social artist, and inform his or her training:

A Planetary Citizen; Comfortable in Many Cultures

The social artist is trained to move comfortably between cultures, capable of understanding and honoring another's belief systems, cultural styles, tribal and national stories, and rituals. Social artists learn to be informed on world issues in the context of different cultures, and not just from the point of view of a particular nation or policy. In terms of human development, this requires high sensory development and a polyphrenic nature, as well as a willingness to keep plunging deeper to tease out the elements of the emerging story. He or she learns to see trends and the emergence of new patterns from apparent chaos.

Polyphrenic Nature

The social artist presents a model for a constantly learning society. Yet he is devoted to the preservation of vital elements in a people's culture, their shared genius, while consistently open to new ideas that can sustain and enrich an emerging planetary culture.

A Seeker of the New Cultural Story

Social artists learn to help members of the culture or organization to preserve the genius of the culture, even as they help move it into the new story. In countries with huge immigrations coming in to upset the given cultural styles, the new story has to do with appreciating the diversity and complexity of the new brew. This means deep appreciation and cross cultural understanding of the stories of the representative cultures. Together, they make for a whole new story. Often the larger picture or story will help movement out of a static reality. This is where consideration of an overarching story or new myth is of great importance. The social artist can hear all the stories and divine the elements of the old that will be needed to help create the new.

The social artist is open to the fact that the membrane between cultures, between worlds, between old and new ways of being, is breaking down. In the past, migrations and diffusions allowed for gradual changes and exchanges between cultures and identities. Now, nothing is gradual; in certain regions of the world we are watching a fast-forward movie of strange multicultural mitosis and its stranger spawn. Cultures once thousands of miles apart are now meeting, a new genesis is occurring, occasionally with a melding of genes, but more often, a mingling of previously divided worlds when thrown together, they undergo a sea change into something rich and strange.

What results is not merely a hyphenated amalgam, Afro-Asian rock music or Chicano-cyberpunk art, but rather hybrid sounds for hybrid selves, a malleable, syncretic fusion that is generating its own cultural matrix. For human beings, the complexity of this not-yet-definable culture is providing sufficient stimulus to call forth latencies in the human brain-mind system that were never needed before, like bacteria learning to breathe rather than die when the culture of oxygen came in, or, closer to home, the ways in which children absorb the mysteries of computer wizardry with ease while their parents struggle to master the basics.

This breakdown of the membrane that I am describing is not merely cultural fusion; it is the joining together of geographies of the mind and body that had never touched before, weaving synapses and sensibilities to create people who are fused into the world-mind with its unlimited treasures, its empowering capacities. The social artist is able to relate to this quickening charge of cultural mitosis, and stays abreast of the ways in which it is occurring in food, music, literature, and theater–the very lineaments of culture. Consciousness is remaking itself.

Several years ago, I was working in New Zealand to help empower a more cohesive and coherent national sense within the society in the wake of the government's abdication as the godfather of all social programs (more about this visit later). The country's situation was both static and chaotic. One evening, a group of Maori social activists recreated the story of the creation itself, as Maoris knew it, and also the awakening from dormant flesh of the human female. Most of the Pakehas, the Anglos, had never heard it, and had certainly never seen it performed. It stirred something vivid and deep within all of us. An essence of Maori legend reveals the power of desire, focus, and the potent, precise, intense, laser-like expenditure of energy required in order to bring something worthwhile to life. It was an immense teaching for all of us, and ignited a pilot light within the participants yearning to create their new society.

It was not the details of the myth that mattered; it was the energy of creation that these Maoris were willing to tap into and reveal that both taught and inspired us. Suddenly we Anglos got a glimpse of the Mana, the spiritual power that is available to the Maoris in the rocks, soil, sea, and air of this country. That immediately enhanced the feeling of profound respect for the Maori, and set up a deep desire to help preserve their culture. Herein was the possible origin of a new story, with the potential for increasing gender equality and creating partnership programs and mutual cooperation.

A Paradigm Pioneer

As a paradigm pioneer, the social artist is able to see trends, and the emergence of new patterns out of apparent chaos. He or she demonstrates that different times require a revolution in management styles. For example, the social artist has the tools to help people to work in collaborative networks and move away from hierarchies and power structures. He or she is one who helps cultures and organizations move from patriarchy to partnership, from dominance by one economic culture or group to circular investedness, sharing, and partnership.

The social artist shows even the most hierarchical and bureaucratically-based organization that the inevitable movement in a world as complex as ours is to circular organizations. The successful new or renewed organizations will look like a series of circles everywhere. Often the most potent training for this partnership pattern comes from indigenous societies where members work together in a webbed network to find creative solutions. The understanding is that while each person knows how to work a problem, seeking the answers together, with many methods, results in solutions that are not only consensual, but also rich and playful.

The Joys of Lifelong Learning

The social artist presents a model for a constantly learning society, consistently open to new ideas that can sustain and enrich an emerging planetary culture. He or she helps to create, wherever possible, new models of education. In our work around the world, my associates and I have found it necessary to work both intraculturally as well as transculturally. In our transcultural work, we try to speak to that which is potential in every human being regardless of local and cultural conditioning. We speak to the yearning that we find in everyone, whether the one who yearns is peddling a rickshaw in Delhi or running an oil company in Dallas. Thus, we try to offer liberating thoughtways that launch understanding, motivation, and problem solving beyond their constricting cultural, and even instinctual, preconditioning.

Using story, art, and metaphor to frame a complex array of processes and techniques designed to bring forth latent potentials, we show people they have a natural access to capacities like being able to think with many different frames of mind—visual, verbal, kinesthetic, interpersonal, subjective, intuitive, and logical-mathematical. This includes capacities that improve the physical use of the body, and that enhance memory, writing, creative expression, and problem solving. Given education and opportunity, we find that most people are able to make remarkable improvements in their functioning and learn new ways of being in a relatively short period of time.

If anything, this is more true in so-called "developing" countries, where we find that people are closer to their innate potentials because they have not yet been shuttered by education and social objectives that inhibit and coerce their natural capacities into "approved" tracks and templates. Wherever we have worked, we have found the possible human is just beneath the surface crust of local culture, and that consciousness and the possible society are not far behind.

Bangladesh offers an example of this approach. Early in 1993, we responded to an invitation from United Nations Children's Fund (UNICEF) to work with over a thousand leaders in Bangladesh. In the world's eyes, this is considered to be one of the most tragic of countries, a nation relentlessly afflicted by flooding, poverty, illness, and futility. But Bangladesh is also a world of metaphor, of high and low theater, of great poetry and music, and of a people deeply engaged by each of these activities. Talk to a rice farmer, and you find a poet. Get to know a sweeper of streets, and you discover a remarkable singer. In the various meetings and seminars that we gave in Bangladesh, we found that participants were remarkably responsive to our methods of learning new ways of being.

Indeed, many seemed to have this knowledge as a kind of second nature, and spoke to us about how for the first time they were being affirmed in what they had long sensed they already knew. It was as if the "imported" culture from the West (mainly England) had dropped a curtain over their more natural artistic thought process and modes of expression. One fellow told us that he had always felt that in his studies he had been made to operate as if he had his hands tied and his lips taped up. Now he felt free for the first time, and his capacity for thought and ideas was blooming. Certainly the participants were filled with plans for how they could apply this kind of work in human capacities to their various endeavors in health, education, and social welfare, and, with the regular help of my team, setting up new curricula and training in methods calculated to celebrate both the human potential and the culture, rather than further calcify it.

To date, these innovations have affected tens of thousands of schools in Bangladesh, with special emphasis on educating and raising the status of women. I think of the Hindu women singing to us at the sweeper colony, and especially of the young woman there with shining eyes. When we asked her what she wanted most in the world, she replied, "I want an education! I want to learn so many things, and I want my children to learn. I want to spend my life learning." In many other cultures this young woman would be highly educated and pursue a career in lifelong learning.

Why Not in Bangladesh as Well?

In our intracultural and educational work, we try to discover the main stories, myths, legends, and teaching tales that underlie the spirit of the culture in which we are working. Then we present these myths and stories as the backdrop upon which to weave our work in human development. We find that people go further, as well as faster and deeper, if their learning is attached to a story, and, most especially, if that story is a key myth of their culture. Thus, in India we have worked with the Ramayana and the life of Gandhi; in Australia, with certain Aboriginal creation myths; in England, with stories of Percival, Gawain, and the search for the Grail. In Bangladesh, we tried to use the poetry of Tagore and other Bengali poets.

One feels instinctively that a new story is needed, for the old stories no longer speak to the current reality. And yet the old stories seem to rise again and again in fractal waves to give power and portent to the culture. What is needed, then, is for the stories to be re-mythologized and rewoven in the light of today's necessities. This has always been the job of culture, to discover again, and on a deeper level, the meaning and relevancy of the once and future story, for without story, a culture becomes denatured and demoralized.

Evoker of Laughter and Life-Advancement

The social artist knows when to be a humorist, fool, or comedian. He or she can break out of the usual projections and expectations and present a world of the absurd. I have always found it useful to learn half a dozen jokes native to the society in which I am working, as well as those joke or comic stories that are universal. Thus, by encouraging such variety of expression, the social artist encourages people to celebrate the new possibilities in creative expressions and all manner of life affirming, life advancing actions – music, songs, humor, dances, rituals, and myths of possibility to be played out, performed, and celebrated.

Building community in the new millennium requires that we create social theater to tell the New Story of a world in transition. A student of mine in a small town in rural Georgia, where life had grown static and in a state of decline, took the disparate and dissenting stories of her area's local history and created a pageant called Swamp Gravy in which everyone in the community participated. This has had a beneficial effect on the community as a whole, leading to much more cooperation and friendliness between the various groups within the community, where deep polarities and resentments had festered between black and white residents. This idea has been replicated with students of mine in Detroit and Chicago, as well as other places.

The Social Artist is Always a New Kind of Healer

Healing involves the mystery of change, of transformation, and of the fluid nature of our bodies, minds, psyches and, by extension, our societies and cultures. We live in a world that is ripe for healing, and this is ultimately what motivates the social artist to take initiative. What the social artist knows is that we are built for healing. The nature and process of healing, the varieties of the healing experience for both persons and societies is the very condition of our humanity, the training ground for our social unfolding.

The critical issue here, and the one that distinguishes the great and inspired social artist compared to the ordinary run of the mill one, is the mystery as to whether he or she regards healing as redemptive or creative, salvational or evolutional. All manners of fundamentalism are generally redemptive in their philosophies and liturgies. One is always trying to fix the old Adam. Or, if one cannot, then one assures one's followers of ultimate fixing on another plane after death.

In our traditional medical technology and healing practice, the emphasis is almost entirely on the redemptive. This redemptive mode is carried over for jihad and defense, or even pre-emptive strikes. One has to take on all manner of extreme measures to fix the fallen state. The evolutional state works on an entirely different basis for life and for healing. There is somewhere to go, there is something to become. Even illness contains within itself the notion of deconstruction leading to a higher reconstruction, chaos leading to cosmos. Healing is *wholing*, the move from a limited condition felt in a most painful way through a process leading to the creation on a new level of a higher order of mind, emotions, and physical being. Something changes, the wounds in the body or of the society are experienced as doorways to higher consciousness and more evolved forms.

The Social Artist Respects the Uniqueness of Each Person

The social artist as healer is the one who helps people to ignite their potential, and does not take the credit for the ignition. The really good social artist-healer is an evocateur who shows people how to access their inner wisdom and knowledge. It is a form of healing that moves us beyond the polarities of left versus right, or us against them, and promotes cooperation, understanding, and networks of mutual aid. It is, above all, compassionate listening, a major training for the social artist.

The Social Artist as Contemplative Creator

The social artist is one who participates in the art of new creation. We are called to explore the mystery of the interface between engagement with external realities and embrace of the inner journey. Creative social artistry is contemplative, a vital synergy between inner and outer realities necessary to transform organizations, institutions, and paths of possibility, as well as visionary endeavors, and, in so doing, unleashing the human spirit of both those who compose the endeavor and those who are served by it. It is an activity of extraordinary balance, a tension in repose. It is a space of exquisite silence and of extraordinary service. In such a state, one has access to remarkable creative ideas, and world making patterns. Beneath the surface crust of consciousness, creative ideas and solutions are always there, ready to bloom into consciousness.

Entry Level Consideration: A Model from New Zealand

The first comprehensive program in Social Artistry was developed and shared in five locations throughout New Zealand. The participants were self-selected leaders, coming from all levels of society, seeking to devise programs for their individual townships and areas. They fell into several discrete groupings: people from civil society seeking to address inequities in education and health care; business leaders wishing to find a means of increasing and supporting the well-being and ongoing learning of their stakeholders; government officials at both local and national levels finding ways of moving from a welfare state to a more cohesive system of shared-governance; Maori elders and spokespersons striving to awaken deeper understanding of their conditions in their European neighbors, while preserving and sharing the profound richness of their culture.

As people spoke together, both in shared interest circles, and in circles across lines of interest, their first hurdle was to surpass the initial goal of getting a huge government or corporate grant to do the work they wanted. Facing the realization that neither government nor corporate society was capable of funding their work, participants then realized that they needed on-going communities of teaching and learning to achieve both greater simplicity and greater effectiveness in their projects.

Many participants realized that this series of meetings marked the first occasion of their learning what others were doing, often in the same field, though in a different city or community. In many cases, this streamlined their efforts, as a group interested in education was able to assist the health and healing sector with educational modes, while the discoveries of those working in health helped the educators see how essential good health is to education. Therefore, offering opportunities to share information about individual and group projects, goals, and discoveries across many interest lines becomes a primary necessity for sharing ethical responsibility and working toward solutions that will achieve millennium goals as well as goals that stretch beyond 2015.

After that initial meeting, it becomes a matter of training in the capacities to move across cultural lines; to become comfortable developing and implementing innovative strategies; to focus on larger, more inclusive, evolving patterns; to generate continuing enthusiasm for the group's endeavors; and to continue working, unfailingly, on one's own physical, psychological, cultural, and spiritual condition. Ongoing meetings with the avowed goal of learning and teaching more intensively are similarly required if burnout is to be prevented and new skills developed. Ancient Greeks called this process *Politeia*: in which an active, engaged, and informed citizenry creates and maintains its community and mutually fulfills societal goals.

The primary entry levels of the Practice of Politeia are individual neighborhoods, communities, and organizations, though many such practices have spawned others to create the sense of Politeia at broader, more populous levels. Social artists working in decentralized governance are trained to provide the model for a Politeia of Participation, offering means by which all members of a community are invited to enjoy the opportunity to influence the political and economic institutions that affect their lives, while simultaneously growing a sense of personal responsibility for fulfilling their needs and wants.

A Politeia of Rediscovery seeks to rekindle spontaneous generosity, with special emphasis on honoring the capacities of others. Here, the social artist is invited to notice the life patterns of things too often taken for granted, and to pay attention to those concerns which call her to high service. She seeks to encourage awareness of the human life stream, finding ways of equalizing access to the best possibilities in gestation, birth, and parenting practices; nutrition, health, and fitness; community life, education, arts, and sciences; as well as our ways of growing old and dying well.

A Politeia of Creativity activates the artistic process as a means of recharging imagination and expressing communal dreams. A Politeia of Healing seeks to move beyond polarities, competitiveness, and confrontation into modes of cooperation and understanding, providing networks of mutual assistance through compassionate listening and dialogues of fairness to address critical issues and move beyond wounding and hatred. A Politeia of Celebration encourages music, songs, humor, dances, ritual, and mythic enactment to be played, performed, and celebrated. Building community today requires a social theater to demonstrate the New Story of the world in transition.

A Politeia of Hope encourages the ever-refreshing attitudes of wonder and gratitude, recognizing the power of even devastating problems to reveal depths of compassion and connection that can astonish us into remembering what is possible. Within the Politeia of Hope, the social artist is trained to never give up, to maintain belief in the infinite creativity of the human being, and to persist in taking positive action no matter what the forces of entropy and negation may proclaim.

In all of these ways and more, the world server as social artist become servant, friend, and co-participant in the great game of life, turning followers into world servers and creating the field in which one can abandon oneself to the strength of others. The social artist helps remove the obstacles that prevent people from being all they can be, thereby enabling them to realize their potential. The world server as social artist is a listener, and listens to the ideas, needs, aspirations, and wishes of others, and then helps them achieve those desires. To this end, participants of the present social artistry program, now in its second year, are invited to create and fulfill a specific project, one that will focus their human development while enabling social change. Ultimately, it is about all of us together co-creating the ever unfolding reality as stated so powerfully by Teilhard de Chardin:

> "The outcome of the world, the gates of the future, the entry into the superhuman, these are not thrown open to a few of the privileged nor to one chosen people to the exclusion of all others. They will open to an advance of all together, in a direction in which all together can join and find completion in a spiritual renovation of the earth."

We are at that stage where the real work of humanity begins. This is the time and place where we partner Creation in the recreation of ourselves, in the restoration of the biosphere, and in the assuming of a new kind of culture – what we might term a culture of kindness – where we live daily life in such a way as to be re-connected and charged and intelligenced, to become liberated in inventiveness, and very engaged in our world and our tasks.

Now there is a quickening, an almost desperate sense of need, for this possible human in us all to help create the possible society and the possible world if we are to survive our own personal and planetary odyssey. Today, community participation and the empowering of grass roots development are essential to transforming the quality of life in societies everywhere. It is through work at these local levels that hope is generated for new and effective ways of shared governance As Kofi A. Annan has said, "*Good governance is perhaps the single most important factor in eradicating poverty and promoting development.*" It has been my experience that working in social artistry at both local and individual levels lays the foundation for good governance.

37

Career and Life Management Skills for Success

Bala S. Prasanna, MS

IBM Corporation, Middletown, New Jersey, USA

"For the lack of a nail, the shoe was lost.
For the lack of a shoe, the horse was lost.
For the lack of the horse, the message was lost.
For the lack of the message, the battle was lost.
For the lack of the battle, the war was lost.
For the lack of the war, the kingdom was lost.
All for the lack of a horseshoe nail."

—*Benjamin Franklin*

"Great things are not done by impulse, but by a series of small things brought together."

—*Vincent Van Gogh*

Workplace habits and expectations have changed significantly over the last few years. Advances in technology, economic liberalization programs throughout the world, social changes, globalization, and business/customer needs, just to name a few, have made carrying the burdens of work place demands round-the-clock become the norm. While the opportunities presented in this present-day workplace environment can be anybody's dream come true, thriving in such an environment has become more challenging than ever. You must be able to grasp the broader picture and adapt, while being aware and honing some essential skills that are vital to career management in particular, and life management in general.

Tying the notion of thriving in the modern work place environment with the notion of success is curiously interesting. In this chapter, we will explore some simple and practical skills that have the potential to yield useful results when practiced consistently. Keep in mind, not all skills suit all personalities, so pick and choose, and master the skills that suit you. Be open and willing to experiment and push yourself into unfamiliar parts of your personality and behavior, but be honest with yourself about your potential and limitations.

G. Madhavan et al. (eds.), *Career Development in Bioengineering and Biotechnology*,
DOI: 10.1007/978-0-387-76495-5_37, © Springer Science+Business Media, LLC 2008

We will now discuss "Career and Life Management Skills for Success" in eight broad categories.

1. Stress Management
2. Effective Communication
3. Networking
4. Conflict Management
5. Leadership Traits
6. Office Politics and Personal Management
7. Volunteerism
8. Over the Top

Stress Management

It is all in the balance! In the work environment, you will have to deal with work load, deadlines, emergencies, overtime, and training. In family or social life, you may deal with health issues regarding either yourself or those in your family, as well as issues regarding vacation, community connections, and hobbies. Juggling all of this causes stress. While a no stress life can be boring, having the right amount of stress at the right time, and managing stress to your advantage is possible. Consider the following:

- Before that important meeting with your boss or customers, or a similarly stressful situation, allow yourself a few minutes in private, if possible. Take a few deep breaths coupled with moderate exercise. Endorphins, sometimes called "feel good hormones" are naturally released chemicals from your body due to exercise. These are known to relieve pain, reduce stress and, overall, to produce a feeling of well-being.
- Be willing to work toward your own behavior modification. For example, you may be comfortable at working on technical issues glued by yourself to a computer. You may shun all the association with other people, and be shy when speaking to an audience. You need to work with yourself to feel comfortable working with others. By doing so, you will carry a lighter burden on your shoulders, and hence be less stressed.
- Remember the adage "The show must go on." We are all dispensable in one way or the other, and for us to think and act that we are indispensable will only cause grief and put more stress on us. If you train a back up for what you do, you will be less stressed, and you can enjoy your vacations better without your manager calling in the middle of the night or during your child's concert, it's a good idea.
- While you are always busy to allow yourself time and space to gain control of your situation, and to change from being a captive of the situation to being able to drive them. Find your own place to think. It may be that quiet room in your office building, or the library, or a park. Make time to think about your issues and ideas beforehand, rather than flying from the seat of your pants.

In a larger context, work your life's "balance muscles."

Effective Communication

Effective communication involves

- *Listening to understand* entails accepting what is heard even when you do not agree, and making efforts to gain clarity about what you have heard. By taking time to listen, you will also have a chance to grasp encrypted information. Sometimes by just letting people speak, they will do your hard work if you happen to be struggling to convey unpleasant or tough-to-sell ideas.
- *Conveying your ideas with purpose and intent:* If the tendency is to mask the real intent, sooner or later you will be exposed, and it will be an uphill task to correct perceptions later.
- *Different Communication Styles:* Individual styles are as individualistic as they come. There are story tellers (get to the point, please!), and people who can only mumble a few words (give me the context, give me more so that I can orient myself to what you are saying!), and everything in between. While there is no ideal style for all times and all people, be flexible to get the most out of people whose style is just different from yours.
- *Speaking up*: Watch your language! Do you have a tendency to use words like "maybe, sort of, kind of," words that seem to comfort you by lessening the responsibility or ownership that comes with what you are saying? When you are speaking to a manager will, and say "I will do it," the manager walk away with confidence that the task will be completed, as opposed to when you say, "I may be able to do it." Just by being careful and using the right language, you are taking ownership. Of course, do not forget to follow through and do what you said you would do.

Further, read between lines and watch for behavioral language. Communication is as much unspoken words and body language as the spoken words. While we all would like and expect others to mean what they just said, it just is not so for human beings. Develop skills and relationships with people so that you can fill in what is said with what was not said, or what was implied. Pay attention to body language (for example, did you roll your eyes when you said it?)

Learning to say NO without feeling guilty may be a hard task for some. You have to say NO when your instincts and circumstances are such, but you can always say NO in positive ways to minimize the impact with your boss or customer or peers ("You want me to do this by next Monday. Is it alright if I do by Friday?" is effectively saying NO, but with a much better aftertaste).

Networking

Remember the saying "*Dig your well before you are thirsty.*" One can never say enough about the virtues of networking. Picture a person walking on stilts. His vision is larger and his strides wider than a person without the equipment. Networking is the "stilt" you need to be more effective, productive, and innovative. Create your own networks– in a family context, with uncles, aunts, siblings, cousins, grandparents; in a friends context, with friends who share your interests, who inspire you, and who draw the best out of you; in a workplace context, with colleagues and managers; in a religious and spiritual context, with clergy or people whose religious and spiritual beliefs and guidance will do you good. The following guidelines should be considered when networking:

- *Why, with whom, and what will I be able to contribute or gain, when and where (note what you will be able to do before you seek to get something out of the network connections).*
- *If you help others, they will be more likely to help you.*
- *Follow up with leads, and acknowledge sources.*
- *Thank contacts.*
- *Do what you promised to do.*
- *Keep your networking "antenna" active.*

Conflict Management

Conflicts are inevitable, and they are a part of organizational life. A conflict can be healthy if it brings out new ideas. Remember, you need to have friction to produce a spark. A win-lose conflict creates distrust and anger. Negotiation is the key to a win-win resolution. Do your homework; use your connections to understand your options and some possible solutions before you begin negotiations. Always approach a negotiation with confidence. Remember, as a negotiating partner, if you think you have power, you have power. If you do not think you have power, you do not have power.

The following are some negotiating styles:

- *Competing*–assert your position without opposing viewpoints.
- *Aggressive*–assert your position and do not accept any other views (my way, or no way).
- *Avoiding*–stall, ignore, sense of timing.
- *Collaborating*–fully satisfying both sides (relationship is important).
- *Accommodating*–forgoing your concerns (selflessness).
- *Compromising*–negotiating, finding middle ground.

Each of the above styles has its own advantages and may be suited for a particular situation. While you may have a dominant style of your own that maps into one of the above styles, you need to train yourself in other styles and avoid depending on one style for all situations.

For example, if you are aggressive all the time, even if you have the material to back you up, you will deny yourself a chance to learn something new just by not listening to other views. People may also find it futile to present their ideas and options and simply not cooperate. Sometimes, it just becomes necessary to put your foot down and go your way. To be effective, you need to develop acumen about it. If your style is accommodating, and if you overuse it, you will simply become a doormat, and everyone will walk over you. You have to develop the backbone to stand up and assert yourself for your ideas and the value that you bring to the table.

Finally, do not let differences harden into anger, resentment, and hate. Do something about it. Negotiate!

Leadership Traits

Leadership is not about position, it is about action. While volumes upon volumes have been written about this topic by the experts, the following can be used as a guide to understanding and cultivating leadership.

- Live and work with paradoxes: This requires openness, accepting that everybody has limitations and virtues, and that you are trying to create an aggregate as a leader by working with these seemingly conflicting aspects of people and things.
- Push the frontiers rather than following.
- Promote innovation: Innovation is the key to success. You do not and cannot be a sole proprietor of innovations all the time. But if you develop the smarts to spot and promote, and integrate in your vision or scope of what you are trying to lead, hats off to you, Leader!
- Influence and guide. Act as a catalyst and champion causes.
- Have that invisible "something" that attracts followers – this could include simple things, such as how you speak of people. Being genuine and using positive language will bring rich rewards and makes your leadership more effective.
- Articulate a vision and make it happen.
- Focus on your stakeholders.
- Focus on real work, and not the mundane things.
- Deal with ambiguities! Rationalize fuzzy data.
- Be transcendent. Realize that your presence as a leader is not to serve your interests for personal gain, rather to serve larger interests that are beyond you. Leaders are expected to have higher levels of confidence.

Office Politics and Personal Management

The Oxford English Dictionary defines office politics as "*activities aimed at gaining power within an organization.*" You always have to play politics. Remember, there are good politics and bad politics. Let us consider some elements of "good politics."

- *Managing your boss:* At first thought, managing your boss sounds counter-intuitive. But taking time to understand your boss's likes or dislikes for everyday communications, will make your relationship with your boss that much more pleasant, or at least worry-free. (Does he or she like email, instant messaging, or calls on his or her cell phone at particular times). Remember, your manager has managers and peers, hence there are demands and expectations placed on him or her.

As a corollary to the above, you should also consider "managing your peers" along similar lines. If not handled properly relations with co-workers can be trickier than with managers.

- *Who are your angels?* Everybody needs angels every day just to help sail through the day. Perhaps also to get choice assignments or promotions. You create your angels. To do so, you need to be actively building positive relationships with peers, or with your manager's peers. If you ask yourself what you have done for them and come out ahead, you have succeeded in creating angels. Without angels, your advancement may just be a mirage.
- *Everyone expects to be paid back:* If you are perceived to have succeeded in either getting that choice assignment, or winning that million dollar proposal, and you had depended on others for your success, then everyone who is party to this expects to be paid back. It could be as simple as a thank you note to the person, or an acknowledgement of the contribution to the person's manager, or some kind of a larger pay back. (*I stuck my neck out for you when you were behind in your assignment, now it is your turn*!). So factoring this in ahead of time and responding as needed will position you better in the long run.

  As a corollary to this, success can create opposition. While it cannot be helped in some situations, on your part, consider how success was achieved. If you were fair and positioned yourself in win-win ways, you can at least minimize the opposition. Anticipate the aftermath of success. If you are perceived as having played win-lose, even if you succeed this time, they will be wishing for your failure in future.
- *Disillusionment at work:* There will be occasions when you will discover your manager was not honest, your colleagues are working behind your back and undercutting you, or your employer changed some benefits that were important to you. To help live through such situations, you need to see what choices you have. In the end, it also helps to be philosophical and derive strength through your religious or spiritual upbringing.

Volunteerism

There are only twenty-four hours in a day, and between family, friends, and work responsibilities, there seems to be no time to do something that brings meaning to your life, and that has the potential to change lives of others. But if you make the time and become a volunteer at your professional society, at your local SPCA, or at Big Brothers and Big Sisters, the personal reward of joy and meaning you reap far exceeds what you gave.

The movie *Sophie's Choice*, the query: "*At Auschwitz, tell me, where was God?*" The response: "*Where was man?*"

What more eloquent plea could sound for volunteerism?

You have a role to play in this world, and you have a reason for being here. But it is up to you to find your part and direct your future. You and you alone determine your destiny through your own efforts. Accept this responsibility not just for yourself, but for us all. You have the power to change your life and the lives of others. Do not back away from the exercise of this power or wait for some one else to act. You can get what you want, but part of what you want should be to help others along the way.

The good life is not a passive existence where you live and let live. It is one of involvement where you *live and help live.*

Over the Top

In this section we will discuss a few miscellaneous topics that complement and supplement our earlier topics.

- *Maintenance free employee:* Managers love employees who are deadline passionate, require little or no supervision, have good writing skills, and can read the boss's mind. (Nobody said it was easy, but it's worth a try!)
- *Expectations:* Always set expectations and work to overachieve them. Expectations are the benchmarks. Meeting expectations makes you an average employee.
- *Brand yourself:* While you cannot be all things to all people, recognize your strengths and create a positive brand for yourself. You may be a good systems engineer, you may be a good architect, you may be blessed with the ability to give great presentations–find what it is you are good at, and offer yourself as a person with that brand.
- *Anger Management:* Properly channeled anger has its place and brings desirable results. But anger, if not managed properly, can be disastrous. You may be able to patch up relations when you get angry and act out in a family situation. But a similar action may produce irreversible effects in a work place environment. You boss may tolerate or forgive you for an honest mistake you made that cost the company several thousand dollars, but he or she may not tolerate what you said about your boss in anger to your colleagues when that makes its way to him or her.
- *Professional Development:* In this day and age, when people switch jobs every few years, managing your career is not your employer's responsibility. You own your career and so you need to work on it, while taking advantage of whatever opportunities you have. Choose whether you want to pursue a technical or management career. Take advantage of employer or professional society-sponsored professional development classes. Remember, in order for you to be promoted, it is not enough to be the best at your current level, you need to demonstrate that you can work at the higher level that you are aspiring to be promoted to.
- *Making the Right Choices:* What a wonderful life, we are given so many choices to make! Choosing friends, choosing how you think and feel, handling criticism and feedback, spending time, training, networking, nurturing relationships. It is up to you to find friends who inspire you, who you can count on. You can

benefit yourself and them. Bad company is a drag, and can be harmful to your aspirations. Thinking optimistically and constructively is a choice that you make. While there is some time and place to feel down, sad, and angry, get over it fast! It always helps to be a bit thick-skinned and able to receive criticism. Not that being sensitive is wrong, it saps energy, and is a choice you make. Remember, when people criticize you, allow yourself to see whether they are exposing your blind side and thus helping to fix some gaps that you had, or whether they are simply having fun at your expense (always ignore the recreational critics!)

Suggested Reading

1. Cohen, H., *You Can Negotiate Anything*, Bantam (1982)

38

Perspectives on Ethical Development: Reflections from Life and Profession

Jerry C. Collins, PhD

Department of Biomedical Engineering, School of Engineering, Vanderbilt University, Nashville, Tennessee, USA

I have been asked to give, as a bioengineer and as an individual, a perspective of my own odyssey in ethics. I claim no special gift, perspective or accomplishment that qualifies me for this task. But that may not be such a bad thing. I hope you see in my ordinary experience an encouragement to explore your own, to see how you arrived at where you are now. The exercise may be worth the effort.

Early Formative Experiences

I was born and grew up in Nashville, Tennessee, in the 1940s. Nashville was a politically conservative, historically racially segregated city. My family was in a Christian conservative religious movement. They were loving but strong-willed; some might say controlling.

I was an early reader and precocious. At the age of five, when I spoke to my grandmother (with whom our family shared a house) about my ability to read and do math, she said "Do not brag!" I stopped what I was saying. I was a pleaser, but parts of me were rebellious and combative.

When I was six, I took a small comic book stuck to a cereal box in our neighborhood grocery from the box, hid it in my pocket, and took it home. My father found it, asked where I got it, and made me take it back to the grocer and apologize.

These and similar instances awakened or helped develop in me a sense of right and wrong. I am still coming to terms with how my early history affected my ethical sensitivities.

In those early years, racial awareness formed a large part of my ethical awakening. I was aware that my household included Mertella, an African-American woman who worked as a maid during the day and helped to care for my great-grandmother. I was also aware that our church included quarters for a

G. Madhavan et al. (eds.), *Career Development in Bioengineering and Biotechnology*, DOI: 10.1007/978-0-387-76495-5_38, © Springer Science+Business Media, LLC 2008

janitor and his family. Robert would prepare communion for our church, then drive across town to an African-American church where he and his family could participate. Conversations within my group of friends included occasional racial allusions, banter, and inappropriate jokes. I was somewhat annoyed, but the pleaser in me did not object.

College and Graduate School

I was awarded a scholarship to Vanderbilt University in the fall of 1958. During the summers of 1958 and 1959, I earned money for school by working as a thread packer in a garment-industry factory. Our work area was not air-conditioned, ostensibly because reducing the humidity would adversely affect the thread-waxing process. The pay for virtually every worker in the plant was minimum wage, $1 per hour. The work was boring but not onerous for me. I relieved the monotony by calculating trivia. For example, at typical rates of production, I could pack enough thread to reach to the moon twice and back in a single summer. I also calculated that at minimum wage for 13 weeks, I could earn 80 percent of university tuition for the academic year.

I also realized that most of the workers were African-American women, some with children. Some had worked at the factory for years, but nearly all made minimum wage. I was not aware of any worker except the manager, the plant repair person, and two employees in shipping who made more than $1.10 per hour. My economic future was bright; in contrast, theirs was limited and circumscribed.

Our family made little money during that time, so I lived at home and commuted with a friend to school. I was not a part of the Vanderbilt social scene, and except for classes, not a part of the intellectual scene either. Nashville at that time was a town targeted by some students for integration of public eating facilities. The Board of Trust of Vanderbilt contained members who were openly critical of and opposed to integration of either lunch counters or the university. The School of Divinity admitted a student, James Lawson, who was later forced to leave because of his leading role in the desegregation movement. (He has recently been welcomed back to Vanderbilt, 45 years later, as a Distinguished University Professor.) Public schools were integrated, but the process was punctuated with anger, violence, and the dynamiting of a distinguished African-American lawyer's house. I was somewhat aware of what was going on, but not moved to take part, and certainly not encouraged to by my family. Since that time, I have reviewed Nashville history from different eyes, including a read of *The Children* [1] by the late David Halberstam, and become convicted that momentous events were happening in Nashville in 1960–62 and that I was ethically naive for not having been aware of them and taken part.

When I graduated from Vanderbilt with a degree in electrical engineering, I accepted a summer position at the US Naval Ordnance Laboratory (NOL) in White Oak, MD. I enjoyed my work, helping design a sonar preamplifier for a submarine; the fields of engineering design and communication theory were fascinating. However, in April 1963, the USS Thresher sank, killing all 129 on board.

When I returned in June for my second summer, I found that my supervisor, Bill, had been on the gangplank the summer before, waiting to board, when his supervisor Don had taken his place to check on some details. The near-miss of Bill and the loss of Don caused me to think: Should I not use my engineering expertise for something more directly supportive of human life?

During my first year in graduate school at Purdue, I discovered the library section of John Steinbeck novels and reinforced my awareness of the plight of the worker. I also discovered that an attractive, brilliant West Lafayette English high school teacher was interested enough in me to spend time with me. We were married in August 1964. Much of the development of my ethical sense has come from and through Sandra.

In August 1963, my second summer at NOL, I received a phone call from my mother. She said that trouble was set to occur, as it had in several places in the South, on the Washington Mall the next week. I should quit work at the Naval Ordnance Laboratory a week early and come back to Nashville for the rest of the summer to avoid it. I did so, and missed the chance to hear "I Have a Dream" as a result.

I returned to Purdue, finished my MSEE in early 1965, and then arranged to transfer to Duke for my doctoral studies. Duke was just beginning their biomedical engineering educational program, and I was one of its first participants, although my degree was still technically in electrical engineering. I decided that I would study with my advisor, the late Theo Pilkington, when I heard him say one Monday "I think we saved a child's life this weekend." He was working with pediatric cardiologists on an electrocardiography modeling program. I spent two years working under Theo's tutelage in the laboratory of Dr. Knut Schmidt-Nielsen, identifying and describing a nasal energy and fluid exchange mechanism. I learned from both of these mentors not only how to conduct a research project, but also how to treat and train student mentees. Knut, his colleagues, his students, and I shared a high regard for nature, and this made me aware of ecological and environmental issues.

In the spring of 1968, my wife and I were placed on the planning committee for an annual regional church lectureship. The committee decided to change the practice of inviting only Caucasians to the lectureship, and to invite and assign people for discussions regardless of color. We had read *The Death of God* by Thomas J. J. Altizer [2] and thought it contained some ideas that our group of experts would consider interesting and relevant. To my surprise, the person assigned to review chose not to deal substantively with the issues raised by Altizer, but to deliver a shallow diatribe. This reinforced a process of critical review of the beliefs and practices of our church fellowship which has continued ever since. Our shallow syllogisms and reasoning changed into a deep contemplative approach, a realization that what might be more important lay elsewhere.

The regional meeting that we were planning occurred the week after the shooting death of Dr. Martin Luther King, Jr. At the time of the shooting, a friend and I were in an assembly of more than a thousand of the world's leading theologians at Duke, listening to a lecture entitled "The Eschatology of Hope" by Jurgen

Möltmann. Möltmann has written extensively on that subject, notably in *Theology of Hope.* [3] As news of the shooting and subsequent death of Dr. King was reported to the audience, anguished cries and shouts of "Barbarians!" echoed through the hall.

I have reflected on that night and the events of that week many times since. As I learned more about Möltmann, I realized that his life experience and his lecture that night were particularly appropriate in the context of Dr. King's murder. Born in 1926, Möltmann, as a 17-year-old German military survived a nearby explosion, and surrendered to the British as a 19-year-old soldier in 1945. He became a Christian theologian after his prisoner-of-war experience in post-war years. Möltmann's life-long thesis is that the hope that the expectation of the future kingdom of God in the Christian "liberates us to refuse to be reconciled (to) that kingdom's contradiction in the present orders of injustice, and thus to conduct our lives here in resistance to the reigning forces of evil." [4] As one who has come to see the resistance to the integration of African America into the full promise of American and world society as a "reigning force of evil," I attach profound significance to the events of that night.

Our church lectureship later that week was interrupted by violence after Dr. King's death. Parts of Durham were on fire, and the city was under martial law. We had five guests in our house, one African-American. As the military vehicles patrolled our deserted street beneath our window, our discussions and devotional periods were unscripted, intense, and personal.

I left Duke in August 1968, aware that my scientific and conservative religious worlds were at odds. I felt no particular tension over that apparent conflict, and never have since. Of greater concern to me was that I had seen injustice in the world, and had been in a position to react constructively (albeit at personal risk) and had failed to do so.

Early and Later Employment

As I began employment at the University of Kentucky in the fall of 1968, I found myself in a department with no female faculty, and very few female and no African-American students. The decade, which had begun as one of peaceful protest focused on racial injustice, turned violent as frustration about the Vietnam war mounted. [5] In 1969, the University of Kentucky Armory was burned to the ground. I was drafted as a non-combatant but my department chair wrote in my behalf as an educator and I was deferred.

The University of Kentucky basketball team was nationally renowned, but was remarkable for not having been integrated. That would only occur several years later; when they won the national championship in 1978, several of their key players were African-American.

Our church fellowship was also predominantly Caucasian. My wife began to volunteer and interact with guests at the local Florence Crittenden Home for unwed mothers. We had several mothers stay with us for lengths of time. Mary, a young African-American, lived with us for several months. We met her family, who lived in a small nearby town, and came to love her gentle, cheerful spirit. We also saw the personal and financial burden her compliant nature and two children placed on her, and were delighted when a good man agreed to marry her and care for her and the children.

When we moved to Nashville in 1977, the incidence of females and ethnic minorities in my division in the School of Medicine was somewhat higher, but was still low. My wife and I met a wonderful African-American family who directed a tutoring program in an inner-city community center. Volunteer after-school tutoring by our family, particularly by Sandra and our children, caused us to realize that tutoring, although helpful, did not address adequately many of the problems inner-city students were having. We became part of a group that funded and started a preschool child care center in the poorest demographic district in Nashville. That center has been in operation for a decade and has the state's highest rating. It certainly is a step in the right direction.

For my first seven years at Vanderbilt, I participated in NIH-sponsored animal research. Our area of study was in microvascular (capillary) transport in the heart and lungs during and following induced myocardial ischemia. I had held a similar position in the Department of Cardiothoracic Surgery at the University of Kentucky, where our interest was preservation of the myocardium during and after cardiopulmonary bypass. In both cases, we used acute canine models. I was at peace with our research, but the day my two older children came to my laboratory and realized that our end was to help human health, but our means was to study heart and lung function through *post mortem* studies in dogs and sheep, my work was difficult to explain. I hoped to minimize acute studies, and eventually realized that precise, detailed anatomical and physiological mathematical models and intelligent statistical design can yield valuable information while reducing the number of animals that must be used in experimental studies.

From 1984 to 1997, I was the director of computer systems in the Clinical Research Center (CRC) at Vanderbilt. I became aware of many ethical issues that I had not been forced to confront when my focus was on animal studies and mathematical models of research, including:

1. The wastefulness of a poorly designed experiment;
2. The potential conflict of interest of an investigator who has a financial interest in a company who supplies drugs or provides financial support for a clinical study;
3. The risk of patient death or morbidity as the result of inadequate or poor testing;
4. The possible deleterious effects of drug-drug interactions;
5. The systematic exclusion of women from studies on the pretext of added variability attributable to female reproductive cycles;
6. The lack of recognition of, or accounting for, the existence of ethnic phenotypes in study design;
7. In the early years, marginalization of AIDS as a "homosexual" problem.

I also realized that institutional review boards set up to review the ethics and efficacy of human research protocols were relatively recent entities, and that the CRC Advisory Boards, first established in 1960, were in many cases antecedent to Institutional Review Boards (IRB).

Professional Activity

My serious involvement in professional, not honorary, society activities began in the 1980s, when I joined the Biomedical Engineering Society and the American Physiological Society. My multiple responsibilities with the Biomedical Engineering Society (BMES), beginning in 1990, included editing the Society Bulletin (1990–2001), serving on the Membership Committee and the Board of Directors of the BMES, and first chairing, then serving as a member of, the Interface with Industry committee. These activities and responsibilities made me acutely aware of disconnects between academic and industrial environments of bioengineering, and the need to improve communication and interaction between the two. In 1999, I became industrial liaison for the NSF-funded VaNTH Engineering Research Center for bioengineering educational technologies, and held that position throughout its tenure.

BMES Code of Ethics

In 2002, I agreed to serve as founding chair of the Ethics Committee of BMES. We formulated a Code of Ethics [6] that was ratified by the membership in 2004. The Code as it was formulated has at least two questionable issues: (1) The Code is aspirational, not punitive; the Society decided on this course in order to limit financial liability, (2) The Code appears to favor compliance to employers and agencies, even in the face of questionable practices. Whistle blowing or nonviolent disobedience, for example, are not options that appear in the wording of the Code. We have recruited committee members from a variety of professions to bring perspective to the activities of the committee.

Several international experiences in the 1990s are worth noting. In 1993 I visited Kiev with a medical team. We found the medical supply system in shambles after the 1991 breakup of the Soviet Union. We were able to arrange transfers of modest gifts of equipment, supplies, and drugs to pediatrician friends in the Ukraine, but by and large, their medical system could not provide even rudimentary diagnosis and treatment. For example, no hospital in the Ukraine had access to a working MRI analysis system for the general public. (My county had dozens.) Parents of children exposed to the nuclear catastrophe of Chernobyl were helplessly watching their children die of the accumulated results of radiation exposure. (Chernobyl exploded on April 26, 1986, the fifteenth birthday of my son. The public watched on state television as boys his age were sentenced to death by radiation exposure, despite the denial by official sources.)

Two years later, I revisited the Ukraine and one of my friends, Dr. V. Malyuk. His laboratory had been asked to measure the release of radioactive strontium 90 accumulated in the bones of children exposed to the aftermath of the explosion. I was able to buy a personal computer to help him in his task. I thought *Here is a scientist as well educated as I who makes less than one percent of what I do.*

During that visit, I addressed a group of 50 pediatricians in a university room out of whose windows I could see Babi Yar, where some say as many as 200,000

Jews and sympathizers were shot, more than 100,000 in a three-day period in 1941. Their remains were bulldozed into mass graves. The Siberian poet Yevtushenko wrote:

> "No monument stands over Babi Yar.
> A steep cliff only, like the rudest headstone.
> I am afraid.
> Today, I am as old
> As the entire Jewish race itself. [7]"

Shostakovich's 13th Symphony commemorates in part the site and the event. As I stood on the Babi Yar site after my lecture, I could say with President Kennedy "Ich bin ein Jude."

In 1997, I was invited to attend an integrative science meeting in St. Petersburg. As I walked the streets of that beautiful city, built on more than 40 islands interlaced with canals, I realized that more than a million people, and 20 million Russians in all, had died during the three-year Leningrad siege of World War II. Living conditions were still difficult and public infrastructure—unsafe streetcars, collapsing building facades—carried great risk. I developed a sense that the citizens of St. Petersburg were more than just survivors, they were heroes. I also realized that some of the survivors of Leningrad might have participated in the anti-Semitism of Babi Yar.

Recent Experiences

The primary purpose of the Tennessee Biotechnology Association (TBA), a state affiliate of the Biotechnology Industry Organization (BIO), is to promote biotechnology industry within the state. As co-chair of the Education Committee of TBA, I have become acutely aware of social issues associated with the growth of biotechnology industry, and the ethical imperative of improving community life through the creation of better jobs. In several states, the average tourism salary is near poverty level for families of two or three. Average biotechnology salaries are two to four times as large. I am also a member of the Center for Bioethics and Culture, a group promoting serious public dialogue on bioethics issues raised by the development of biotechnology.

I had teaching responsibilities for a number of years in a sophomore-level course in biomedical engineering thermodynamics, in which we developed ethics modules to complement instruction in scientific principles. For example, when discussing the principles of mass, energy, and species conservation applicable in a dialysis machine, we used small group role-play to consider whether a company manufacturing dialysis equipment should develop a home dialysis machine. Answering this question involved students in addressing a great number of issues, including marketing, safety, risk and cost/benefit ratio, and reliability. This activity has been documented elsewhere. [8]

Biomedical Engineering Ethics Course

With the assistance of student Emily Mowry and colleague Dr. Sean Brophy, [9] I established an undergraduate course, Biomedical Engineering Ethics, in 2004. This course has been offered four times, with a total enrollment of more than 70 junior and senior students. The class met for two hours on Tuesdays, in which the class viewed a film or engaged in an extended discussion with an expert or a panel, and an hour on Thursdays. Through exposure to classical ethics models and practices, classic and contemporary cases such as the Tuskegee syphilis study, the German medical atrocities during World War II, and the early ignorance and avoidance of study and treatment of the human immunodeficiency virus and AIDS, to issues of the Human Genome Project, and to a number of cases of alleged professional and corporate malfeasance, students developed proficiency in ethical decision making in the following respects: to be **perceptive** about issues, to be **conditional** in point of view (to be able to investigate multiple points of view), to be **discriminatory** in their approach to ethical issues, and, finally, to become **competent** in identifying and approaching ethical issues. The text we chose for the class was Tom Beauchamp's and James Childress's *Biomedical Ethics.* [10]

This text develops, through the presentation of a number of case studies, a list of four tenets of biomedical ethics: beneficence (good will), nonmalificence (absence of ill will or malice), justice (equal treatment), and respect for autonomy (informed consent). Are these tenets sufficient for a platform of ethical behavior? I posed this question to a team of five students from different religious and cultural backgrounds: a Hindu, a Buddhist, a Muslim, a Christian, and an atheist. Each developed a list of tenets characteristic of her/his culture. Then the team identified common elements and eliminated others. They selected the four tenets listed above and added a fifth: personal responsibility. This commonality may indicate that there is a common morality underlying ethical behavior. It is quite possible that the similarity of student experiences negates cultural differences among the students. I will address this point more personally in concluding this article.

Toward a Common Morality and Experience

These ethical tenets are literally written in blood: in the blood of millions of victims of the Holocaust; of hundreds of African-American syphilitic men in Tuskegee, Alabama, unjustly sentenced to death through non-treatment of syphilis when treatment was available; of millions of AIDS victims sentenced to death through an early purposeful avoidance of study and treatment; of thousands of women and ethnic minorities excluded from clinical studies because gender and ethnic differences made studies more difficult, expensive, and involved. In many, perhaps most, of these cases, a disdain for the omitted or ignored constituency has played a role in the mistakes or deficiencies of clinical studies. In ethics, including biomedical and bioengineering ethics, we learn from the mistakes and sins of the past.

There is a widely-held conception that the progression of science, fashioned according to the scientific method, is an orderly, impersonal process, in contrast to the rancorous arguments that frequently impede or prevent cultural or religious resolutions. In fact, nothing could be further from the truth. Scientific differences are often accompanied by personal disagreements. Differences of opinion are expressed. Egos are at stake. Voices are raised.

I had an epiphany while going down the escalator to the Exhibit Hall at a conference in 2004. There were hundreds of posters on the floor, each presented by a scientist convinced of the truth of her/his poster and eager to present her/his point of view to colleagues who would listen. Each scientist was a preacher and his poster was his church. Differences of opinion would be expressed and arguments would ensue, and probably spill over into the scientific literature. The initial uncertainty of knowledge in the scientific process would eventually be resolved, and life would go on, a little better for having gone through the process.

In 2002, a discussion of the human tragedy of the war in Afghanistan was occurring in the NIH Clinical Center between three people: a student of mine; a relative of President Karzai of Afghanistan; the mentor of President Karzai's relative, who had grown up on a kibbutz in Israel; and myself. We discussed the possibility of bringing humanitarian aid to the children of Afghanistan, which a friend of mine, a bioengineer at Harvard called to active duty, was trying to do in his time off. My student paused suddenly and held up his hand for attention. "Here we are: a Muslim, a Christian, and a Jew, agreeing on this common ground that we should try to help these poor children." Another epiphany for me.

In 2004, I had the honor of interviewing Earl Bakken, co-founder and first CEO of Medtronic, the pacemaker of pacemaker companies. He told me about the first portable pacemaker he designed, from a metronome circuit published in a popular electronics magazine. His surgeon partner, Walton Lillehei, had asked for it because his children recovering from heart surgery would arrest and die after losing rhythm. They tried it first on a dog, with success. The next day, Dr. Bakken was dismayed to find his device on a child, not a dog, as they had previously agreed. When he asked Dr. Lillehei about it, his response was: "The dog was doing all right and the child was dying." This story reminded me of still another epiphany of the nonviolent rebellion of the Nashville students and of Dr. King in the 1960s. Sometimes in ethics or bioengineering or life, it's better to do the right thing than to follow the rules.

In 2005, *Time* magazine selected the unlikely trio of Bill and Melinda Gates and Bono as People of the Year. The especially remarkable thing about the selection of these three was that none was selected for having a primary expertise in music or software or management. They were selected for their altruism and compassion. The selection resonated with people around the world, from every possible religious and cultural persuasion. We are tired of war and of systematic deprivation of the less fortunate. We must for our own sake give ourselves to the greater good.

From where does altruism originate? Is there a gene? Does it come from a religious culture or cultures, a voice of "deep magic" that we all listen for? Evolutionary biologists believe there is a process, or processes, that naturally select altruistic behavior. I am coming to realize that the important thing is not to understand the mechanism so much as to act upon the reality. The voices of Bono and the Gateses are louder and clearer to me than the voice of corporate culture that says that the research budgets of some large pharmaceutical companies are larger than that of the entire National Science Foundation, but they have one goal: to make money for the shareholder. I won't divest myself of my diverse stock portfolio, but there is room under my tent for additional goals: developing a technology infrastructure from which all can benefit, and mustering the political and social will to implement it.

I have told you in some detail about the limitations my cultural, personal, and religious background, and my own personal choices, presented to my full participation in redemptive experience, in order to lead you to this final place. I apologize for using my own religious language to ask myself the question: is my Kingdom of God large enough to transcend religious, cultural, and scientific differences and allow me to welcome you to work together for the greater good? Is yours? Like Jurgen Möltmann, can we refuse to give in to the present orders of injustice, and thus conduct our lives here in resistance to the reigning forces of evil? Can you and I welcome others, with whom we might disagree in particulars, to join us in that work?

References

1. David Halberstam, *The Children*, Fawcett, 1998.
2. Thomas J. J. Altizer, *The Death of God*, Bobbs-Merrill Company, 1966.
3. Jurgen Möltmann, *Theology of Hope: On the Ground and the Implications of a Christian Eschatology*, Augsburg Fortress, 1993.
4. www.adamsteward.wordpress.com/2007/08/18/jurgen-moltmanns-theology-of-hope/
5. David Halberstam, *The Best and the Brightest.* Ballantine Books, 1993.
6. See: www.bmes.org/pdf/2004ApprovedCodeofEthicsShortForm.pdf
7. *The Collected Poems 1952–1990* by Yevgeny Yevtushenko. Edited by Albert C. Todd with the author and James Ragan, Henry Holt and Company, 1991. Used by permission of the author.
8. Collins J, Mathieson, C. Case studies in economics and ethics in an early biomedical engineering class. *Proceedings*, ASEE Annual Conference, Montreal, June 2002.
9. Mowry, E., Collins, J., Brophy, S. Creation of a bioethics course for the undergraduate biomedical engineering curriculum. *Proceedings*, ASEE Annual Conference, Nashville, June 2003.
10. Beauchamp, Tom L., Childress, James F. *Principles of Biomedical Ethics (5th* Ed.). Oxford University Press, 2001.

Suggested Reading

1. Altman, Lawrence. *Who Goes First? The Story of Self-Experimentation in Medicine.* Random House, New York, 1986.
2. Angell, Marcia. *The Truth About the Drug Companies: How They Deceive Us and What to Do About It..* Random House, 2004.
3. Avorn, Jerry. *Powerful Medicines.* Knopf, 2004.
4. Bakken, Earl E. *One Man's Full Life.* Medtronic, Inc., Minneapolis, Minnesota, 1999.

5. Beauchamp, Tom L., Childress, James F. *Principles of Biomedical Ethics (5th Ed.).* Oxford University Press, 2001.
6. Bellah, Robert N., Madsen, Richard, Sullivan, William M., Swidler, Ann, Tipton, Steven M. *Habits of the Heart: Individualism and Commitment in American Life.* & Row Publishers, New York, 1985.
7. Beyleveld, Deryck. *Human Dignity in Bioethics and Biolaw.* Oxford University Press, New York, New York, 2001.
8. Biagoli, Mario, Galison, Peter. *Scientific Authorship: Credit and Intellectual Property in Science.* Routledge, 2003.
9. Buranen, Lise and Roy, Alice M. (eds.) *Perspectives on Plagiarism and Intellectual Property in a Postmodern World.* State University of New York Press, 1999.
10. Chapman, Audrey. *Designing Our Descendants: The Promises and Perils of Genetic Modifications.* Johns Hopkins University Press, Baltimore, Maryland, 2003.
11. "Committee on the Biological and Biomedical Applications of Stem Cell Research." *Stem Cells and the Future of Regenerative Medicine.* National Academy Press, Washington, D.C., 2002. Also available with WEBACCESS.
12. "Committee on Public Information, Section on Intellectual Property Law, American Bar Association." *Intellectual Property: A Guide for Engineers.* ASME Press, New York, 2001.
13. "Council on Ethical and Judicial Affairs." *Code of Medical Ethics.* 1996–1997 Edition. American Medical Association, Chicago, 1997.
14. Davies, Kevin. *Cracking the Genome: Inside the Race to Unlock Human DNA.* The Free Press, New York, 2001.
15. Devettere, Raymond J. *Practical Decision Making in Health Care Ethics.* Georgetown University Press, Washington, D.C., 1995.
16. Dhanda, Rahul. *Guiding Icarus: Merging Bioethics with Corporate Interests.* John Wiley & Sons, New York, New York, 2002. Available via WEBACCESS.
17. Drane, James. *More Humane Medicine: A Liberal Catholic Bioethics.* Edinboro University Press, Edinboro, Pennsylvania, 2003.
18. Florman, Samuel C. *The Existential Pleasures of Engineering.* 2nd Edition. St. Martin's Griffin, New York, 1994.
19. Galbraith, John K. *The Culture of Contentment.* Houghton Mifflin Company, New York, 1992.
20. George, Bill. *Authentic Leadership: Rediscovering the Secrets to Creating Lasting Value.* Jossey-Bass, San Francisco, 2003.
21. Gladwell, Malcolm. *The Tipping Point: How Little Things Can Make a Big Difference.* Back Bay Books, Boston, 2002.
22. Gluck, John P., Dipasquale, Tony, Orlans, F. Barbara. *Applied Ethics in Animal Research: Philosophy, Regulation, and Laboratory Applications.* Purdue University Press, 2000.
23. Goozner, Merrill. *The $800 Million Pill: The Truth behind the Cost of New Drugs.* University of California Press, 2004.
24. Greenberg, Daniel. *Science, Money and Politics: Political Triumph and Ethical Erosion.* University of Chicago Press, 2001.
25. Guerin, Wilfred L. (Ed.) *A Handbook of Critical Approaches to Literature.* Oxford University Press, 1999.
26. Haynes, D. The integrity of research published by Stephen E. Breuning. *Bull Med Libr Assoc* July; 76(3): 272, 1988.
27. Hilts, Philip J. *Protecting America's Health: The FDA, Business, and One Hundred Years of Regulation.* Borzoi Books, 2003.
28. Jones, James H. *Bad Blood: The Tuskegee Syphilis Experiment*, The Free Press, New York, 1993.

29. Kass, Leon R. *Life, Liberty and the Defense of Dignity: The Challenge of Bioethics*, counter Books, San Francisco, 2002.
30. LaFollette, Marcel C. *Stealing Into Print: Fraud, Plagiarism and Misconduct in Scientific Publishing*, University of California Press, Berkeley, California, 1992.
31. Lambrecht, Bill. *Dinner at the New Gene Cafe: How Genetic Engineering is Changing What We Eat, How We live, and the Global Politics of Food.* Thomas Dunne Books, St. Martin's Press, 2001.
32. Lundberg, George D. with Stacey, James. *Severed Trust: Why American Medicine Hasn't Been Fixed*, Basic Books, New York, 2000.
33. McKibben, Bill. *Enough: Staying Human in an Engineered Age.* Times Books, New York, New York, 2003.
34. Moss, Lenny. *What Genes Can't Do.* MIT Press, Cambridge, 2003.
35. Mowery, David and Rosenberg, Nathan. *Technology and the Pursuit of Economic Growth.* Cambridge University Press, New York, 1989.
36. Naumes W, Naumes MJ. *The Art and Craft of Case Writing.* Thousand Oaks, California: Sage Publications; 1999.
37. O'Neill, Onora. *Autonomy and Trust in Bioethics.* Cambridge University Press, New York, New York, 2002.
38. Polanyi, Michael. *Faith and Society.* University of Chicago Press, 1946.
39. Postrel, Virginia. *The Future and Its Enemies: The Growing Conflict Over Creativity, Enterprise, and Progress.* Touchstone, New York, 1998.
40. President's Council on Bioethics. *Beyond Therapy: Biotechnology and the Pursuit of Happiness/ a Report of the President's Council on Bioethics.* The President's Council on Bioethics, Washington, D.C., 2003. Also available through WEBACCESS.
41. President's Council on Bioethics (foreword by Kass, Leon R.). *Human Cloning and Human Dignity: The Report of the President's Council on Bioethics.* Public Affairs, New York, 2002.
42. Relman, A. S. Lessons from the Darsee affair. *N Engl J Med* 208:1415–7, 1983.
43. Schrag B, ed. *Research Ethics: Cases and Commentaries*, Vol. I – IV (revised), Bloomington, Indiana: GREE; 2000.
44. Schulze, Quentin J. *Habits of the High-Tech Heart: Living Virtuously in the Information Age*, Baker Academic, Grand Rapids, Michigan, 2002.
45. Spears, Larry C. (ed.) *Insights on Leadership: Service, Stewardship, Spirit and Servant Leadership*, John Wiley and Sons, Inc., New York, 1998.
46. Steneck, Nicholas H. *Introduction to the Responsible Conduct of Research.* Office of Research Integrity, Washington, District of Columbia, 2004.
47. Veatch, Robert. *The Basics of Bioethics.* 2nd Edition. Prentice Hall, Upper Saddle River, New Jersey, 2003.
48. Waters, Brent and Cole-Turner, Ronald, eds. *God and the Embryo: Religious Voices on Stem Cells and Cloning*, Georgetown University Press, Washington, District of Columbia, 2003.
49. Zweiger, Gary. *Transducing the Genome: Information, Anarchy, and Revolution in the Biomedical Sciences.* McGraw-Hill, New York, 2001.

Part V

Growth and Responsibilities Beyond the Profession

39

Technology Development and Citizen Engagement

Joseph O. Malo, PhD

University of Nairobi and the Kenya National Academy of Sciences, Nairobi, Kenya

Modern fields such as bioengineering and biotechnology have evolved out of the solid foundations of physics, chemistry, and, more importantly, biology. Despite the trailblazing advancements, society still experiences an imbalance in the scientific and technological literacy of citizens. This has serious implications for public policy formulation, especially for developing countries.

Citizen engagement, especially in the realm of bioengineering and biotechnology, is vital to optimally respond to the many challenges of international development. Science and technology development might best be viewed as a "bottom-up" participating process where citizens play a central role. Instead of citizens being viewed as passive beneficiaries of a trickle-down effect in technology transfer, they should be considered as knowledgeable, active, and centrally involved in both upstream design of technologies and downstream delivery and regulation.

I urge bioengineers, biotechnologists, and life scientists to consider the following key representative issues during their science and technology development endeavors:

1. What role can technologies play in the future of people in the developing world?
2. How can poor people become more involved in shaping their own technological futures?
3. What makes science and technology work for the poor?
4. What forces will be involved in shaping science and technology for the developing world?
5. How can those who work in science and technology development assist the poor and those in the developing world?
6. How and where should research and development funds be directed to best help those with the least means (at least at present) to help themselves?

One of the main bottlenecks of acceptance of new technologies is the regulation of risk and uncertainty arising from technological applications. In this connection, those frequently marginalized individuals who would ultimately be

G. Madhavan et al. (eds.), *Career Development in Bioengineering and Biotechnology*,
DOI: 10.1007/978-0-387-76495-5_39, © Springer Science+Business Media, LLC 2008

involved in the use of innovation and technology should also be involved in the initial decision-making processes. Thus, it is time to adopt a new vision of citizenship that goes beyond public engagement, encompassing how science and technology agendas are framed, the social purpose they serve, and who stands to gain or lose from them.

Advances in bioengineering and biotechnology should not ignore the perspectives and concerns of poor citizens in a trickle-down framework. For these technologies to provide wellbeing for the citizens, innovation must be rooted in the local realities. Hence, a more participatory approach is needed, where innovations are seen as part of broader system of governance and markets that extend from local to regional, national, and international levels. No doubt this will increase access and indeed ownership for sustainability, and this is where bioengineers, biotechnologists, and life scientists have a crucial role to play in the present and the future.

40

In Defense of Science and Technology

Elizabeth M. Whelan, ScD, MPH

The American Council on Science and Health, New York, New York, USA

Modern society benefits from a spectacular range of scientific and technological advances that have provided us with improvements in every sphere of life—from food safety to health care to transportation. We have reaped these benefits in improved length and quality of life, and in health. But according to many media outlets, our well-being is threatened, rather than improved, by scientific advances.

What creates this paradox? Part of the reason is that science has become more and more complex, requiring a greater degree of sophistication than was true in the past. Ninety years ago, it was simple to explain that heating milk in the process of pasteurization would kill pathogenic bacteria and prevent illness. Today, it is more difficult to explain to the non-scientist exactly what recombinant bovine somatotropin (rBST) is, and how it works to increase cows' milk production. Similarly, when the non-scientist hears about cloning animals to improve the stock of meat and milk producers, he or she is vaguely uneasy about consuming such products—because there is little understanding of what the processes are and the effects they might have on either the animals concerned or the humans at the end of the food chain.

Developed nations, such as the USA, today have had little experience with the diseases that science has all but eradicated from our society—such as polio, measles, whooping cough, and rubella. Thus, among some segments of society, parents are reluctant to vaccinate their children for fear of possible hypothetical effects of various vaccines. Added to a simple lack of information about vaccines and diseases, is the activity of various special interest groups that pump up fears about these major life-saving advances. For example, many parents believe, based on a misunderstanding of the scientific process, that the preservative used in vaccines is causally linked to autism—a hypothesis that has been disproved time and time again.

In no other sphere has the lack of sophisticated scientific understanding been more damaging than in the response of many to the concepts of biotechnology. Although gene-splicing can result in the decreased use of pesticides on many crops,

G. Madhavan et al. (eds.), *Career Development in Bioengineering and Biotechnology*,
DOI: 10.1007/978-0-387-76495-5_40, © Springer Science+Business Media, LLC 2008

and improve crops' ability to provide nutrients and withstand pests and droughts, many view the techniques involved as unnatural, and thus to be avoided. Partly because societies are becoming predominantly urban, people have little contact with the methods of food production, and do not realize that agriculture itself is a man-made, not a natural system. They thus are prey to bogus speculations about the deleterious effects that gene-splicing might have.

Science—today and everyday—is under assault. The assailants are members of the media, trial lawyers, self-appointed consumer-activists, and environmentalists. The science being mutilated pertains to a wide spectrum of health topics—including "facts" on the purported health hazards around us, including acrylamide (a chemical formed in cooking high-carbohydrate foods), breast implants, polychlorinated biphenyls, phthalates (plasticizers), aspartame (Nutrasweet), and Olestra (Procter & Gamble's doomed fat substitute).

In these instances—and so many more—outright blatant misrepresentations of the available science are made, health hazards that do not exist are claimed and picked up by the news media, and ultimately by lawyers intoxicated with the possibility of a cash reward in court from a corporate deep pocket.

The challenge for scientists—and only scientists can do this—is to educate the public about the realities of scientific processes. Why are scientists not accepting the challenge and countering the misinformation that is so common today? There are several reasons:

First, most scientists feel more comfortable in labs and classrooms than on op-ed pages and TV studios—and they have no real clue about how to go about challenging what they read and see. Second, in virtually 100% of cases where scientists have stepped forward to debunk the "carcinogen scare *du jour*," they have been subject to *ad hominum* attacks and labeled "paid liars" for industry. That threat of humiliation is enough to cause many to bite their tongues. Third, science these days has become so very specialized that the overwhelming portion of our country's scientists has a very narrow area of expertise. Those with a Ph.D., for example, in entomology, biology, veterinary medicine, or physics might possibly be as duped as the average citizen when activists orate about the supposed health threats from vaccines or trace levels of environmental chemicals.

These obstacles must be addressed and overcome because the consequences of the silence of the scientific community (interpreted as assent) are profound. The assault on science not only distorts health risks, but it threatens innovation and jobs—and our country's enviable high standard of living. Only scientists can effectively counter scientific misinformation. Scientists have a duty and obligation to do so, especially in the context of health sciences, bioengineering, and biotechnology.

41

Science, Ethics, and Human Destiny

John C. Polanyi, PhD

Nobel Prize Laureate, Department of Chemistry, University of Toronto, Ontario, Canada

Science, as you know, used to be termed "natural philosophy." It was a good description. We scientists philosophize about nature. We have been doing this to such effect in this century that we have transformed the accepted view of matter, energy, space, life, death, and almost the universe. Coincidentally, we have reshaped the world we live in, extending and enriching human life and, regrettably, furnishing the machinery for mega-death.

Though we do science so well—some would say too well—we do not explicitly know how we do it. We have no instruction booklets to offer. That is why we have our students work alongside us in our laboratories. They are required to imitate somebody who has the skill that they want to acquire, and they should imitate us on the occasions that we do it right.

This last statement implies that, though we have no rules for making discoveries, we do have some inkling of when we have made them. Nothing too surprising about that, you may say, it is paradoxical but not surprising. At least, nothing uniquely surprising. It is also true of composing music, returning a tennis ball, or flying an airplane.

For all of these skills, rule books, though they may exist, are absolutely inadequate; learning is by apprenticeship, and what is crucial in that apprenticeship is to be able to distinguish success from failure. It is also what is most mysterious, since it suggests that one must already have a skill in order to exercise it. It is a circular argument. It is a paradox. I have no answer for it. It worried Plato a lot. It should also worry infants, who are the most prodigious acquirers of skills, but the paradox does not worry them at all.

The same is true of my colleagues, who have a lot in common with infants. They spend almost no time puzzling over the fact that they cannot explain, nor therefore logically justify, the methods that they use. Most, if cornered, would say that the proof of the pudding is in the eating; science works. However, real philosophers, trained, as they are, not to understand things, find this answer unpersuasive.

G. Madhavan et al. (eds.), *Career Development in Bioengineering and Biotechnology*,
DOI: 10.1007/978-0-387-76495-5_41, © Springer Science+Business Media, LLC 2008

Nobody doubts that science works, they say; what they question is its claim to be *scientia*—knowledge. Just because it tastes like pudding does not prove that it is pudding. This contemporary skepticism deserves to be taken seriously. I believe that it says things that are true, but derives from them things that are false.

Science, say the deconstructionists, is accepted because eminent personages, who accord themselves the title 'scientist,' claim it to be true. This is incontrovertible; no one is qualified to check all the laws of science. We accept the major part of science that is beyond our expertise as a "social construct." In this, the deconstructionists are right.

They are also right when they go on to say that logic alone does not compel us to accept the truth of science. This is a way of saying that unequivocal proof does not exist. There is always room for equivocation. It is at this point in their argument, however, that the deconstructionists go off the rails. They conclude that, in the absence of proof, the proper attitude to science is "absolute skepticism." This is absolute nonsense.

Before I elaborate, let me say that what we are encountering here is a widely touted 'post-modern' view of science. Its proponents hope that you will accord it the dubious compliment of finding it "scientific." I decline. However, a wrecker's ball should not be taken lightly.

What is the flaw in the deconstructionists' belief in disbelief? There is a hint of it in my question. But let us consider more closely what it is that scientists do.

Usually, scientists make observations of a numerical sort. They plot these on a computer screen. The points that appear on the screen may suggest a pattern. But they cannot, as the deconstructionists rightly insist, demand a particular pattern. Only an infinite number of points can make a continuous curve and persuade you of a shape.

We are not going to wait for infinite time. What we do is get together with our colleagues and peers, and we agree that we have seen a pattern. This consensus is a product of logic informed by the conventions of the day. It is indeed a social construct. The post-modernist has nothing against social constructs, of course. Nor have I. In a moment, I shall be so bold as to hold up those of science as the best hope of mankind. What the post-modernist objects to is that these social constructs are dignified as science.

What has a social construct, they ask, to do with truth? The answer is everything.

First of all, let us note that this question would not be asked if the questioner did not believe in knowable truth. That I share this belief was evident in my earlier claim that in the pursuit of scientific truth, we depend upon our ability to distinguish success from failure. Clearly, I believe that success in achieving understanding is possible; what is truer can indeed be distinguished from what is less true.

Of course, we make mistakes. When we distinguish truth from falsehood we do so fallibly. This is important, and it is part of the post-modern thesis. The truth of science, we have already noted, is not guaranteed by logic. It is fallible. Put differently, seeing requires an instrument, which is a pair of eyes. But it requires more than a pair of eyes, and yet, we do see. This is why, in addition to the word *see*, we have the word *vision*. It describes what the human mind does with the points of light on the retina. It also describes what the scientist does with the points on the computer screen; through his vision he links them, one with another.

The marvel of science is precisely the aspect that makes the post-modernists despair; the fact that our vision is informed not merely by data, but by our experience of the world. To widen that experience, we call upon the experience of others—their ideas of what is fitting or differently stated, what is beautiful.

It is, in fact, the triumph of science that it is a social construct; not any old social construct, such as that stemming from feudalism or tyranny, but the most democratic social construct, called consensus. For the achievement of consensus implies respect for the experience of others. If—and this is our shared premise—the truth is knowable, it is hard to conceive of a better means for approaching it than this process of casting our net beyond logic to beauty.

It may be objected that a consensus, however wise, is a product of its time, whereas the truth is eternal. The objection is valid. It has not been my purpose to argue that what at some time in history is regarded as "science," is the truth for all time. It is a station on the way to truth.

Looking back over the route science has taken, I would not even wish to say that each way-station is closer to truth than the last. I would only say that the accumulation of experience has brought with it, in the long run, more profound insights. It is this same conviction that causes hundreds of thousands to devote themselves passionately to the enterprise of science. Their belief in that enterprise is the very antithesis of the absolute skepticism enjoined on us by the post-modernists. If that were accepted, it would be the death of science.

These remarks about the social construct that we call science have implications for the two other topics of this meeting: ethics and human destiny. A social construct of the sort that I have been describing implies a society of constructors. What makes this particular society, the society of science, so extraordinary is not only the fact that for centuries past, it has been an international society, though that is significant. Much more remarkable is its ethic, which requires that a shared goal, which is this approach to truth, be put ahead of personal advantage.

So "*does science, scientific knowledge, imply an ethical position, or is the knowledge ethical without values incorporated?*" That is a very difficult question. I am talking about the ethics that make possible discovery that makes possible the society of science. What do I mean by putting a shared goal against personal advantage? A scientist who did not believe that objectivity, so far as it can be achieved, takes precedence over self-advancement, would not belong in science. If he or she put these unethical ideas into action by, for example, falsifying data, such an individual would be banished from the community of science forever. The same, of course, is true not only of science, but of any scholarly pursuit.

That ethic is pretty rare, and that commitment to truth is at the same time a commitment to the fundamental tenets that we call "human rights." The truth is no monopoly of one race, religion, nationality or gender. The devotion to truth, ahead of self interest, implies tolerance of dissent, freedom of speech, and willingness to listen–all central virtues of a democratic society. And as we are beginning to realize, from such anguished situations as Kosovo and Darfur, from the acknowledgment of human rights, there should flow a sense of responsibility to safeguard those rights. Just as individual responsibility has flowered in society at large over the past decades, so it has among scientists.

It is no longer considered ethical to don a white lab coat and lead a life of monastic devotion to one's specialty. Scientists are citizens. Better yet, they are global citizens. Conspicuously, though too infrequently, they have acted as such. In their citizenship, they have to be clear about what it is that they have to offer. They are numerate and literate in science. They belong to an international community with a commitment to objectivity—so far as that lies within them—and with bonds of trust that stem from that shared commitment. But they do not have a unique path to truth. They are not obviously all-wise on any subject.

What do scientists have to contribute, then, in the shaping of human destiny? Their humanity, in the first place, as I have stressed. There have been conspicuous instances, in Nazi Germany for example, where their humanity has failed them, but it is more remarkable for resisting corruption, even in a climate of tyranny. You probably were first introduced to science as some impersonal and implacable machinery of proof. That is a fiction tyrants love because they want to co-opt that imaginary non-existent machinery.

The truth of the matter is, however, that science is a human activity; a social construct rather than a machine driven by undeniable logic. Scientists are people reasoning fallibly, as people do. Sometimes scientists make brilliant use of symbols to gather human experience into patterns. Many of the patterns are superficial, some illusory, all provisional; for it is the lot of every scientist to be proven wrong in some respect by another scientist. Yet each scientist, if he has genuinely sought the truth, has, in his stumbling progress, helped to find it.

That is a statement about the human condition, more than about "human destiny." What fundamental alterations in our circumstances could science bring? One thinks of longer life and greater leisure. They are, increasingly, a reality. But could there not also be in science a model for a more *decent* life and more *meaningful* leisure? Surely there could.

So far, I have not dwelt sufficiently on the miracle of the society of science even existing, a society in which virtually every member feels him or herself a participant. It is a real society, with leaders, laws, fellowship, and history. Amazingly, it has held together for centuries without formal government, without inherited privilege, and without violence, police, or prisons. It is sufficiently tolerant to actually invite dissent. Its heretics are not burned at the stake, but hailed as heroes.

It is not, of course, a society of angels. Personal ambition is a major driving force. But to a high degree, and that is the model that I thought of in the context of this wonderful phrase of human destiny which we have been offered, to a high degree, this force has been harnessed to a common goal. That goal is not venal or cruel, but the humane one of understanding.

If the society of science could, through its existence, its example, give us this as our common destiny, it would have made its greatest gift to mankind.

42

Motives, Ethics, and Responsibility in Research and Technology Development

Subrata Saha, PhD

Departments of Orthopaedic Surgery & Rehabilitation Medicine; Neurosurgery; Physiology & Pharmacology, State University of New York – Downstate Medical Center, Brooklyn, New York, USA

Pamela Saha, MD

Department of Psychiatry, State University of New York – Downstate Medical Center, Brooklyn, New York, USA

Bioengineering and biotechnology present exciting career opportunities that will have a profound impact on human life. These technologies allow us to transform our environment, improving significantly our prospects for health and longevity. We may even be able to transform ourselves, making us super human, enhancing our physical attributes and our intelligence. But this begs the question—do we want to do all of the things we may be capable of doing?

We may step back and look at the motives that lie behind our goals, desires, and dreams for our future. What is the best motive behind a decision to devote one's time and resources to bioengineering and biotechnology? We often claim altruism as a motive. This includes the desire to enhance the lives and health of others, and to improve the environment. It also includes the desire to improve the well-being of many people—not just the elite few, and to extend benefits and resources across the globe as economically as possible. Ultimately, altruism appears to be a good motive for seeking a career in bioengineering and biotechnology.

However, other motives can bring similar beneficial results. These motives can include the desire for profit, renown, or respect—or even the simple desire to experience the exhilaration of being first to make an important discovery. Interestingly, today, whether a person's motive is altruism or self-advancement, the question of whether the ends justify the means often arises. This problem is clear in the current debates over stem cells, where researchers claim the altruistic motive of curing Parkinson's disease, paralysis, diabetes, or other chronic illnesses.

G. Madhavan et al. (eds.), *Career Development in Bioengineering and Biotechnology*, DOI: 10.1007/978-0-387-76495-5_42, © Springer Science+Business Media, LLC 2008

What about the motivation to learn all that can be learned? This is the "need to know" motive that drives scientists and engineers to seek knowledge regardless of its usefulness. Today, most of us experience an overload of information. This overload is resulting in an existential crisis, leading to concerns over where science and technology are taking us. Neuroscientists, for example, suggest that the mind and the brain are one and the same, and that our thoughts, feelings, hopes, and dreams can all be reduced to biochemical reactions within neural circuits. While we desire gains from science and technology, we fear the meaning of their discoveries and outcomes, and wish to control their dissemination in our public schools. Creationists want evolution removed from the text books or to have a creationist version of human beginnings taught in science and engineering classes on par with the theory of evolution.

Bioengineers and biotechnologists may decide to determine their own identity. They may, for example, decide to lead the public to recognize the value of science and technology education, and take a strong position on how education should be designed and carried out. For instance, they may state clearly that creationism is not appropriate as material for science education, as it will undermine well-accepted principles of science.

Bioengineers and biotechnologists may also set goals in terms of achieving certain virtues. Multifaceted statesman Benjamin Franklin had at one time made a list of virtues that he intended to achieve, and kept a record as to his progress achieving them. Along similar lines, bioengineers and biotechnologists may develop and use a code of ethics to achieve virtues such as placing patients and society above their own interests.

Many pharmaceutical technologies and medical devices are too expensive for most of the world's population. We should strive to reach the higher goal of not only serving our employer or country, but serving the world as a whole to increase the quality of life in underdeveloped countries.

We want laws prohibiting certain kinds of research, and not just because the research violates human or animal rights or safety. We are uncomfortable with research that combines human genetic material with those of animals, even if the resulting tissues would be destroyed before any evidence of awareness would occur.

Should the "need to know" motive be restricted by what society can bear? Are the individual rights of the scientists to be bound by a societal need to slow the discovery of new information that challenges human identity, perceptions of animal rights, or ways the legal system should investigate or deal with criminal behavior? There has always been controversy over whether or not the mentally ill, brain damaged, or mentally retarded should be held fully accountable for criminal behavior. Now there is controversy related to new research indicating that certain brain scanning techniques might eventually determine criminal intent. This type of crime prevention in advance was visualized in Steven Spielberg's movie *Minority Report*. But should we continue our research in this area because we can? Are there things that we should deliberately decide to remain ignorant about?

Much has been written on the rationale behind moral decision making. Curiously, neuroscientific research has discovered that much of our moral decision making is emotional rather than rational. Our elaborate systems of ethical decision making are actually our brain's effort to confabulate an explanation for our choices after the fact. Research has indicated there are innate circuits in the brain that lead to automatic moral decision-making before we understand the reasons behind them. Not only is this an advancement in understanding ourselves as humans, but it likewise sheds light on how we may go about navigating our future judgments as they are applied to bioengineering and biotechnology.

Our innate gifts of moral decision making may require closer examination, as these are based on human reactions designed for life before today's technology existed. Today's humans are unique in that we not only have the opportunity to determine our future in terms of material things, but we even can determine the future of our very character, as we have the tools to understand ourselves more deeply than we have ever before. Bioengineers and biotechnologists of the future are choosing a career and development pathway that can mean so much more than simply building a better mouse trap. But how far we should go with the possibilities available to us requires us to adopt more than a utilitarian point of view with focus on material gains.

Science and Technology Policy for Social Development

Semahat S. Demir, PhD

National Science Foundation, Arlington, Virginia, USA

"If you need to understand it to make policy, you should turn first to people who are scientists and engineers for factual information."

—*Vinton Cerf*

Bioengineering and biotechnology are multidisciplinary fields where collaborative partnerships strengthen the technological outcomes and development. The fields, as such, have many of open-ended problems, and hence, opportunities. These problems can be complex, and sometimes better solved with groups of people from different disciplines, with different perspectives. Policy formulation, analysis, and implementation are such areas, where working at the interface of government, physical sciences, medical sciences, life sciences, law, and engineering, can be extremely beneficial for social development. Areas where bioengineers and biotechnologists have already been making an impact through governmental policymaking include science, technology, engineering, and mathematics (STEM) education; healthcare delivery; regulatory issues related to health care services; medical technologies, methods and devices; drugs and drug delivery; patent and intellectual property issues related to medical methods and devices; immigration issues; export and import issues related to medical and healthcare products; research and research funding administration for medical and life sciences; and research and development tax codes.

Government services can motivate engineers and life scientists to be life-long learners, to integrate technical knowledge at different disciplines, to explore and challenge their potentials, and to become the policy developers. It is hence an essential responsibility for all engineers interested in policy-making and government relations to focus on professional development and interpersonal "soft" skill development, most notably negotiation techniques, inter-cultural dynamics, and refined thinking. Instances where all of these skills play together include legislative expert witnessing, testimonies, and congressional/parliamentary communications. Though many processes such as legislative bills and resolutions appear

G. Madhavan et al. (eds.), *Career Development in Bioengineering and Biotechnology*, DOI: 10.1007/978-0-387-76495-5_43, © Springer Science+Business Media, LLC 2008

linear, their formulation, analysis, and implementation require non-linear thinking that can be developed through consistent practice. In short, the readership of this book is recommended to assume a proactive role in their career development, and to ultimately help enhance social processes, a responsibility that we all share as fellow professionals.

44

Medicine and Health Safety

Richard A. Baird, PhD

National Institute for Biomedical Imaging and Bioengineering, National Institutes of Health, US Department of Health & Human Services, Bethesda, Maryland, USA

Roderic I. Pettigrew, PhD, MD

National Institute for Biomedical Imaging and Bioengineering, National Institutes of Health, US Department of Health & Human Services, Bethesda, Maryland, USA

As medicine and healthcare have become more sophisticated, bioengineers have been called upon to play an increasingly important role in developing and improving technologies to predict, diagnose, and treat disease, and in accelerating the translation of these technologies to the clinical setting. These technologies have included the development of imaging and image contrast agents, minimally invasive image-guided interventions, and the *in vivo* and *in vitro* application of many imaging modalities, including magnetic resonance, nuclear, optical, ultrasound, and X-ray. Bioengineers have also developed many implantable devices and sensors for monitoring, stimulating, and improving organ function, as well as many naturally derived and synthetic materials for drug and gene delivery, tissue scaffolds, and tissue repair, regeneration, and replacement.

The increasing sophistication of medical technology has also dramatized the need for a greater understanding of the basic science underlying this technology, including the effects of electromagnetic and ionizing radiation on living tissues; the toxicity of drugs, probes, and imaging agents; the biocompatibility between materials and their biological substrates; and the long-term interactions between implantable devices and the biological systems into which they are introduced. In addition, a broad range of health safety concerns, while obviously important and beneficial for patients, have made the clinical translation of medical technology more difficult. These concerns place increasing design constraints on hardware, devices, and biomaterials intended for clinical usage. They have also spurred the development of new clinical safety standards and increased the regulation of medical technology.

Developments in basic and clinical science, particularly in genomics and molecular biology, are transforming the medical landscape and ushering in a new era of

G. Madhavan et al. (eds.), *Career Development in Bioengineering and Biotechnology*,
DOI: 10.1007/978-0-387-76495-5_44, © Springer Science+Business Media, LLC 2008

medicine. In this new era, healthcare is shifting from a curative paradigm, where the emphasis is on managing disease, to a preemptive paradigm, where the emphasis is on predicting and preempting disease using genomic information and molecular technologies. In this preemptive paradigm, identification of an individual's genetic inheritance and the molecular bases of disease are beginning to allow us to personalize medical care by predicting the disease predisposition of individual patients, tailoring medical treatment to these predispositions, and customizing disease management to achieve optimum medical outcomes. This new paradigm will ultimately reduce the burden of disease, as well as the personal and societal costs of healthcare, by detecting and treating disease at earlier stages and preempting more serious consequences. It will also help to address the challenges of health disparities by allowing diagnoses at the point of initial physician contact, and by tailoring therapy to the individual patient.

This transformation of the medical landscape will require the improvement of many existing medical technologies, and the further development of many new and emerging ones. Many biomarkers, measurable indicators of the physiological state of cells, tissues, or organs, have already been identified, and have been shown to be early indicators of health or disease. However, for medicine to become truly preemptive, there is a profound need to identify and validate new, molecular-based biomarkers that can be used to identify disease and monitor specific biologic processes and therapeutic interventions. Some of these biomarkers are present in blood and other bodily fluids. One example would be C-reactive protein, which is a marker for acute inflammation. However, in many cases, markers observed in bodily fluids do not indicate the site of inflammation in the body. This requires imaging agents and biomarkers of the target disease. One class of emerging imaging biomarkers are magnetically labeled macrophages. The migration of these "tagged" immune cells to different regions in the body (e.g., inflammatory sites in transplanted hearts undergoing rejection) can be followed using magnetic resonance imaging and used to detect inflammation or transplant rejection at an early stage. Recent studies have also demonstrated that many imaging biomarkers can be linked with therapeutic agents, allowing physicians to combine both diagnostic and therapeutic technologies.

The need to safely monitor molecular biomarkers in healthy individuals to anticipate and prevent disease, will require further development of quantitative analytical techniques, as well as substantial improvements in imaging technology. Current examples of bedside, or point-of-care (POC) devices, are often based on simple technical designs appropriate for performing single analyte analyses, such as measuring blood glucose. To fully realize the potential of biomarkers for preemptive medicine and personalized care, POC devices will need to have multiplexing capabilities; i.e., the ability to simultaneously test for the presence of large panels of biomarkers in a range of biological samples. Progress is being made with respect to devices capable of detecting a limited number of analytes, such as with cardiac markers. For example, POC devices are now available that can quantitatively measure creatine kinase-MB, myoglobin, and troponin I to aid in the diagnosis of acute myocardial infarction.

Other currently available POC devices are able to rapidly and accurately detect protozoan species that cause diarrheal diseases. The next generation of these devices, incorporating the latest advances in nanotechnology, biosensors, and microfluidics, is currently under development, and will be able to simultaneously detect many analytes, including drugs, small metabolites, and infectious agents. Although this technology is still primarily laboratory-based, the future transfer of this technology into clinical tools will play a critical role in realizing the vision of molecular medicine. But many challenges remain in developing platforms capable of pushing the limits of multiplexing with the necessary integration to allow for full and varied sample processing. One significant challenge is the development of a device robust enough for use in the home by the patient.

Empowering physicians to make bedside, or point-of-care, decisions using portable, self-contained diagnostic and monitoring devices for near-patient testing has the potential to profoundly impact healthcare delivery. As medicine becomes more personalized, with greater consumer participation in timely diagnostics and therapeutic delivery, the focus of medical care will inevitably shift from the large hospital-based setting to the clinic-based, and ultimately home-based, settings. To a greater and greater extent, we expect biomarker monitoring to occur remotely, with patients teleconsulting with their physicians to rapidly detect and diagnose disease or modify their course of therapy following a biomarker change. This shift to smaller settings will increase patient involvement and remove initial diagnosis and therapeutic planning from the traditional clinical and clinical support team. This will inevitably create further health safety challenges for bioengineers, as diagnostic and monitoring devices will need to be able to work in nonclinical environments without presenting a danger to the patient, yet maintain a simple interface for use by non-specialists or untrained personnel.

Finally, advances in medical technology and health safety concerns have placed increasing demands on both clinical and bioengineering education and research training programs. This has led to a revamping of the medical school curriculum, and restructuring of residency requirements to allow more time for training and research. Bioengineering departments are also working closely with medical schools to co-train bioengineers and medical students, and to partner bioengineers and physicians in multidisciplinary teams. This approach, which allows bioengineers to shadow physicians, gives both groups a greater appreciation of the unique challenges of the clinical setting. Bioengineering departments are also restructuring their own curriculum to provide not only a rigorous engineering education, but to train a new generation of researchers at the interface of the biological and physical sciences. These interdisciplinary scientists are being trained to communicate and work collaboratively with scientists in both the biological and physical sciences, and will have the knowledge and skills needed to effectively conduct research across these scientific disciplines. Collectively, these changes in engineering and medical education will help to train a new cadre of bioengineers better equipped to meet the challenges of the 21st century.

We commend these emerging challenges, opportunities, and paradigm shifts in medicine, and the corresponding new health safety challenges required to address them, to the attention of today's bioengineers. We hope that by fully embracing the above training opportunities, today's bioengineers will be ready to tackle these important clinical problems and to play a pivotal role in their solutions.

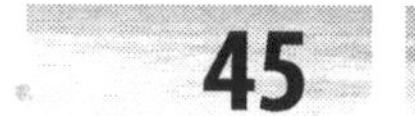

Patient Safety: Building a Triangle of Importance

T.K. Partha Sarathy, MD

Sri Ramachandra Medical University and Harvard Medical International Alliance, Chennai, India; Division of Academic Affairs, Gulf Medical College, Ajman, United Arab Emirates

Healthcare systems across the board have become increasingly complex over the years, and the expectations of healthcare providers, management, and administrators, not to mention patients, keep growing. While we appreciate the increasing levels of consumer consciousness, the numbers of medical malpractice-based litigations have reached staggering heights, with sometimes unreasonable, jury awards. Perhaps for this reason, and also because of the leaps made in the technology of medicine and care, the cost of healthcare has sky rocketed. But this has marginalized a good segment of the population from even seeking the basic levels of standard and quality care.

One might wonder why we should even bother to address this problem when the big governments are themselves unable to handle it. Unless this problem is addressed in every forum, and unless everyone in the healthcare industry, most notably engineers and technologists, also becomes involved in bringing healthy solutions, it will continue to elude us. Looking at this scenario, the three areas that one can immediately attend to include the *safety* of patients in healthcare systems, the *cost control and affordability* of medical care, and conformity to *bioethical standards.* This triangle of importance is certainly a demanding imposition on the healthcare facilities and the care givers. Obviously it is a lifelong, inter-generational, multidisciplinary team effort, and the engineers and technologists have a lot to offer in this game.

For instance, care givers' errors in judgment or medications, inappropriate practices, lack of effective communication with patients and their relatives, and failure to seek expert opinions will continue to generate litigation and increases in healthcare costs, and superficially it may appear that engineering and technology staff can do so little to help in these areas. But in the lines of the old adage, "an ounce of prevention is worth a pound of cure," engineers and technologists have enormous responsibilities in terms of appropriate design, development, and

G. Madhavan et al. (eds.), *Career Development in Bioengineering and Biotechnology*,
DOI: 10.1007/978-0-387-76495-5_45, © Springer Science+Business Media, LLC 2008

maintenance of biomedical diagnostics and therapeutics. Effective project management from the engineering and technology side will drastically help the healthcare system, in terms of reducing hospital down-time, the cost of operation, and patient spending. In this context, information technology and systems engineering have a lot to offer in developing cost-efficient, user-friendly, easy-access solutions to many of the day-to-day problems and challenges that the healthcare professionals and management personnel face.

Today, there are a multitude of software solutions available to the healthcare industry. But a good many of such solutions are expensive, and the operating costs add to the financial strain rather than reducing it. Where high-volume turnover is possible in big facilities, their use is quite efficient. But it adds considerable financial strain to smaller institutions working at the local community level. Several small, community-based sixty to seventy bed hospitals exist in almost every district or county, and most are not financially healthy. Every year we hear that many of these hospitals are folding due to financial crisis. Information technology and systems engineering have to carefully study this scenario and prepare exclusive technology solutions to suit such hospitals, keeping in mind that technical support may not be adequate or ideal. Engineers and technologists must help find and offer solutions to keep these financially strained hospitals afloat and help them cut down waste and improve productivity.

Further, patients come to a hospital to be cured of or helped with a disease or condition – not to substitute other health-related infections or transmissions. To this end, environmental engineering and safety is a vital area that is open for engineering innovation, starting with an appropriate waste management program and including facilitation of proper disposal of biodegradable and non-degradable wastes. Engineers and technologists should assume leadership in developing innovative, environmentally sound, and patient-friendly systems for assisting the healthcare system. The bottom line is patient safety, and the engineers and technologists involved should be consciously aware of the demands of the profession.

Modern bioengineers and biotechnologists work very closely with physicians, surgeons, and medical scientists in vital technology developments. In developing and using such technologies, they should pay attention not only to patient safety, but to the ethical soundness of such trials. The controversy surrounding embryonic stem-cell research is only too well-known. Although animal cloning has been successful, and the technical know-how of human cloning may well be known, engineers and technologists should be extremely careful and aware of the ethical issues.

I place these issues for the consideration and practice of engineers and technologists, with the hope that the healthcare system will further significantly benefit from their long-lasting contributions and innovations within the triangle of importance.

Design of Appropriate Medical Technologies: Engineering Social Responsibility and Awareness

Nigel H. Lovell, PhD

Graduate School of Biomedical Engineering, University of New South Wales, Sydney, and National Information and Communications Technology Australia, Eveleigh, New South Wales, Australia

Professionals undergo extensive training to perform a particular set of tasks. Engineers are trained to design, develop, and build. Bioengineers will typically develop technologies for application in the health and medical domains. It could be argued that the natural gravitation of an engineer is to over-design, and build into a product a rich feature set. However, if one examines what truly comprises a visionary engineering design, the key concepts are *simplicity* and *appropriateness.* The technology itself sub-serves and grows from identification and deep understanding of the medical problem–the solution is an approach that best meets social, economic, and health needs.

The technology, therefore, is a secondary consideration after identification of the problem. It follows that not all problems can or should be solved with engineering-based technology solutions. In this chapter, the concepts of broader professional responsibilities are explored as an integral part of the nexus between medical need, technology, and solution.

In exploring this nexus, it should be understood that the definition of health technology goes well beyond the traditional concept of an instrument or piece of equipment, but also encompasses any method used to promote health, or any approach used to prevent and treat disease and improve rehabilitation or long-term care. Technologies in this broad context could also include drug discovery and procedures, settings of care, and screening programs.

The concept of design of appropriate technology to meet a medical need is by no means new. Indeed, departments of health and other agencies world-wide have formed groups to manage and assess health technology. No matter the technology or region of the world, the four basic issues that relate to such technology assessments comprise need, effectiveness, safety, and cost. The participants that

G. Madhavan et al. (eds.), *Career Development in Bioengineering and Biotechnology*, DOI: 10.1007/978-0-387-76495-5_46, © Springer Science+Business Media, LLC 2008

must be engaged when performing such evaluations include researchers, engineers and developers, practitioners, administrators, patients and their support networks, and policy decision-makers.

One may think that the sole focus of an appropriate technology orientation would be on developing countries. Indeed, from a social equality viewpoint, this is the area that deserves primary attention. However, technology developers, assessors, care providers, legislators, and policy-makers from both developing and developed countries have huge responsibilities, no matter what the economic status of a nation.

In developing countries, it is self-evident that expensive, low patient throughput medical diagnostic and therapeutic equipment is of very low priority in comparison with rudimentary devices, tools, and techniques. Such issues as the containment of infectious disease through the provision of clean drinking water, sanitation, and access to essential medications for immunization, must naturally be the focus. However, regardless of the economic development status of a nation, there will remain the necessity for provision of primary care through community or regional health centers.

In developed countries, naturally, the critical issue of delivering quality health care to all citizens remains. The focus, however, moves away from managing acute and infectious diseases to managing chronic and complex diseases, such as congestive heart failure, chronic obstructive pulmonary disease, and diabetes. However, even in the most economically advanced countries, the need to deliver quality health care in an economically viable and sustainable fashion is of concern, as is the issue of continuum of care across the various health services. With increasingly aging populations, the demands on limited health resources will escalate over the next decade.

The US spends over 15% of its gross domestic product (GDP) on health. At a cursory level, this would appear to be a laudable outcome. However, in a somewhat ironic consequence of excessive health resource stratification, a large population within the US does not have access to the full spectrum of medical services. Other developed countries spend 7–12% of their GDP on health, but can provide socialized medical services to their entire population. In a very simplistic analysis, the reason such countries can provide crucial health services to their entire population is the tight intersection of triaged health service delivery models with appropriate technologies. Whether these systems will remain sustainable with an ageing population is another issue, and is open for debate.

While at the government level the US is now actively refocusing health care programs on primary care and on the promotion of point-of-care diagnostics, telehealth, and telecare initiatives for the remote delivery of health care and for the promotion of self-care, it remains the remit of the engineer to design the appropriate technologies for delivery to this health care sector. There are complexities involved in assessing and evaluating such technologies and health service delivery models. Do the approaches improve the quality of life of the patient? Are they cost effective? Are the approaches economically sustainable? Just because the questions are difficult (and, in some situations, potentially unanswerable until extensively trialed), it is by no means an excuse for the engineer to remove him or herself from the debate.

Telehealth services is just one example of a particular approach to care in which engineers can not only be engaged in the design of appropriate technology, but also do so in a way that influences and governs health delivery.

The path from birth, wellness, sickness, and to eventual death are essentially the human condition–there is no alternate route, and the path is traveled by all. Along the way, we want to experience the healthiest existence for ourselves and those who are close to us. A more altruistic view would be to achieve the same for society as a whole. Here is where the engineer plays a critical role. The engineers' remit must be to understand and appreciate the economic and social implications of the technology they are designing. This understanding must be made in the context of where the technology is planned to be deployed (which country, setting, or scenario). With this awareness, the engineer does not just design a product, but must engineer a social awareness and can indeed be one of the key players in enacting both health and social reform.

47

Ubiquitous Healthcare: A Fundamental Right in the Civilized World

Pradeep Ray, PhD

Asia-Pacific Ubiquitous Healthcare Research Centre, University of New South Wales, Australia

Dhanjoo Ghista, PhD

School of Mechanical and Aerospace Engineering, Nanyang Technological University, Singapore

Ubiquitous healthcare (u-Health) is the next step to electronic health (e-Health) that is becoming a major agenda in the goals of international development to help bridge the gap between developed and developing countries. Until recently, approximately 40% of the population in developing countries have not had basic healthcare. In this domain, computer communication technologies could play a major role, as per the recent e-Health communiqué from the World Health Organization (WHO) [1].

Today, e-Health, (understood to mean use of information and communication technologies locally and at a distance), presents a unique opportunity for the development of public health of uniform levels across communities. The strengthening of healthcare systems through e-Health may even contribute to the implementation of fundamental human rights by improving equity, solidarity, quality-of-life, and quality-of-care. WHO member states and groups of member states are drafting their own strategies for e-Health, and other organizations of the United Nations system have drawn up strategies for information and communication technologies in their domains.

Unlike many other e-Business areas, healthcare offers major challenges due to its high complexity involving the interplay of disciplines such as nursing, radiology, pathology, cardiology, and oncology. Hence, there is a strong need to develop an integrated e-Health architecture (covering primary to tertiary care), that involves a multi-disciplinary approach comprising the following:

- Clinical data management
- Decision support systems

G. Madhavan et al. (eds.), *Career Development in Bioengineering and Biotechnology*, DOI: 10.1007/978-0-387-76495-5_47, © Springer Science+Business Media, LLC 2008

- Technical, hardware, and network issues, including telemedicine
- Database structure and constraints
- Autonomous smart devices
- Standards for communication between healthcare providers
- Data exchange standards for healthcare devices
- Legal and ethical considerations
- Patient-oriented computing

There have been substantial developments in e-Health in the developed world to provide healthcare to remote communities such as the outback in Australia, and the Alaskan region in North America. Indeed, e-Health has the potential to alleviate, to some extent, the problems of healthcare related to an aging population and increasing percentage of chronic patients (especially in the areas of cardiology, diabetes, cancer, and HIV/AIDS) who require close attention. While these are some of the challenges in the developed world, it is interesting to see how technological developments in e-Health in developed countries can help under-developed nations.

E-Health is now making remarkable progress in some of the fast growing countries in the Asian region thanks to the already established broadband communication infrastructure that reaches nearly 75% of the population. These countries are now pioneering the paradigm of u-Health that makes healthcare facilities available to everyone in the country through broadband wired and wireless mobile communication infrastructures. Such a paradigm is also very useful in disaster situations that require the rapid establishment of healthcare services, despite the breakdown of infrastructure at the site of disaster.

The motivation and commitment to telemedicine in developing countries (facing problems of infrastructure and growing population) is often backed by a willingness to pay for systems that are expected to improve health outcomes and lower medical costs in the long run. Telemedicine services may be perceived as more of a necessity in developing countries than they are in the developed countries, resulting in a greater willingness among the former to change established methods of doctor-patient interaction and healthcare administration to improve the level and quality of medical services (especially in rural communities) at low costs.

The Telecommunication Development Sector (ITU-D), a UN body advising nations on telecommunication development, has been researching technological developments that have the potential to support telecommunication applications that are commercially viable or sustainable through other transparent financing mechanisms in rural and remote areas of developing countries that exhibit the following[2]:

- Scarcity or absence of reliable electricity supply, water, access roads, and regular transport
- Scarcity of technical personnel
- Difficult topographical conditions, e.g. lakes, rivers, hills, mountains, or deserts, which render the construction of wire telecommunication networks very costly

- Severe climatic conditions that make critical demands on equipment
- Low level of economic activity, which is consequently based on agriculture, fishing, and handicrafts
- Low per-capita income
- Under-developed social infrastructures (health, education, electrical power, transportation, etc.)
- Low population density
- Very high calling-rates per telephone line, reflecting the scarcity of telephone service.

These characteristics make it difficult to provide public telecommunication services of acceptable quality by traditional means at affordable prices while also achieving commercial viability for the service provider. However, it is felt that u-Health is probably one of the most compelling factors motivating political leaders of the world to find ways and means to invest in information and communication technologies.

Technological excellence is growing in low-income countries, which are developing their own expertise. However, for many, the benefits expected of e-Health have not yet materialized. Prerequisites for the successful integration of e-Health into healthcare systems include long-term government commitment based on a strategic plan, national awareness of the benefits of e-Health, and availability of skilled human resources.

Flow of health data no longer has any barriers. It is essential to evaluate and share experiences in order to develop individualized cost-effective models and, in particular, to understand the determinants involved in the adoption and sustainability of e-Health. Although this situation needs regulating, it is also an opportunity for faster and more comprehensive epidemiological surveillance. A global approach to handling data flow will help to promote standardization and pluralistic low-cost services, as evidenced by the establishment of a number of specialized international healthcare networks for the control of healthcare problems such as HIV-AIDS and pandemics. Bioengineers and biotechnologists are strongly recommended to make use of all the global opportunities in this arena for proliferating u-Health and enhancing public health.

References

1. *World Health Organisation (WHO) e-Health Report by Secretariat*, Fifty-Eighth World Health Assembly A58/21, April 2005
2. *ITU-D Focus Group 7 Final Report on New Technologies for Rural Applications*, April, 2005 www.itu.int/ITU-D/fg7/pdf/FG_7-e.pdf

Towards Affordable and Accessible Healthcare Systems

Xiaofei F. Teng, PhD

The Joint Research Center for Biomedical Engineering, Department of Electronic Engineering, The Chinese University of Hong Kong, Shatin, N. T., Hong Kong SAR

Yuan-Ting Zhang, PhD

The Joint Research Center for Biomedical Engineering, Department of Electronic Engineering, The Chinese University of Hong Kong, Shatin, N. T., Hong Kong SAR; Institute of Biomedical and Health Engineering, Shenzhen Institute of Advanced Technology, Chinese Academy of Sciences, China

The aging population is a worldwide phenomenon. In 2000, the estimated amount of population aged 65 or above was 421 million, and it is anticipated that the figure will reach 1.5 billion by 2050 [1]. Associated with this issue, the prevalence of chronic diseases is inevitable. This results in new and long-term demands on healthcare systems. If not cautiously prevented and managed, chronic diseases will be the leading cause of disability by 2020, and hence become the most expensive financial burden on our society. For example, according to the World Health Organization's 2005 global report, China will spend up to $558 billion over the next ten years due to premature deaths caused by heart disease, stroke, and diabetes, [2]. In addition, global health spending is continuing to outpace economic growth. *Per capita* health spending has increased by more than 80 per cent from 1990 to 2005, exceeding the 37 per cent rise in gross domestic product (GDP) per head [3]. Meanwhile, one billion people around the world, mainly in the developing countries and rural areas, demand access to affordable and accessible healthcare systems [4].

Andy Grove, Intel's chairman and former CEO, has used "the mainframe era" in the sixties as a metaphor to describe the current situation of healthcare [5]. As the dominating systems, mainframes in the sixties were localized in a dedicated environment and operated by skilled specialists. They were not affordable for the general public, and their quantity was limited. In terms of quantity and performance, mainframes were rapidly outstripped by personal computers in the eighties; and further

G. Madhavan et al. (eds.), *Career Development in Bioengineering and Biotechnology*, DOI: 10.1007/978-0-387-76495-5_48, © Springer Science+Business Media, LLC 2008

surpassed by personal digital assistants (PDAs) and mobile phones in the nineties by their size and cost. Could we imagine a similar trend for our healthcare systems to evolve from a centralized mainframe era to a distributed personal epoch?

To cope with rapid demographic changes, significant resource challenges, and inadequate accessibility of the current healthcare system, a multi-level, multi-dimensional, and standardized healthcare system has recently been proposed [6]. As illustrated in Figure 48.1, the system has a four-layer structure, namely personal healthcare system (PHS), home healthcare system (HHS), community healthcare system (CHS), and hospital healthcare information system (H^2IS). In contrast to traditional healthcare systems, which respond to a particular instance at clinics and hospitals, the new system features pervasive, preventive, and personalized healthcare. By introducing technologies that do not require labor-intensive operation, this system will allow users to self-monitor or even self-treat anytime and anywhere, despite the potentially shrinking healthcare workforce. The system, which provides a platform for prevention measures, is of particular importance in terms of healthcare cost-reduction for preventable chronic diseases.

Interdisciplinary expertise is necessary to put the system into practice. From the perspective of bioengineers, the focus will be on developing various types of front-end sensors and medical devices, especially miniaturized, integrated,

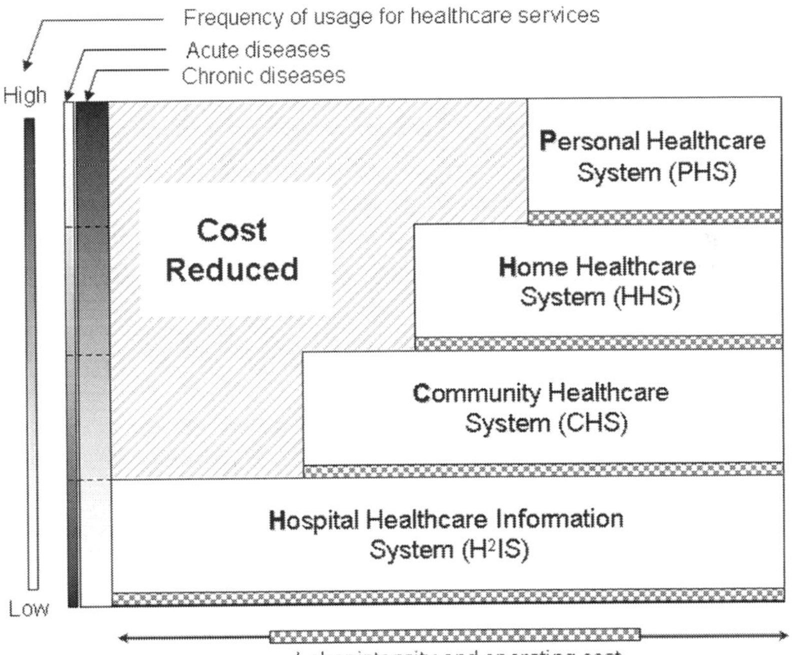

Figure 48.1 The four-layer structure of the proposed healthcare system (modified from [6, 7])

networked, digitized, and standardized (MINDS) medical devices, [6] which are required for the usage at the PHS and HHS levels. The state-of-the-art technologies, including miniaturization of electronics and sensors, lower power computing, e-textile materials fabrication, and information processing [8–11], have paved the way to pervasive and wearable monitoring of physiological conditions, providing the technological platform for personal healthcare. The growing variety of low-cost personal healthcare devices allows more people to monitor their own health conditions, greatly increasing the affordability and the accessibility of personal healthcare services. For example, a novel cuffless and wearable blood pressure measurement method has been developed in recent years [12–17]. The design uses low-cost optoelectronic sensors and application specific integrated circuits (ASIC) to ensure reduced size and the cost of the product [18, 19]. To measure blood pressure continuously without disturbing the user's daily activities, the sensors and circuit can be incorporated into a watch, shirt, or chair [15–16]. An example is shown in Figure 48.2. The image illustrates the use of the health-shirt (h-shirt) with Wearable Intelligent Sensors and Systems for e-Health (WISSH). E-textile materials were used in the h-shirt as electrodes and wires for electrocardiogram

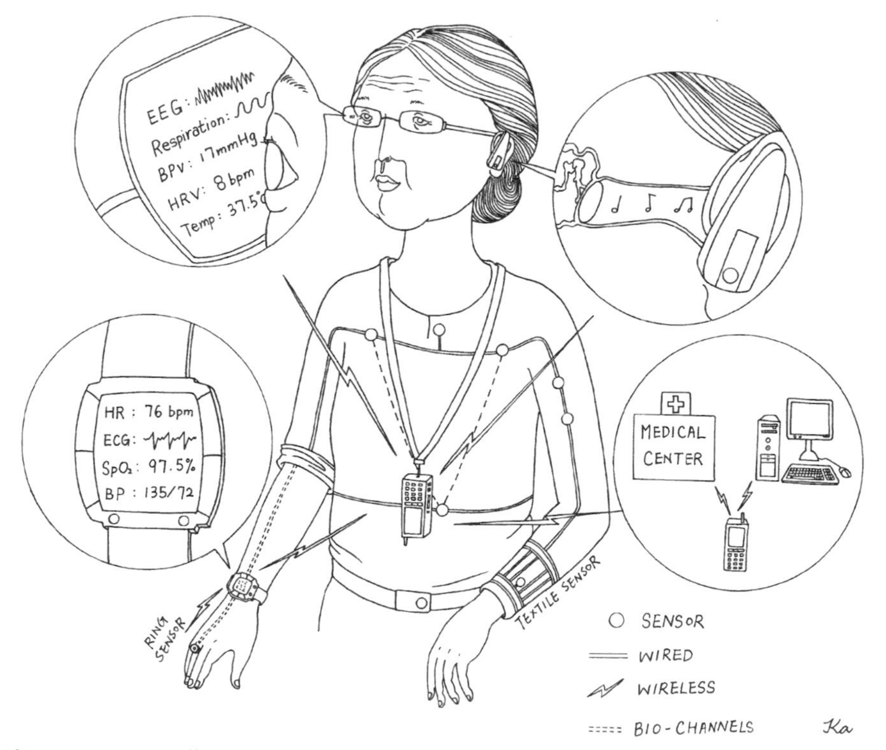

Figure 48.2 An illustration of the Wearable Intelligent Sensors and Systems for e-Health (WISSH) with the h-Shirt[16]. (Artwork: courtesy of Ms. Joey K.Y. Leung, The Chinese University of Hong Kong)

collection and signal transmission. Further development of this h-shirt, namely the Convertible, Universal Health-Suit (CUHS), would include both sensing functions and real-time bio-feedback mechanisms for therapeutic treatment and health management [16].

Moreover, significant advances in mobile computing, wireless communications, and network technologies, integrated with parallel advances in pervasive and wearable systems, have had a radical impact on modern healthcare systems. Mobile health (m-health) is propelling the evolution of e-health systems from traditional desktop "telemedicine" platforms to wireless and mobile configurations [20].

Healthcare is undergoing a paradigm shift, from a symptom treatment system to a preventive, early risk detection, and early treatment system. Given the complexity and changing nature of the healthcare environment, preventive measures and self-monitoring implemented at the personal level is crucial for building affordable and accessible healthcare systems.

It took us almost forty years to move from the mainframe to personal computing, but we cannot afford to wait that long for a similar transformation of the healthcare system. After all, we are talking not about mainframes, but about life and death.

Acknowledgement The authors would like to thank for the financial support from the Hong Kong Innovation and Technology Fund (ITF) and the industrial sponsors: Golden Meditech Company Ltd., Bird International Ltd., and Bright Step Corporation.

References

1. United Nations, Population Division, "World population prospects: the 2006 revision population database," www.esa.un.org/unpp, 2006.
2. World Health Organization, Global Report, *Preventing Chronic Diseases: Vital Investment*, 2005.
3. *Financial Times*, www.ftchinese.com/sc/story_english.jsp?id = 001012849, 2007.
4. A. Shah, "Global health overview," *Risk Group LLC Newsletter*, 2005. www.globalissues.org/health/overview/
5. B. Schlender, "Intel's Andy Grove on the next battles in tech: the IT visionary says tech needs to learn to think bigger," *Fortune*, May 12, 2003.
6. Y. T. Zhang, Y. S. Yan, C. C. Y. Poon, "Some perspectives on affordable healthcare systems in China," in *Proceedings of the 29th Annual International Conference of the IEEE Engineering in Medicine and Biology Society*, France, 2007.
7. C. C. Y. Poon, Y. T. Zhang, and Y. S. Yan, "Some perspectives on high technologies for low-cost healthcare: Chinese Scenario", manuscript prepared for *IEEE EMB Magazine* (special issue), 2007.
8. A. Lymberis, "Smart wearables for remote health monitoring, from prevention to rehabilitation: current R&D, future challenges," in *Proceedings of the 4th Annual International IEEE Conference on Information Technology Applications in Biomedicine*, UK, pp.272–275, 2002.
9. K. Hung and Y. T. Zhang, "Implementation of a WAP-based telemedicine system for patient monitoring," *IEEE Transactions on Information Technology in Biomedicine*, Vol. 7, no. 2, pp. 101–107, Jun. 2003.

10. G. Troster, “The agenda of wearable healthcare,” *IMIA Yearbook of Medical Informatics*, pp. 125–138, 2005.
11. J. Luprano, J. Sola, S. Dasen, J. M. Koller, O. Chetelat, “Combination of body sensor networks and on-body signal processing algorithms: the practical case of MyHeart project,” in *Proceedings of the 3rd International Workshop on Wearable and Implantable Body Sensor Networks*, pp. 76–79, USA, 2006.
12. Y. T. Zhang, C. C. Y. Poon, S. D. Bao, and K. C. Liu, “A wearable and cuffless device for the continuous measurement of arterial blood pressure,” in *Proceedings of the International Ubiquitous-Healthcare Conference*, Korea, 2004.
13. C. C.Y. Poon and Y. T. Zhang, “Cuff-less and noninvasive measurements of arterial blood pressure by pulse transit time,” in *Proceedings of the 27th Annual International Conference of the IEEE Engineering in Medicine and Biology Society*, China, pp. 5877–5880, 2005.
14. C. C. Y. Poon, Y. T. Zhang and Y. B. Liu, “Modeling of pulse transit time under the effects of hydrostatic pressure for cuffless blood pressure measurements,” in *Proceedings of the 3rd IEEE-EMBS International Summer School and Symposium on Medical Devices and Biosensors*, U.S.A., pp. 65–68, 2006.
15. C. C. Y. Poon, Y. M. Wong and Y. T. Zhang, “M-Health: the development of cuff-less and wearable blood pressure meters for body sensor networks,” in *Proceedings of the IEEE/NLM Life Science Systems & Applications Workshop*, NLM/NIH, USA, 2006.
16. Y. T. Zhang, C. C. Y. Poon, C. H. Chan, M. W. W. Tsang, and K. F. Wu, “A health-shirt using e-textile materials for the continuous and cuffless monitoring of arterial blood pressure,” in *Proceedings of the 3rd IEEE-EMBS International Summer School and Symposium on Medical Devices and Biosensors*, USA., pp. 86–89, 2006.
17. X. F. Teng and Y. T. Zhang, “Theoretical study on the effect of sensor contact force on pulse transit time,” *IEEE Transactions on Biomedical Engineering*, Vol. 54, no. 8, pp. 1490–1498, Aug. 2007.
18. A. Wong, K. P. Pun, Y. T. Zhang, and K. Hung, “A near-infrared heart rate measurement IC with very low cutoff frequency using current steering technique,” *IEEE Transactions on Circuits and Systems I*, Reg. Papers, Vol.52, pp. 2642–2647, Dec. 2005.
19. A. Wong, K. P. Pun, Y. T. Zhang and K. N. Leung, “A NIR CMOS preamplifier with DC photocurrent rejection for pulsed light source,” in *Proceedings of the IEEE Biomedical Circuits and Systems Conference*, U.K., 2006.
20. R. Istepanian, E. Jovanov, and Y. T. Zhang, “Introduction to the special section on m-health: beyond seamless mobility and global wireless health-care connectivity,” *IEEE Transactions on Information Technology in Biomedicine*, Vol. 8, pp. 405–414, 2004.

49

From War to Law via Science

John C. Polanyi, PhD

Nobel Prize Laureate, Department of Chemistry, University of Toronto, Ontario, Canada

In mythical times, a minor God by the name of Prometheus is supposed to have stolen fire from his boss, Zeus, giving it as a gift to man. Zeus, not to be outdone, sent another disruptive gift to Earth: a gorgeous woman named Pandora. She brought her famous box, from which emerged the welter of human creativity we call science and the arts.

The Greeks of 3,000 years ago who told this tale saw the peril that would come from such gifts. They also saw mankind's chief hope of survival: then, as now, it lay in imagination. Not too long ago, (in terms of millennia), in 1945, the God Prometheus returned to Earth, bearing a new gift of fire. This time, it did not merely illuminate the dark of our caves. Through the instrument of atomic fission, and the still greater one of atomic fusion, it liberated the ultimate energy: the energy that binds matter together. The caves themselves became combustible.

With the caves gone, there was nowhere left to hide. That is the central truth about our world. We have no place that is not visible to our enemies, so rather than hide from them, we had better engage them—talk and listen to them. Our hope lies in Pandora's box. Through science and the arts, humankind can learn.

In the very same year of 1945, in the aftermath of their agony, the nations of the world joined in signing the Charter of the United Nations (UN), one of history's greatest acts of imagination. Today, that charter, in principle and ultimately in practice, is law for all 192 nations. At its core lies the provision that "all states refrain from the threat or use of force" against others. A state can only use force for self-defense under circumstances defined and authorized by the UN.

Iraq, let me say, did not meet those conditions. But even if the world does not change overnight, the right to use force has devolved, under the agreed terms of the UN Charter, from individual nations to the collectivity that now speaks, however imperfectly, for mankind: the United Nations. For the first time in human history, war had been made subject to law. This epoch-making change has taken place only in principle. The great ship that is history changes course slowly. But it cannot fail to respond to its rudder, the freely expressed will of humankind. The

G. Madhavan et al. (eds.), *Career Development in Bioengineering and Biotechnology*,
DOI: 10.1007/978-0-387-76495-5_49, © Springer Science+Business Media, LLC 2008

reason is that the principle underlying the UN Charter is clear and correct. It is based on respect for the most fundamental of human rights, the freedom to live without fear.

Physics is not such a vastly different system of thought from that which informs the UN Charter. It, too, blends logic with aesthetics—what is sensible with what is thought to be right. When the greats of physics meet to sign a new declaration in favor, for example, of relativity or quantum mechanics, that too is only a change in principle. Decades of debate and insistent pressure from the new generation are needed before practice catches up with theory. But none among scientists discounts the importance of new principles. The same applies to the world leaders contemplating a transformed world. Principles matter, since ultimately those who govern need the consent of the governed, who care deeply about principles.

Declarations of principle, when they resonate in many minds, shape history. King John of England did not realize when he signed the Magna Carta in 1215 that he was doing anything different from earlier kings—Henry I, Stephen, and Henry II—who also signed charters that made promises. But the Magna Carta's time had come. Its promise of the fundamental right to freedom from arbitrary arrest has been the battle cry of citizens ever since.

But why are you hearing this from a scientist? Because science embodies the values we seek to build a world governed by law. I do not mean to claim that scientists are paragons of virtue. I am only saying that science is a civilized pursuit. And what is that pursuit? What is science? Not the collection of facts, but the establishment, through open debate, of new principles that command wide acceptance. This may seem to be a peculiar description of science. After all, you might say, scientists do not have to formulate the laws of nature, they merely discover them. But that is a superficial view.

Force equals mass times acceleration was a proposition made by Isaac Newton, not by God. The proof that the Creator did not use that equation is that, on close examination, it is wrong. It only applies to objects of large mass. It is altogether misleading when applied to subatomic particles. Why is it so useful then? Because most of us can go through an entire day without consciously encountering a subatomic particle. Newton's equation works superbly for everyday things, like apples and planets. How, then, did Newton's equation gain acceptance? New ideas have a hard time being accepted. Conservatism—fear of the new—applies in science as it does in politics. The science establishment, abetted in the 17th century by the Church, had a vested interest in pre-Newtonian mechanics.

It is here that science has something to teach us. The beauty of the new insight, explaining so much so concisely, captured scientists' imagination. Speaking figuratively, an international jury of scientists voted overwhelmingly in favor of the new view, overruling the powers that be. The great ship of science altered course.

I revert to this earlier image intentionally, to stress that scientists, in common with the rest of society, adopt new views through a gradual process of consensus. There is no magical moment of "proof" when all embrace a new orthodoxy. There is always room for dissent. Science is exemplary is in its handling of dissent. Since every accepted view in science was at one time a minority view, scientists feel obliged to treat minorities with respect. Dissent, they know, is vital to the development of thought, so they go so far as to encourage it.

That, to put it simply, is why we have scientific conferences. At our meetings, we judge the worth of dissenting arguments, not on the basis of the rank or race, color or creed of the dissenter, but on universal principles of good sense and beauty. Argument in science is passionate—a lifetime's effort may be at stake—but civility prevails. Though feelings run high, the resort to force is unthinkable, since the battle is for people's minds. So it is, if we would but recognize it, in the world at large. This profound re-consideration of the role of force in human affairs should not lead us to reject the use of force under all circumstances. Nor does the Charter of the UN outlaw force. It insists, however, that the use of force be sanctioned by the world community, and it requires that force be used demonstrably as a last resort, and to the minimum possible extent. What, then, should we be asking a future regime of international law to do?

Before addressing that question, we should be aware how much we are already asking international law to do, through such instruments as the UN, the International Monetary Fund, the World Trade Organization, the International Atomic Energy Agency, the Atmospheric Test-Ban Treaty, the Nuclear Non-Proliferation Treaty, the Convention on the Prevention and Punishment of Genocide, the Convention Against Torture, the International Tribunals on Yugoslavia and Rwanda, the International Criminal Court, the Kyoto Protocol on Climate Change, and so on. This is a formidable list. You would think that the world was a highly civilized place if you did not know that not all of these agreements are accepted by all countries, nor are all honored by those nominally accepting them. But it remains true that a body of international law is emerging; that none of it existed before 1945; and that a huge amount remains to be done, inescapably and inspiringly, to build upon this basis. For it is the only direction that offers hope.

There is no limit to how far we can go in building international law; it is within human capability to go as far as needed. This takes me back to where I began: ancient Greece. The Greeks were well acquainted with killing and revenge. In their plays, they symbolized the recourse to force as arbiter of disagreement. It ran through play after play—the Greek tragedies—as it has run through human history. But there came a pivotal moment in a particular Greek play, *The Oresteia* by Aeschylus, in which the cycle was definitively broken.

The stage was awash with blood (think of the year 1945), and on it stood Orestes, who had, in the name of justice, killed his mother, since she in turn had killed his father. Next in line to be killed was Orestes himself, now a murderer. Force served unendingly as an invitation to further force. At this point, the Goddess of wisdom, Athena, intervened. She did something without precedent: She appointed a court of citizens to make a decision that all would abide by. In the play, that court was the audience. At this point, the responsibility for what took place on the stage shifted dramatically to the audience; the collectivity in place of the individual.

Athena speaks (as translated by Ted Hughes, the late British poet laureate):

"Citizens of Athens!
This is the first case of homicide
To be tried in the court I have established.
This court is yours,
From today every homicide
Shall be tried before this jury
Of twelve Athenians.
Here my laws shall stand
Unchanged through the hours of the days.
And awe, that humbles the heart
Shall keep the pride of Athenians in check.
Protect this court
Which will protect you all
From the headstrong license of any man's will.
In this court you have a fortress
Over the peace of men and their families."

With that, the audience ceased to be witnesses to a play, becoming instead players. That is the most hopeful thing that we can do. It is this collectivity that gives force to law. It is this collectivity alone that can break the endless cycle of grievance and revenge. Individual will becomes subject to collective will; war becomes subject to law.

50

Science and Technology for Sustainable Well Being

Rajendra K. Pachauri, PhD

Intergovernmental Panel on Climate Change, Geneva, Switzerland and The Energy and Resources Institute, New Delhi, India*

Economic progress achieved since the advent of industrialization has resulted largely from advances in science and technology (S&T). Yet even as society benefits from S&T through choices that we have come to take for granted, decisions on its future are increasingly being questioned and scrutinized. The current path of economic growth deviates from the objectives of sustainable development. It is not only society at large, spearheaded by leaders of public opinion, that is expressing concerns, but also the scientific community itself, which is looking for ways to promote the sustainable well-being of all humanity.

This microscopic analysis of science and its applications emanates from several valid concerns: the role of science in the development and extensive use of lethal weapons; the continued existence of widespread poverty, with over a billion people in the world remaining virtually untouched by the benefits of modern S&T; and the threat of serious environmental externalities from unprecedented levels of production and consumption of goods and services.

A meaningful discussion of S&T solutions to contain war, terrorism, and heinous crime cannot be included in this limited space, but the other two issues deserve elaboration. The distinguished economist Kenneth Boulding, a rare intellectual far ahead of his time, pointed out that two centuries earlier, the difference in average income between the poorest country in the world and the most prosperous was no more than 1:5. When he expressed this concern 30 years ago, he estimated it as being 1:50. Income and wealth disparities are even sharper today. Despite astounding progress globally, the S&T gap between rich and poor nations is, ironically, wider now. If this growing chasm is not bridged, fissiparous tendencies will inhibit and even reverse prospects for enhancing human welfare.

* Nobel Peace Prize Receiving Organization

G. Madhavan et al. (eds.), *Career Development in Bioengineering and Biotechnology*,
DOI: 10.1007/978-0-387-76495-5_50, © Springer Science+Business Media, LLC 2008

Unfortunately, the global community has failed to bring technological opportunities and skills to underprivileged and impoverished communities across the globe.

The challenge of widespread worldwide poverty has typically been addressed through doles and handouts as convenient but largely ineffective palliatives. Seldom have programs in this area created avenues for applying modern S&T to develop local skills and capacity, which alone can generate income and employment on a sustainable basis. Creating opportunities for the productive application of S&T by the most dispossessed communities of the world is a task that engineers, scientists, and policymakers must embrace with urgency.

Among the negative externalities created by human activities, the cumulative emissions of greenhouse gases have had by far the most serious consequence in the form of global climate change. Cuts in emissions of these gases require technological initiatives to stabilize the concentration of greenhouse gases. Because the impact of climate change will continue for centuries, adaptation measures will also require the timely application of S&T. However, these will not take place in a policy vacuum. Regulatory and fiscal measures will have to be put in place by governments, facilitated if necessary by multilateral agreements to trigger the development and application of appropriate technological solutions.

The agenda for the global scientific and engineering community is very clear. We must recognize and evaluate the most critical impediments to the sustainable welfare of human society, including various threats to human life and global peace, disruptions in the delicate balance of Earth's natural systems, and the growing gap between rich and poor. These three sets of conditions are intimately interlinked, requiring a coordinated approach to solve them. Scientists and engineers must work with decision-makers to devise rational policy measures that mobilize desirable responses in the form of development and deployment of suitable S&T solutions in these areas.

51

Nonviolence for Technocrats

Arun M. Gandhi

M. K. Gandhi Institute for Nonviolence and University of Rochester, Rochester, New York, USA

When I was asked to write some words of wisdom for a book on technology, my first reaction was, "How can the philosophy of nonviolence interest technocrats?" Then it struck me, why not? Technologists are human too, right? For a split second, I found myself falling into the common trap of looking at nonviolence simply as a tool for the peaceful resolution of conflict. I quickly reminded myself that nonviolence is much more than that: it has grave implications for all aspects of human life.

In a very important way, nonviolence is like an iceberg. It hides more than it reveals. We assume that because we do not fight we are nonviolent, but this is not true. Our vision of nonviolence is limited by our narrow perspective of violence, and since both are correlated, we can understand the depth of nonviolence only after we have plumbed the depths of violence. This requires a personal commitment to conduct an honest introspection. As a young boy of twelve, what my grandfather, Mohandas Karamchand "Mahatma" Gandhi, made me do was to build a genealogical tree of violence with two branches – *physical* and *passive*. Physical, of course, is where we exert active, physical force upon others, while passive is the kind of violence that is committed quietly without any force being exerted. I was asked to use a moral yardstick – *will the actions or the inactions that I contemplate taking harm or hurt someone in any way directly or indirectly*? If the answer is yes, then your action is violent. It could be anything from wasting resources to over-consuming resources. This form of candid introspection blew my mind when I saw that within a few months, I filled a whole wall in my room with acts of passive violence. Until then, like most others, I lived in the illusion that I was nonviolent.

It was then explained to me that the more insidious passive violence, committed indiscriminately by all of us against each other, ignites physical violence because it generates anger in the victim. This lesson, like no other, made me understand the depth and breadth of violence. Since passive violence causes the victim to become angry, the second lesson in my initiation into the philosophy was to understand the power of anger, and how to use this energy effectively and constructively. I did have a lot of pent-up anger seething within me because of the

G. Madhavan et al. (eds.), *Career Development in Bioengineering and Biotechnology*, DOI: 10.1007/978-0-387-76495-5_51, © Springer Science+Business Media, LLC 2008

humiliation and prejudices [passive violence] that I suffered during the South African racial apartheid. Even at that young age, I saw visions of eye-for-an-eye justice. Grandfather had a unique response to anger management too.

Most psychiatrists and psychologists today emphasize the need to get anger out of one's system – yell at nature, punch pillows, or simply walk it off. This, grandfather believed, was temporary relief. If one does not address the issues that caused the anger in a manner that would satisfy everyone concerned, then, very soon, we would tire of letting off steam and blow up. My grandfather used the analogy of electricity to explain the power of anger. First, he said, anger is to human beings what a trip switch is to the electric circuit. When something is wrong, it trips the switch, cutting the flow of electricity to avoid a major disaster. Anger, similarly, is a form of internal trip switch that warns us of impending danger. Whether we take that warning signal to blow up or beat up on someone, or use it as an opportunity to stop and ponder over what an appropriate response would be, depends largely on how dominated one is by the culture of violence. The more one believes that effective control over people can be exerted only through fear, the more one subscribes to the culture of violence. When we shout, scream, or yell at people, or in extreme cases become physical, we seek to control through fear, instead of controlling through love and respect, as one would in a culture of nonviolence. Control through fear is quicker, but it does not last unless one is able to continuously exert fear over the person, whereas control through love and respect is long-lasting and mutually beneficial.

In another sense, anger is also like electrical energy, which is very useful when used intelligently, and deadly and destructive when abused. It is important for us to learn how to channel anger as we do electricity so that we can use the energy generated by anger intelligently and effectively. Grandfather taught me how to write an anger journal with a difference. Instead of simply pouring my anger into the journal, my journal was written with the intention of finding an equitable solution to the problem that caused the anger. I then had to commit myself to finding a solution. When grandfather was physically beaten up by some white rowdies in South Africa, he refused to file charges against them because he said "they will not learn the lesson that this form of violence is evil." He told them, "I am appealing to your conscience to show you that violence destroys the perpetrator more than it can destroy the victim."

The first two steps laid the foundation for understanding the rudiments of nonviolence. Now I want to dwell on two significant aspects of the philosophy which, I believe, would be more relevant and meaningful for professionals and those who deal with the production of material wealth and goods. These two concepts from Mahatma Gandhi's philosophy of nonviolence are referred to as *trusteeship* and *constructive action.*

What does trusteeship mean? In simple terms, it means that the talent that one acquires through education or inheritance is, in capitalist materialist society, considered one's personal possession. We acquire this talent or inherit it and exploit it for our own personal gains. Gandhi looked at life from the moral rather than the material perspective. As a result, he concluded that one really does not own the talent, but that one is simply a trustee of the talent and, therefore, one should be willing to use the talent for others as much as one does for oneself. In other words, instead of being selfish and materialistic, we should be compassionate and moral. One may say that one does share when contributing to charity but, in the Gandhian perspective, there is a distinction between sharing out of pity and sharing out of compassion.

To explain this in simple terms, when we encounter a homeless, hungry person on the streets, our normal response is to give the person a dollar or two to buy some food. What we are saying to the person through our actions is 'take this money and get lost. I hope I never see you again.' That is giving out of pity, because it is quick and requires no commitment. However, the effect on the person receiving the charity is negative. It reinforces the belief that the poor and the destitute are worthless human beings who will never achieve anything. Anyone living in poverty and/or oppression loses his or her self-respect and self-confidence, and society's approach to their problem convinces them they have to live a life of dependence.

On the other hand, when one gives out of compassion, it means getting involved and making a commitment to help the poor regain their self-respect and self-confidence so that they can fend for themselves. In 1970, in the over-crowded city of Mumbai (then, Bombay), India, there were millions who were homeless and destitute and living on the sidewalks. There still are just as many, if not more.

However, my friends and I decided to use Mahatma's concept of trusteeship and constructive action to help, in a small way, find a resolution to this crisis. We had to approach this problem with utmost humility. The tendency for those of us who are successful in life is to believe we have a solution to all of world's problems, and we begin to dictate to the homeless what should be done. The fact is, we have never been homeless, so we really did not know what it meant to be in poverty.

We assembled together about 700 people who were willing to participate in this experiment. We discovered that most of them worked at odd jobs, and some as day laborers, and barely earned enough for a living. While the men lived in the city, their families languished in the village. Once we learned from them about their problem and all agreed that economic stability was the first criteria, we realized that we had just two options. Either we go to a foundation and get a big grant to build an economic base for them, or we challenge them to raise the seed money required for this purpose. We ruled out the first option because clearly that would give them the impression they could ask for whatever they needed, and someone would make it happen. This was dependence of another sort, and went against the principles of nonviolence. We resorted to the second option and challenged them to work harder or make some sacrifices so that they could collectively save a coin every day and raise the money for the economic infrastructure. It was difficult for people who did not know where the next meal was going to come from, but for their own good they had to do it. They readily took up the challenge and surprised themselves as much as they surprised us when, by the end of 1971 – almost 19 months after the program was conceived – they collected the equivalent of $11,000 in Indian rupees.

We had learned from them that they came from arid lands about 200 miles south of Mumbai where agriculture was impossible. For several generations, most of them made a living making cloth on hand operated looms. Then the gigantic textile industry in Mumbai ran them out of business. Since they had some background in the production of cloth, we decided to use this talent by setting up power-operated looms in their village. The money they had collected was just enough to buy ten second-hand power looms and erect a little corrugated iron shed. When this modest factory was ready, we sent about seventy of them back to their village to run this factory under our guidance and make it profitable. We had to guide and train them through planning, production, and marketing. When they felt confident enough, we handed over charge to them. Over the years, they worked hard and expanded the factories so that today they have four units with more than 600 power looms, and all the seven hundred-odd people who contributed to the capital fund are now living in their village earning more money than they did, with a much better lifestyle for their families. The idea of putting aside a coin every day had unexpected consequences. It taught them the value of saving money. Consequently, by 1978,they had enough savings to start a cooperative bank – the Sangli Jilla Kranti Weaver's Cooperative Bank – in Mumbai, which now has seven branch offices and total assets worth $2 million. Through this bank, they help poor people get small loans to start a business of their own.

Mahatma Gandhi's philosophy of nonviolence requires that we recognize that not all people have the same up-bringing or the same opportunities. For one reason or another, some people are marginalized and they do not know how to get up. The capitalistic/materialistic societies around the world are based on individualism. They tend to make people selfish and self-centered. This has seeped so deeply into the human psyche that we prod our children to be successful in life by any means. "Do not think about others. Set your own goals and get to the top," is the kind of attitude we teach them so that they turn into selfish, self-centered adults. The system is unfortunately based on the jungle law "survival of the fittest." This is neither civilization nor the way to build a cohesive community.

The culture of violence requires that all negative attitudes, thoughts, and desires take control of our lives and our relationships. Consequently, hate, prejudice, suspicion, greed, and the like lurk just under our skin, and, when faced with any form of challenge, they take over our reactions. The salvation of humanity lies in replacing this culture of violence with a culture of nonviolence, which simply means allowing positive attitudes and emotions to supersede and suppress the negativity in us. Nonviolence means showing greater, love, respect, compassion, understanding, acceptance, and appreciation for fellow humans and all other forms of life. It is only when we achieve this transformation that society can truly be considered to be civilized and progressive.

52

Humanistic Science and Technology for a Hunger-Free World

M.S. Swaminathan, PhD

World Food Prize Laureate and UNESCO-Cousteau Ecotechnie Chair, M.S. Swaminathan Research Foundation, Chennai; Rajya Sabha, Parliament House, Government of India and National Academy of Agricultural Sciences, New Delhi, India; United Nations Millennium Development Goals, Project Task Force on Hunger, USA; and Pugwash Conference on Science and World Affairs[†]*, Pugwash, Nova Scotia, Canada*

Hunger is the extreme manifestation of poverty. To establish the actuality that lasting peace is possible only if we eliminate hunger, I quote George Hebert:

> "Sweet Peace, where dost thou dwell?
> I humbly crave.
> Let me once know.
> I sought thee in a secret cave.
> And ask'd if Peace was there.
> A hollow wind did seem to answer. No:
> Go seek elsewhere.
> At length I met a rev'rend good old man:
> . . .
> Whom when for Peace
> I did demand, he thus began:
> "Take of this grain, which in my garden grows.
> And grows, for you;
> Make bread of it: and that repose
> And peace, which ev'rywhere
> With so much earnestness you do pursue,
> Is only there."

[†] Nobel Peace Prize Receiving Organization

G. Madhavan et al. (eds.), *Career Development in Bioengineering and Biotechnology*,
DOI: 10.1007/978-0-387-76495-5_52, © Springer Science+Business Media, LLC 2008

The same idea can be found in the following words of the Roman philosopher Seneca, who was convinced at the beginning of the first millennium that where hunger rules, peace cannot prevail, "*A hungry person listens neither to religion nor reason, nor is bent by prayers.*" In spite of the recognition that food and drinking water constitute the first among the hierarchical needs of human beings, currently over a billion children, women, and men go to bed partially hungry or malnourished each night, and access to safe drinking water is increasingly becoming a luxury in most developing countries. International economic inequity has been the prime challenging force that has debilitated or demoralized many past international missions from meeting their set targets in hunger reduction.

Since the onset of the Industrial Revolution in Europe, technology has been a major source of economic inequity among nations and among communities within countries. If technology has been a cause of economic and social inequity in the past, today we have the opportunity for making science and technology a strong ally in the movement of social, gender, and economic inequity. As an example, modern information technology provides this opportunity. Knowledge and skill empowerment can now be achieved at a fast pace. However, the technological and skill empowerment of the poor cannot be achieved through programs designed on the basis of a patronizing top-down approach. The information provided should be demand-driven and need-driven, and should fuel the knowledge revolution for ending economic and gender inequity.

The accomplishment of these tasks requires considerable inter-disciplinary technical, managerial, and financial resources. Scientists at the International Peace Research Institute, Oslo, have studied the causes of armed conflicts in the last fifteen years, and have found that violent conflicts in most cases can be traced to economic rather than ideological differences. They have hence suggested that investing in agriculture and biotechnology, which helps to promote food and livelihood security in many nations, is an effective strategy for preventing future wars, eradicating poverty, diminishing environmental degradation, and reducing violence. Unfortunately, even now, far too high a proportion of the national gross domestic product is being invested on arms and military equipment, as compared to programs designed for poverty eradication and meeting basic needs such as the food, water, and health of underprivileged sections of mankind.

We now have a huge stockpile of scientific discoveries, engineering revolutions, technological innovations, and entrepreneurial track records. This stockpile is more than adequate to help all nations provide every adult human being a healthy and productive life, and every newborn child a happy future. It is therefore a distressing commentary on our political, social, and spiritual value systems that the number of children, women, and men living in hunger and poverty exceeds the entire human population of the year 1900.

To inject some hope into the situation, the formidable power of science and technology can benefit humankind only if we know how to temper it with humanism. We do indeed have the capacity to achieve universal freedom from hunger, if we bring about the requisite fusion among political will and action, humanitarian values, ethics, inter-disciplinary skill sets, and people power. Let us hope that the progress in bioengineering, biotechnology, and information revolution will bring forth a new yield of scientist and technologist leaders who will be able to usher in an era of humanistic science to eradicate hunger and poverty. I conclude with a poem by Wystan Hugh Auden in the hope that we will see the emergence of hunger and poverty fighters in whose dictionary the word "impossible" does not occur, with reference to ensuring that every child, woman, and man has an opportunity and right for a healthy and productive life.

"Hunger allows no choice
To the citizen or the police
We must love one another or die
Defenseless under the night
Our world in stupor lies;
Yet, dotted everywhere,
Ironic points of light
Flash out wherever the Just
Exchange their messages;
Of Eros and of dust
Beleaguered by the same
Negation and despair
Show an affirming flame."

53

Feeding the Hungry

Norman E. Borlaug, PhD

Nobel Prize, Presidential Medal of Freedom, and Congressional Gold Medal Laureate, Department of Soil and Crop Sciences, Texas A&M University, College Station, Texas, USA; International Maize and Wheat Improvement Center, Mexico; Sasakawa Africa Association, The Nippon Foundation, Tokyo, Japan

"The war against hunger is truly mankind's war of liberation."

—*John F. Kennedy*

Global poverty and human development are two important issues of our time that require a collaborative effort to increase awareness, interest, and research. Some 800 million people still experience chronic and transitory hunger each year. Over the next 50 years, we face the daunting job of feeding 3.5 billion additional people, most of whom will begin life in poverty. The battle to alleviate poverty and improve human health and productivity will require dynamic agricultural development.

Breakthroughs in wheat and rice production, which came to be known as the Green Revolution, signaled the dawn of applying agricultural science to the Third World's need for modern techniques. It began in Mexico in the late 1950s, spread to Asia during the 1960s and 1970s, and continued in China in the 1980s and 1990s. Over a 40-year period, the proportion of hungry people in the world declined from about 60% in 1960 to 17% in 2000. The Green Revolution also brought environmental benefits. If the global cereal yields of 1950 still prevailed in 2000, we would have needed nearly 1.2 billion more hectares of the same quality, instead of the 660 million hectares used, to achieve 2000's global harvest. Moreover, had environmentally fragile land been brought into agricultural production, the soil erosion, loss of forests and grasslands, reduction in biodiversity, and extinction of wildlife species would have been disastrous.

Today, nearly two-thirds of the world's hungry people are farmers and pastoralists who live in marginal lands in Asia and Africa, where agro-climatic stresses and/or extreme remoteness make agricultural production especially risky and costly. Africa has been the region of greatest concern. High rates of population growth and little application of improved production technology during the past three decades have resulted in declining per capita food production, escalating food deficits, deteriorating

G. Madhavan et al. (eds.), *Career Development in Bioengineering and Biotechnology*,
DOI: 10.1007/978-0-387-76495-5_53, © Springer Science+Business Media, LLC 2008

nutritional levels among the rural poor, and devastating environmental degradation. There are signs that smallholder food production may be turning around through the application of science and technology to basic food production, but this recovery is still fragile. But African capacity in science and technology needs strengthening, and massive investments in infrastructure are required, especially for roads and transport, potable water, and electricity.

For the foreseeable future, plants—especially the cereals—will continue to supply much of our increased food demand, both for direct human consumption and as livestock feed to satisfy the rapidly growing demand for meat in the newly industrializing countries. The demand for cereals will probably grow by 50% over the next 20 years, and even larger harvests will be needed if more grain is diverted to produce biofuels. Seventy percent of global water withdrawals are for irrigating agricultural lands, which contribute 40% of our global food harvest. Expanding irrigated areas will be critical to meet future food demand, but expansion must be accompanied by greater efficiencies in water management.

Although sizable land areas, such as the cerrados of Brazil, may responsibly be converted to agriculture, most food increases will have to come from lands already in production. Fortunately, productivity improvements in crop management can be made all along the line: in plant breeding, crop management, tillage, fertilization, weed and pest control, harvesting, and water use. Genetically engineered crops are playing an increasingly important role in world agriculture, enabling scientists to reach across genera for useful genes to enhance tolerance to drought, heat, cold, and waterlogging, all likely consequences of global warming. Biotechnologists will be essential to meeting future food, feed, fiber, and biofuel demand.

The battle to ensure food security for hundreds of millions of miserably poor people is far from won. We must increase world food supplies but also recognize the links between population growth, food production, and environmental sustainability. Without a better balance, efforts to halt global poverty will grind to a halt.

54 Environmental Consciousness and Sustainable Engineering Design

Raghav Narayanan, MS

Department of Civil and Environmental Engineering, Carnegie Mellon University, Pittsburgh, Pennsylvania, USA

Ashbindu Singh, PhD

United Nations Environment Programme, Division of Early Warning and Assessment- North America, Washington, District of Columbia, USA

At the United Nations Conference on the Human Environment (Stockholm, 1972), Rene Dubos, advisor to the conference, authored the now popular maxim "*Think Globally, Act Locally.*" It is solutions at a local level that can help resolve the global environmental concerns faced by all of us today.

It is now a well-acknowledged scenario that today's engineers and technologists cannot be isolated and work in a single field, but must adopt an interdisciplinary approach encompassing, among many others, life sciences, sociology, public policy, and business. The focus of product development can no longer be restricted to such issues as functionality, cost and product life. Environmental and social factors need to be considered along with economic constraints. It is necessary to design and develop technology and solutions keeping in mind the *entirety* of the process, rather than just the ends in themselves. The complete life cycle of the material (from material extraction, purification, and processing to manufacture and product use, and finally to waste management) has to be reckoned with. Such a cradle-to-grave analysis must be part of an engineer's responsibilities.

The United Nations Environment Programme (UNEP) uses the term Cleaner Production to refer to a mentality of how goods and services should be produced with the minimum environmental impact under existing technological and economic limits. [1] Cleaner production requires the development of both an environmentally sustainable process and environmentally friendly products. Designing processes and products while keeping the environment in mind, often called "design for the environment," [2] comprises design for recovery and reuse (material and component recovery), disassembly, waste minimization, energy conservation, material conservation, chronic risk reduction, and accident prevention. It should

G. Madhavan et al. (eds.), *Career Development in Bioengineering and Biotechnology*,
DOI: 10.1007/978-0-387-76495-5_54, © Springer Science+Business Media, LLC 2008

also be noted that the role of the perfect design team is not only to follow these guidelines, but also to innovate when under pressure to ensure that neither product quality nor environmental attributes are compromised. [2] This, in fact, could be one of the challenges facing engineers and technologists—to satisfy these designs for the environment conditions while also working under constraints involving economic viability, safety, reliability, and performance.

It is also necessary to manage resource utilization efficiently. Resource management has not just environmental benefits, but also economic benefits. Waste and by-product reuse can both reduce consumption, and increase efficiency of the entire process. The famous case of Kalundborg in Denmark indicates the possible gains of such a system. [3] Kalundborg has a network of organizations where residuals and by-products from one industry are used as raw materials for another. The networked industries include the pharmaceutical plant Novo Nordisk, the enzyme producer Novozymes, the oil refinery Statoil, the soil remediation company Bioteknisk Jordrens, the Energy E2 Asnæs Power Station, the plasterboard factory BPB Gyproc, as well as the waste company Noveren I/S, and finally, the Kalundborg municipality itself. This symbiosis has resulted in significant savings in oil, coal, and water consumption, as well as decreased emissions of sulfur and carbon dioxide. In economic terms, the industries estimate to have thus for saved $160 million. [4]

There are many examples of such possible symbiotic industries; however, most are not viable. The problems are both economic as well as logistic (for example, transportation of flue gases is not feasible for distances larger than a few kilometers). It is thus necessary to concentrate upon developing local ways to reuse waste products.

Bioengineers and biotechnologists, in particular, need to respond to pressures from all fronts—economic (in the form of resource conservation and usage efficiency), political (in the form of more stringent regulations), and social (in the form of growing environmental and health consciousness of the public). The role of such present and future interdisciplinary engineers and technologists, therefore, is not just to develop processes that achieve the required function, but also to symbiotically incorporate principles related to sustainable design.

References

1. United Nations Environment Programme (19th February 2007). In *UNEP Cleaner Production Activities.* Retrieved May 10, 2007, from www.uneptie.org/pc/cp/home.htm
2. Fiksel, Joseph, *Design for Environment: Creating Eco-Efficient Products and Processes*, McGraw-Hill, New York, 1996
3. Ehrenfeld, John R. and Marian R. Chertow, "Industrial Symbiosis: The Legacy of Kalundborg," Chapter 27 in A Handbook of Industrial Ecology, Robert U. Ayres and Leslie W. Ayres, Editors, Edward Elgar, Cheltenham, U.K., p. 334–348.
4. United Nations Environment Programme (20th June 2001). In UNEP Production and Consumption Branch, Environmental Management of Industrial Estates. Retrieved May 10, 2007, from www.uneptie.org/pc/ind-estates/casestudies/kalundborg.htm

Improving Public Health Quality and Equity through Effective Use of Technology

Andrei Issakov, PhD

Health Technology and Facilities Planning, Department for Health System Governance and Service Delivery, World Health Organization, Geneva, Switzerland

S. Yunkap Kwankam, PhD

E-Health, Department of Health Statistics and Informatics, World Health Organization, Geneva, Switzerland

Technology plays a pivotal role in delivering health services to the population and improving people's health. It constitutes the material platform on which the delivery of care rests, the basis for provision of all health interventions, and it equips health professionals with indispensable means to perform their functions more effectively and efficiently. Technological advances are shaping and changing health systems as we know them, and have a tremendous impact on health systems design, operation, and performance. Regardless of national culture, level of country development, and degree of health system sophistication, healthcare everywhere is heavily dependent on health technology, and requires putting together an extraordinary array of properly balanced and managed technological resource inputs. Investments required are enormous and decisions must be made carefully, ensuring the best match between supply of those inputs and the health system's needs, as well as the appropriate balance between capital and recurrent investments.

While technological evolution brings in generally positive changes underpinning the promise of medical and health progress, it also presents health authorities with complex policy choices and challenges, and has far-reaching social, ethical, legal, and economic impact on health systems, going far beyond pure clinical implications. The way in which technology is introduced and managed significantly influences the effective lifetime of health system capital and its impact on service delivery. Old investments quickly become outdated as new

G. Madhavan et al. (eds.), *Career Development in Bioengineering and Biotechnology*,
DOI: 10.1007/978-0-387-76495-5_55, © Springer Science+Business Media, LLC 2008

technologies emerge, or they rapidly deteriorate and become unusable in the case of poor care and maintenance. Investments are wasted and assets underutilized if assessment and planning are insufficient, technology support and user capacity are inadequate, and technology resource inputs are not properly balanced.

The distribution and use of technology is closely linked with equity concern and the gap between the "haves" and "have-nots." In many countries, sophisticated tertiary hospitals heavily concentrated in the largest cities and dominated by indiscriminate use of low-volume, high-cost technology are often available only for those who can afford them. At the same time, there is a severe shortage of essential services of a higher priority at a district level and for primary health care in cities themselves, which makes those who are most in need—the rural population and urban poor—suffer most.

Health technology has become an increasingly visible policy issue, and health technology management strategies have repeatedly come under the spotlight in recent years. Policy- and decision-makers are deeply concerned with the benefits expected from new technology regarding increase in efficiency and effectiveness, weighing those benefits against the potential costs.

Only a small number of countries are producers of the world's health technology. The rest depend on technology transfer, the success of which depends not only on technology itself, but also on the context to which it is transferred. While in industrialized nations, technology assessment and asset management have controlled costs and improve access to quality care, in most developing and transitional countries, rapid technology proliferation has far outpaced the capacity of health systems to track the innovation, and to put in place adequate support systems for evaluation, introduction, and utilization of health technology. This has caused serious imbalances.

Many technical programs in the World Health Organization (WHO) have gone a long way in developing, assessing, and transferring specific appropriate technologies and procedures addressing particular diseases and health conditions. Another prime focus of WHO's work is the broader health systems and services context, in which emerging and existing technologies are applied. This work aims at optimizing health systems' capacity to efficiently acquire and utilize technology so that a health care institution or system as a whole meets the demands for quality care and remains competitive, providing the highest possible level of care to the population.

It should be noted, however, that the role of health technology management to the provision of quality care is understood primarily in a qualitative rather than a quantitative manner. Further work is needed to identify the causal pathway between health technology management and health outcomes. This pathway is yet uncharted, and there is currently a dearth of models (deterministic or statistical)—let alone evidence—linking health technology management to health outcomes. Establishing this link is critical for developing the evidence base and understanding the impact of health care technology on health system performance. To arrive at this understanding, it is important not only to identify technology inputs in the provision of health interventions, but also to formulate a concept of quality in technology management. This should be accomplished together with appropriate methodologies and indicators to measure quality and its impact on the overall quality of care, as well as strategies and tools for its improvement. Measures should also be taken to improve the collection, analysis, and use of information on distribution, availability, utilization, and management of health technology. This should be linked with epidemiological data, disease profiles, priority health needs, and health outcomes data.

Beyond the evidence, poor use of health care technology cannot be fully explained by factors such as lack of competent staff, absence of consumables, and broken down equipment. These are merely proximate determinants of underutilization. The real determinants would probably lie in the broader health systems governance domain. One would need to examine why decision-makers would acquire technology that is subsequently not used. Many of these countries have failed to put in place viable institutions that assure the six domains of stewardship needed for proper deployment and use of health technology in their health systems:

- Generation of intelligence—collection, analysis, and dissemination of information
- Formulation of strategic policy framework for technology
- Ensuring formal mechanisms for policy implementation: powers, incentives, and sanctions
- Building and sustaining partnerships for technology acquisition and utilization
- Creating a fit between policy objectives and organizational structure and culture
- Ensuring accountability, responsibility, answerability to the population, and consumer protection

In addition, health authorities are confronted with the challenge of attracting and retaining competent technologists as, generally, public sector salaries do not compare favorably with those in the private sector. This results in flight of competent technical staff from the public sector to private enterprise, or even out of the country.

At global and regional levels, the way forward includes the development of evidence on the impact of health technology management on health and health systems performance. This will support improved advocacy and establishment at a national level of long-term political commitment. It will also help build an institutional environment for the formulation, implementation, monitoring, and evaluation of appropriately defined and targeted health technology policies and strategies. Clear policy guidance and effective tools for handling complex technology choices are necessary for country decision-makers and managers if they are to adopt efficient healthcare delivery practices in response to health needs and people's expectations. Assistance should be provided to countries to train essential human resources to assess, plan, organize, manage, and operate health technology. Centers of excellence should be identified and strengthened to support national activities, and training programs adapted to national and regional contexts should be developed and implemented.

Bioengineers and biotechnologists are critical in further developing and applying innovative concepts and methodologies to accomplish the following:

- Increase awareness of the need for quality in health technology management as a key contributor to improved patient and public health outcomes
- Work toward modeling the link between quality technology management and quality of public health care
- Discuss the role of implementation of appropriate guidelines and standards as a key link between health care technology management and health outcomes
- Identify ways of instituting collaborative mechanisms and instruments for sustainably carrying forward this important work.

56

Information Sharing in the 21st Century

Vinton G. Cerf, PhD

Google, Inc. Herndon, Virginia, USA

If we have learned anything in the long evolution of civilization, it is that information lies at its heart. It is not information that is power. It is the sharing of information that gives power. When the first hominids who discovered how to make tools, weapons, and fire shared this information with others of their kind, they began civilization's slow and often uneven evolution toward our 21st century.

The invention of writing ignited a bright new stage in the progressive sharing of information that has fueled the evolution of civilization. While the oral tradition allowed us to communicate directly only with the next generation or two, writing allowed us to record and communicate with virtually all future generations. In writing, our species learned to accumulate and accurately share knowledge obtained from many members of our society.

The invention of movable type and the printing press triggered yet another dramatic evolutionary step. With the invention of the transistor, the integrated circuit, and digital technology, our civilization has made yet another dramatic step. Coupled with the invention of networks such as the Internet, computers and digital storage have become tools of information-sharing on a scale never before accomplished in all the millennia that precede our present day.

The scientific method dictates that it is the sharing of information that permits the reproduction of experiments and verification of the theoretical predictions that anticipate their outcomes. The Internet has become a new tool for the capture, discovery, and sharing of information contributed by potentially every member of civilized society. The potential democratizing of information through this medium, and the stunning ability to comb effortlessly through unimaginably large archives of recorded data, confer on our civilization an opportunity for scientific, technological, and social progress on a scale never before imagined or experienced in history.

That the works of every author, and potentially the knowledge of every individual member of global society, could be available to everyone at the touch of a keyboard is beyond historical experience. This gift is not without its potential hazards.

G. Madhavan et al. (eds.), *Career Development in Bioengineering and Biotechnology*, DOI: 10.1007/978-0-387-76495-5_56, © Springer Science+Business Media, LLC 2008

The same technology that delivers knowledge can withhold it. That technology, in fact, can deliver misinformation with the same facility as it delivers codified and verified results. As users of this new medium, we may be faced with the daunting task of distinguishing between the works of genius and the works of countless monkeys typing at their myriad keyboards.

Shared databases are becoming the wellsprings of the scientific method. The human genome databases made it possible for genetic researchers to discover all that is known about particular sequences of human DNA base pairs simply by comparing a sequence with those already recorded in the database. The capture of the genetic sequences of all life forms will confer this power on anyone interested enough to ask the question. Similar scenarios may be painted for all scientific information.

If the essence of the industrial revolution was the harnessing of water, steam, and ultimately electrical power to magnify human or animal mechanical power, then information and computers are their analogs in our globally networked world. Digital processing magnifies our human ability to remember, analyze, and predict. Having reached this important plateau in our ability to capture, harness, and share information, there is, however, no guarantee that the process will be uniformly cumulative. Information in digital form may be among the most fragile of media. Who now can reliably read the information on an 8" floppy word processing diskette or a 7 track magnetic tape? And even if these fragile bits can be read, do we still have access to the software that can interpret their meaning and render the information useful?

Those of us who dream of a modern day Prospero's Library available to everyone have found challenges in abundance. We must preserve not only the literal bits of digital information, but their means interpretation. That we shall overcome these obstacles is not guaranteed. But if the history of our civilization is any guide, it seems fair to predict that we shall make our way forward, at times unevenly, towards a future in which the knowledge of mankind is available to everyone in equal and abundant measure.

Energy and Sustainability in the 21st Century

John P. Holdren, PhD

American Association for Advancement of Science, Washington, District of Columbia; Woods Hole Research Center, Falmouth, Massachusetts; Program on Science, Technology, and Public Policy, Belfer Center for Science and International Affairs, John F. Kennedy School of Government, Harvard University, Cambridge, Massachusetts; and Lawrence Livermore National Laboratory, Livermore, California, USA

The problem of energy looms as a central element in the 21st century. Well-being has environmental, sociopolitical, and cultural as well as economic dimensions. The goal of sustainable well-being entails improving all these dimensions in ways and to end points that are consistent with maintaining the improvements indefinitely. This challenge includes not only sustainably improving the standard of living in developing countries, but also converting to a sustainable basis the currently unsustainable practices supporting the standard of living in industrialized ones.

Civilization's ability to meet this immense challenge clearly depends on our strengths in natural science and engineering. However, it also depends on our strengths in the social sciences, and in "social technology" in the form of business, government, and law, as well as on the societal wit and will to integrate all of these elements in pursuit of the goal of sustainable well-being.

No part of this challenge is more complex or more demanding than the energy dimension. This is so, in part, because energy supply is tightly intertwined with national and international security, and with many of the most damaging and dangerous environmental problems–from indoor air quality to global climate change–as well as with the capacity to meet basic human needs and fuel economic growth.

The multiplicity and importance of these linkages would make energy a vexing issue even in a world where energy demand was constant. But that is not the world we live in. Continuing population growth and rapidly rising affluence in many parts of the globe are driving a rate of increase in energy use that has staggering implications. Even if the energy efficiency of the world economy–gross world product per unit of energy–were to continue to increase at the long-term

G. Madhavan et al. (eds.), *Career Development in Bioengineering and Biotechnology*,
DOI: 10.1007/978-0-387-76495-5_57, © Springer Science+Business Media, LLC 2008

historical rate of about 1% per year, the realization of middle-of-the-road population and economic projections would entail quadrupling world energy use in this century.

In today's world, where one-third of primary energy comes from oil (two-thirds of the remaining high-quality supplies of which probably lie under the volatile Middle East), that middle-of-the-road energy trajectory cannot be managed simply by expanding what we are already doing. Such a path is not merely unsustainable; it is a prescription for disaster. After all, 80% of primary energy comes from oil, coal, and natural gas combined, and virtually all of the carbon dioxide from the combustion goes straight into the atmosphere.

The perils of oil dependence and climate change, coupled with the demand for large increases in the *per-capita* availability of energy services, compel an early transition to a different path. Its requirements include a reduction in global population growth (achievable, fortunately, by means that are desirable in their own right), and a sharply increased emphasis on improving the efficiency of energy conversion and end use; aiming to improve the energy efficiency of the world economy not by 1% per year, but by 2% per year or more.

Also required is a several-fold increase in public and private investments to improve the technologies of energy supply. We need to know whether and how the carbon dioxide from fossil-fuel use can be affordably and reliably sequestered from the atmosphere; whether and how nuclear energy can be made safe enough and proliferation-resistant enough to be substantially expanded worldwide; and to what extent biofuel production can be increased without intolerable impacts on food supply or ecosystem services. Ultimately, we need to improve the affordability of the direct harnessing of sunlight for society's energy needs.

I hoist this issue to the serious attention of the bioengineering, biotechnology, and biomedical sciences community. Nothing is more important to the human condition in the 21st century than rising to this set of challenges.

Health and Human Rights: A Global Mandate

Sarah Hall Gueldner, RN, DSN

Frances Payne Bolton School of Nursing, Case Western Reserve University, Cleveland, Ohio; and School of Nursing, College of Health and Human Development, The Pennsylvania State University, University Park, Pennsylvania, USA

> "Peace, in the sense of the absence of war, is of little value to someone who is dying of hunger or cold. It will not remove the pain of torture inflicted on a prisoner of conscience. It does not comfort those who have lost their loved ones in floods caused by senseless deforestation in a neighboring country. Peace can only last where human rights are respected, where the people are fed, and where individuals and nations are free."
>
> —*His Holiness the XIV Dalai Lama Tenzin Gyatso*

Health, like peace, of any nation is intimately linked with the economic philosophy and politics associated with the procurement and distribution of wealth. There is general agreement that the overall well-being of a country can be predicted by its overall economic status. Unfortunately, the financial destiny and distribution of a country's economic resources is generally determined by the elite few, who also have the greatest political clout. There is power in having money, and those who are well-off financially also tend to be the ones who have the power to control the distribution and dynamics of money.

This economic inequity leads to separation and fragmentation of citizens into segments. We only know our own life, and are not likely to come to know in a tangible way the life circumstances of others, especially the impoverished. In this context, while many wealthy individuals and institutions make substantial personal donations to the poor, one-on-one solo efforts will never be sufficient to make a difference in the social conditions associated with poverty and disease. Rather, system-wide changes through inter-disciplinary strategies have to be set in place to transform the wide social disparities between those who have enough and those who do not. This requires that the politics of each society be based on a strong commitment to the basic human right of every citizen to equal access to the resources they need, including education and healthcare.

G. Madhavan et al. (eds.), *Career Development in Bioengineering and Biotechnology*,
DOI: 10.1007/978-0-387-76495-5_58, © Springer Science+Business Media, LLC 2008

Achieving that high level of commitment requires a significant portion of a country's innovative social wealth be spent on reforms that can gradually transform poverty, illiteracy, and healthcare disparities into work opportunities and social stability to improve the life of the disadvantaged. There is an enormous burden to mount the political and social movement to strengthen a sense of community and civic responsibility to reach out to the vulnerable individuals and groups within the community, not just out of kindness or sympathy, but with awareness of the systemic good of the whole.

To achieve this lofty humanitarian goal, commitment and solutions must come from the global community, and especially from inter-disciplinary engineering, technology, and scientific professionals, in collaboration with nursing, medicine, and both profit and non-profit business communities. The power to eradicate many social systemic issues will be leveraged by this professional cooperation. It is particularly important to cooperatively develop and implement socio-educational programs to diminish international healthcare disparity by extracting the best of the above disciplines for the betterment of human health. Utilization and maximization of new affordable global health and educational technologies to improve the need of the impoverished is one major avenue to help global community achieve its best. Strategies like this will require leaders with a *vision* that extends beyond today's primarily profit-driven and wealth accumulating mentality to a keen social attentiveness to innovation.

Good health is a basic human right, and it is our shared responsibility to ensure the common good in healthcare for all. However, the provision of this human right, like many others, including educational literacy, empowerment, humane society, shelter, food, social equity, and safety from crime, violence, and drugs, are so closely linked with the world economy that it is becoming difficult to implement. This is one of the greatest challenges facing us, as inter-disciplinary professionals and citizens, in our persevering attempts to become ardent advocates and catalytic practitioners of worldwide health and human rights reform. I present this solemn appeal to the inter-disciplinary readership of this volume, with a sincere hope for action.

"Never doubt that a small group of thoughtful, committed citizens can change the world: Indeed, it's the only thing that ever has."

—*Margaret Mead*

Gender Equality: Progress and Challenges

Yunfeng Wu

School of Information Engineering, Beijing University of Posts and Telecommunications, Beijing, P.R. China

Yachao Zhou

Department of Computer Science and Technology, Tsinghua University, Beijing, P.R. China

Metin Akay, PhD

Harrington Department of Bioengineering, Fulton School of Engineering, Arizona State University, Tempe, Arizona, USA

"Fully committed to implementing international agreements that foster women's rights, we will continue to support advocacy for women's empowerment and build their capacity–for a life free from violence, poverty and discrimination."

—*Noeleen Heyzer, Executive Director, United Nations Development Fund for Women*

Today it is widely acknowledged that gender equality is not only essential for social justice and poverty reduction, but it is also crucial for the promotion of fundamental human rights and women's empowerment, and the improvement of basic health and well-being.

Thanks to the many efforts of international agencies and organizations, considerable progress has been achieved during the past three decades in narrowing the gender gap in some spheres of society. Barriers to women's entry to the labor force have been eliminated. The blooming civil industry and prosperous global economy provide more career opportunities for women, which helps raise household income and lower the disproportionate levels of poverty. In the dimension of educational attainment, which is considered to be a prerequisite for sustainable social and economic development, many developing countries launched initiatives to accelerate literacy rates and ameliorate gender disparities. For example, the compulsory

G. Madhavan et al. (eds.), *Career Development in Bioengineering and Biotechnology*,
DOI: 10.1007/978-0-387-76495-5_59, © Springer Science+Business Media, LLC 2008

education system carried out since the 1980s by China's government has strongly increasing the enrollment rates of women in primary and secondary education. Additionally, a large number of children from the countryside benefit from the recent tuition-free policy. At the tertiary education level, more and more female students enroll in a variety of degree programs, particularly in the emerging fields of bioengineering and biological sciences. According to the American Society for Engineering Education, 39% of biomedical engineering bachelor's degrees in the year 2000 were awarded to females, while at graduate levels, 34% of masters and 32% of doctoral degrees were conferred to women. The leadership of women in community and society service is also considerable. From 1990–2007, the IEEE Engineering in Medicine and Biology Society, the largest bioengineering society in the world, has had four female presidents. Additionally, from 2000–2006, four female students were successively elected to serve as the student representative on the society's administrative committee. This empowered leadership enhances a stronger voice of women in the formulation of policies influencing their societies.

In spite of past progress, significant challenges still lie ahead. The steadily increasing economic participation of women do not mean they receive equal pay. In developed countries, women may obtain offers with relative ease, but their employment appears to be subject to a glass ceiling that impedes their upward mobility and opportunity for advancement. Women who are in managerial positions often face a painful choice between career success and family. Gender-biased violence and deficient maternity rights protection become severe obstacles that affect women's physical security and bodily integrity, nutrition, healthcare, and well-being.

Progress toward greater gender equality is slow, since it challenges regional economic inequity and deeply entrenched cultural attitudes. Legislation ensuring equal rights for all women and men is critical, and must also be accompanied by effective implementation. The example of female pioneers can inspire and encourage the younger generations to realize and unleash their potential. Improved gender equity will bring women a strengthening of economic capacity, broader access to allocable resources, an intensified sense of self-worth and dignity, integrated infrastructure for the promotion of health and well-being, and equal opportunity to achieve political influence and community leadership.

60

Complexity: Mastering the Interdependence of Biology, Engineering and Health

Yaneer Bar-Yam, PhD

New England Complex Systems Institute, Cambridge, Massachusetts, USA

The opportunities for new approaches at the interface of engineering, biology, and health are being dramatically enhanced by the study of complex systems. This field provides general frameworks in which the engineering aspects of biology and the biological inspiration of engineering, and their relation to health, can be better recognized. Key aspects of the field of complex systems include self-organization, pattern formation, networks of influence, evolutionary improvement, and multi-scale characterization.

Complex systems science provides new insights that are at the core of a revolution in approaching science, engineering, management, and medicine. Studying collective properties of a system is not the same as the studying component properties. The implications of this perspective are specifically relevant to the relationship of biology and engineering and the improvement of health. The greatest opportunities arise from informing research with universal principles, including characteristics of collective behaviors, evolutionary improvement of complex systems, and the relationship of structure and function.

Since this brief paper cannot do justice to introducing complex systems concepts, we limit the discussion to a few notes about insights for engineering concepts in support of biological science and biologically inspired engineering systems, brief discussions of how the understanding of evolution is changing large engineering projects, and how multi-scale analysis of structure and function is impacting on our understanding of the delivery of health services. The suggested readings contain citations to the work done by others in the field.

The divergence between biology and engineering is often understood in terms of the substance of its components. However, of greater importance is the nature of how functional capabilities are understood. Since function arises from the interaction of multiple components, how we frame the collective behaviors of a system defines our understanding of functional capabilities. In traditional biology, this connection has been difficult because of the limits of understanding the

G. Madhavan et al. (eds.), *Career Development in Bioengineering and Biotechnology*,
DOI: 10.1007/978-0-387-76495-5_60, © Springer Science+Business Media, LLC 2008

components, while in engineering, the relationship of components and function is logically derived and explicitly specified. Principles that underlie the behavior of systems can enable us to generalize across multiple systems, both biological and engineering.

Today, the list of biological components is much better established through genomic knowledge and related high-throughput data technologies. The basic notion of functional purpose as a key engineering concept, therefore, should be increasingly applied in describing biological systems. Still, it is our understanding that the relationship of collective behaviors to components does not follow the traditional engineering paradigm, and should be built instead upon a more general multi-scale representation. This approach enables us to understand collective properties of cells, organisms, or collections of organisms without describing a "prescriptive" structure of components. Conversely, a growing impact of complex systems on engineering can be understood as biological inspiration. Biological inspiration can take many forms; after all, the logic of computers was originally derived from considering an aspect of human thought. Today, complex systems insights about biological collective behaviors, for example, the collective organization of insects, flocking, wolf-packs, and other distributed systems in which functional behaviors are not explicitly described, but arise from the aggregate behavioral properties of components, are inspiring many engineered systems. Self-organization is also embraced in major "engineered" systems that couple human activity with engineering infrastructures such as the World Wide Web and Web 2.0 systems like Wikipedia. Our analysis of multiscale properties of complex systems implies that the functional distinctions between engineering and biological systems must diminish as engineering systems act in real time response to a large variety of conditions that require both robustness and sensitivity. Such capabilities cannot be met by traditional functional specifications. Theorems prove this assertion.

These issues become increasingly manifest when we consider the most complex engineering projects. Traditional engineering uses a problem solving approach. While this is the usual approach for engineering projects involving one person, it is also the general approach to projects that encompass many thousands in the design and operation of large infrastructure systems like the air traffic control system. Such engineering projects often follow a paradigm akin to the Manhattan Project and the space program. The project is framed by an understanding of the underlying principles (the relationship between energy and mass or Newtonian mechanics) and a specific goal. To achieve this goal, the task is divided into components and assigned to groups, then to individuals. The results are then combined to form the final whole. Those two historical projects may be rightly considered monumental achievements. However, today, large engineering projects are frequently not successful. In the worst cases, projects that cost up to billions of dollars have been abandoned as failures.

What this approach fails to appreciate is the role of networks and interdependence of components. Such dependencies lie at the root of the study of complex systems and the intersection between seemingly disparate disciplines, including biological sciences and engineering. Better understanding of the properties of dependencies enables us to address the kind of unanticipated indirect effects that arise in highly planned projects that lead to their breakdown. While the discipline of systems engineering has developed many methods for addressing component dependencies, including abstraction, modularity, layers, and hierarchy, these methods ultimately can only be as successful as the underlying framework for decomposition of the task allows. Beyond a certain level of dependency, the traditional approach of decomposition is insufficient because, in a technical sense, the complexity of the dependencies can exceed the complexity of the components.

The insights we have gained from recent years of considering this limitation have turned the efforts of the systems engineering community to the opportunities that complex systems bring, from the understanding of biological development of complex systems to these tasks. These insights include the understanding of networks of dependency and evolution applied to an engineering context. Evolution as a biological process results in a functional complex system through processes that do consider the independent functional role of components, but only their role in context.

The complex systems analysis of function of systems extends also to social systems. Perhaps it is better to consider that the field of complex systems does not create artificial distinctions between social, engineered, and biological systems. This is important when we consider the functional effectiveness (health) of the healthcare system in the same manner as we consider the functional effectiveness of a biological or engineered system.

While medical advances are often considered to be tied to our understanding of biology, many of the problems that we face in providing effective healthcare have to do with the mechanisms of the delivery of care. The existing difficulties with the healthcare system, including medical errors, poor quality, and low measures of overall effectiveness, can be understood in relation to the limitations of the underlying structure of the system to provide the services that are otherwise known. Thus, an effective understanding of the system of healthcare is necessary. This understanding is similar to understanding the functional behavior of engineered or biological systems. We have reported on the essential structure of the healthcare system, and have performed studies of the informational aspects of medical errors. These studies point to key aspects of the system that must be changed in order for the system to provide effective care. Among the key insights is that there is a need to separate the delivery of different kinds of care according to the categories "large scale" and "highly complex" to organizations that are designed to better provide each of the kinds of services. This insight arises from the recognition that an organization that effectively provides highly repetitive large scale services is organized quite differently from an organization that provides highly complex services. Requiring one to do the task of the other would only result in ineffectiveness, and this is inherent in the structure of the existing system. We also pointed out key weaknesses in the structure of communication for prescriptions, and more general aspects of information in the healthcare system that are related to medical errors.

Quite generally, our studies suggest that scientific analysis based upon complex systems principles can provide effective guidance to improving the delivery of health services.

In this brief paper we have pointed to some developments in the understanding of complex systems principles that can be used to advance bioengineering and biotechnology and essential health services. These and related convergences should result in improved understanding and capabilities of systems that we all rely upon.

Suggested Readings

1. Y. Bar-Yam, *Dynamics of Complex Systems*, Perseus Press, Reading, 1997.
2. Y. Bar-Yam, *Making Things Work: Solving Complex Problems in a Complex World*, Knowledge Press, 2004.

61

Enhancing Humanity

Raymond C. Tallis, D. Litt

The University of Manchester, Hope Hospital Clinical Academic Group, Manchester and Salford Royal Hospitals National Health Services Trust, Salford, UK

"Tereza is staring at herself in the mirror. She wonders what would happen if her nose were to grow a millimeter longer each day. How much time would it take for her face to become unrecognizable? And if her face no longer looked like Tereza, would Tereza still be Tereza?"

—*The Unbearable Lightness of Being, Milan Kundera.*

There is increasing concern among a wide range of commentators that human nature is in the process of being irrevocably changed by technological advances which either have already been achieved or are in the pipeline. According to a multitude of op-ed writers, cultural critics, social scientists, and philosophers, we have not faced up to the grave implications of what is happening. We are sleep-walking, and need to wake up. Human life is being so radically transformed that our very essence as human beings is under threat.

Of course, apocalypse sells product, and one should not regard the epidemiology of panic as a guide to social or any other kind of reality. The fact that many quoted panickers about the future have gotten both the past and the present wrong, should itself be reassurance enough (see for example *The End of History*, by Francis Fukuyama). Nevertheless, it is still worthwhile to challenge the assumptions of those panickers who are trying to persuade us to be queasy about the consequences of the various technologies that have brought about enhancement of human possibility. Indeed, some of these challengers want to call a halt to certain lines of inquiry, notably in biotechnology.

The most often repeated claim is that we are on the verge of technological breakthroughs – in genetic engineering, in pharmacotherapy, and in the replacement of biological tissues (either by cultured tissues or by electronic prostheses) – that will dramatically transform our sense of what we are, and will thereby threaten our humanity. A bit of history may be all that is necessary to pour cool water on fevered imaginations. In 1960, leading computer scientists, headed by the mighty Marvin Minsky, predicted that by 1990 we would have developed computers so smart that they would not even treat us with the respect due to household pets.

G. Madhavan et al. (eds.), *Career Development in Bioengineering and Biotechnology*,
DOI: 10.1007/978-0-387-76495-5_61, © Springer Science+Business Media, LLC 2008

Our status would be consequently diminished. Has anyone seen any of those? Smart drugs that would transform our consciousness have been expected for 50 years, but nothing yet has matched the impact of alcohol, peyote, cocaine, opiates, or amphetamines, which have been around a rather long time.

It was expected that advances in the understanding of the neurochemistry of dementia in the 1970s would permit us not only to restore cognitive function in people with Alzheimer's disease, but also to artificially boost the intelligence of people without brain illness. The results have been disappointing, as we are reminded by the United Kingdom's National Institute of Clinical Excellence's judgment that anti-dementia drugs have only modest benefits. Gene therapy that was supposed to deliver so much in the 1980s is still waiting to deliver now, two decades later.

So do not hold your breath; you will die of anoxia. Of course, changes will come about eventually. But it is the *pace* of change that matters. We can individually and collectively adapt to gradual technological changes; that is why they never quite present the insuperable challenges some doomsayers and dystopians anticipate. In Victorian times, it was anticipated that going through a dark tunnel in a train at high speed (30 mph) would be such a shocking experience that people would come out the other side irreversibly damaged. In one of his last poems, published in 1850, Wordsworth opined that the infantility of illustrated newspapers – the first tentative steps towards the multimedia of today – would drive us back to "caverned life's first rude career" ('Illustrated Books and Newspapers').

Railway journeys and tabloid newspapers have not had the dire effects that were predicted. Even the most radically transformative technologies have not had the impact that we might have expected. The dramatic electronification of everyday life that has taken place over the last few decades has not fundamentally altered the way we relate to each other. Love, jealousy, kindness, anxiety, hatred, ambition, bitterness, and joy still seem to bear a remarkable resemblance to the emotions people had in the 1930s. Teenagers communicating by mobile phones and texts and chat rooms and webcams still seem more like teenagers than nodes in an electronic network. I have to admit a little concern at what we might call the e-ttenuation of life, where people find it increasingly difficult to "be here now" rather than dissipating themselves into an endless electronic elsewhere. But inner absence and wool-gathering is not entirely new, even if it is now electronically orchestrated. It just becomes more visible. What's more, there is something reassuring about electronic technology. Because it is widely and cheaply available – and so smart – it allows us to be dumb. This compresses the differences between people.

Of course, people are worried about more invasive innovations; in particular, the direct transformation of the human body. This is where the gradualness of change is important, because as individuals who have gone through the experience of growing up through puberty into adulthood, we have a track record of coping with changes without falling apart or losing our sense of self entirely. After all, throughout our lives we have all been engaged in creating a stable sense of our identity out of whatever is thrown at us. This idea is worth dwelling on.

We humans are unique among the animals in having a coherent sense of self, and this begins with our appropriating our own bodies as our own. This is our most fundamental human achievement: that of transforming our pre-personal bodies – with their blood, muscles, snot and worse – into the ground floor of our personal identity (see *The Kingdom of Infinite Space: A Fantastical Journey Around Your Head*, Atlantic Books by Raymond Tallis). Looked at objectively, our bodies beneath the skin are not terribly *human*; indeed, they are less human than our human technologies. There is very little in my purely organic body that I could say is *me.* Most of the meat of which I am made and which I assume as myself is pretty alien: "our flesh/Surrounds us with its own decisions" as Philip Larkin said in 'Ignorance' in *The Whitsun Weddings.* On the whole, those decisions are not very pleasant.

At the root of humanity is what in *I Am: A Philosophical Inquiry into First-Person Being* I have called 'the Existential Intuition' – the sense that 'I am this'; our appropriation of our own bodies as persons who participate in a collective culture. Even at a bodily level, this intuition withstands quite radical changes. By this I don't just mean coping with a wooden leg or a heart transplant, or being able to reassume ourselves and our responsibilities each morning when we wake up or when we come round from a knock-out blow. I mean something more fundamental – namely, normal development. We grow from something about a foot long and weighing about 7 pounds, to something about 6 feet long and weighing about 150 pounds, and for the greater part of that period, we feel that we are the same thing. We assimilate these changes into an evolving and continuous sense of our own identity.

This is possible because change happens gradually and because it happens to all of us. Gradualness ensures continuity of memory alongside an imperceptible change in our bodies and the configuration of the world in which we live. That is why my earlier reassurances emphasized the gradualness of technological advance. If I look at myself objectively, I see that I am the remote descendent of the 10-year-old I once was, and yet my *metamorphosis* is quite unlike that of Kafka's man who turns into a beetle. My dramatic personal growth and development is neither sudden nor solitary; and this will also be true of the changes that take place in human identity in the world of changing technologies.

Yes, we shall change; but the essence of human identity lies in this continuing self-redefinition. If we remember that our identity and our freedom lie in the intersection between our impersonal but unique bodies, our personal individual memories, and our shared cultural awareness, it becomes easier not to worry about the erosion of either our identity or our freedom by technological advance.

If, as I believe, the distinctive genius of humanity is to establish an identity that lies at an ever-increasing distance from our organic nature, we should rejoice in the expression of human possibility in ever-advancing technology. After all, the organic world is one in which life is nasty, brutish, short, and dominated by experiences which are inhumanly unpleasant. Human technology is less alien to us than nature (compare: bitter cold with central heating; being lost without global positioning system and being found with it; dying of parasitic infestation or spraying with pesticides). Anyone who considers the new technologies to be inhuman, or a threat to our humanity, should consider these points. Better still, they should spend five uninterrupted minutes imagining the impact of a major stroke, severe Parkinson's disease, or Alzheimer's disease on their ability to express their humanity. Those such as Fukuyama, who dislike biotechnology, do not seem to realize that the forms of 'post-humanity' served up by the natural processes in our bodies are a thousand times more radical, more terrifying, and more dehumanizing than anything arising out of our attempts to enhance human beings and their lives. Self-transformation is the essence of humanity, and our humanity is defined by our ever-widening distance from the material and organic world of which we are a part, and from which we are apart.

L'homme passe infiniment l'homme. (Blaise Pascal, *Pensées*)

In short, do not be afraid.

62

Translational Research

Gail D. Baura, PhD

Keck Graduate Institute of Applied Life Sciences, Claremont, California, USA

Translational research is generating much ink these days. Taking research from the bench to the bedside and back again is the National Institute of Health's (NIH) mandate in recently launching a consortium of twelve academic health centers for multidisciplinary investigation. [1] The Alfred E. Mann Foundation for Biomedical Engineering is making individual grants of over $100 million each to create institutes directed solely at developing new medical products. According to Foundation President and Chief Executive Officer A. Stephen Dahms, medical device investigators chosen within each institute to pursue product development of their research must understand the design control process and other Food and Drug Administration (FDA) good manufacturing practices that are required for market-released devices. [2] The Wallace H. Coulter Foundation is funding nearly 100 projects through its two translational research grant programs for bioengineering. [3]

Is translational research as rare as NIH describes? According to its website, NIH recognized that a "broad re-engineering effort" for research investigation was needed, after discussions with deans of academic health centers, recommendations from the Institute of Medicine, and meetings with the research community. Its new "Roadmap for Medical Research" is intended to "synergize multidisciplinary and interdisciplinary clinical and translational research and researchers to catalyze the application of new knowledge and techniques to clinical practice at the front lines of patient care." [1] The American Association for Cancer Research depicts the translational research cycle as a process with the following steps: (1) idea generation, (2) discovery, (3) synthesis with medical knowledge, (4) application to a medical problem, and (5) movement into clinical testing. [4] A strong case can be made that bioengineering researchers in the medical device industry have always practiced translational research. This device implementation of translational research may serve as a model for translational research in other areas.

In the early 1950s, patients with complete heart block began to be resuscitated with electrical stimulation. Pacemakers quickly progressed from Paul Zoll's large, line-powered, external pacemaker to Earl Bakken's handheld, battery-powered, version. Zoll was a cardiologist who was essentially also a bioengineer. Bioengineer Bakken, who co-founded Medtronic, worked with heart surgeon C. Walton Lillehei.

G. Madhavan et al. (eds.), *Career Development in Bioengineering and Biotechnology*,
DOI: 10.1007/978-0-387-76495-5_62, © Springer Science+Business Media, LLC 2008

The goal of these and other researchers was to develop a totally implantable pacemaker. Bioengineer Wilson Greatbatch and heart surgeon William Chardack achieved this milestone with their first successful human implant in 1960. Greatbatch used then-novel transistors and an epoxy-encased Ruben-Mallory zinc mercuric oxide battery in his design. [5]

Greatbatch licensed his design to Medtronic. Medtronic later developed the first implantable, programmable neurostimulator for control of Parkinsonion tremor in response to neurosurgeon Alim-Louis Benabid's request in the late 1980s for a brain pulse generator for experimentation. [6, 7] This deep brain stimulation technique is generally considered the second revolutionary therapy for treatment of Parkinson's disease, the first being administration of levodopa.

The list of devices gleaned from translational research does not stop with Medtronic. As entrepreneurs attempt to start their device companies, the first question venture capitalists ask before funding is, "What is the market size?" While market size translates into potential profits for the venture capitalists it translates into clinical need for bioengineers. If the market size is sufficient for funding, the company then attempts to deliver a product that will win FDA approval. A key component for premarket approval of a new type of device is substantial clinical data that support the proposed indication for use. Once approval is obtained, the next hurdle is again related to clinical need. To obtain Medicare reimbursement, a company must demonstrate, through further substantial clinical studies, the need for its device. Without Medicare reimbursement, most devices will not be able to succeed in the marketplace, as insurers generally follow Medicare's lead.

The Medical Device Excellence Awards provide an excellent showcase for many innovative devices that are the result of translational research. One of the best examples is Olympic Medical Corporation's Cool-Cap system, which is the only FDA-approved treatment for hypoxic-ischemic encephalopathy (HIE). (Olympic is now a unit of Natus Medical Inc.) HIE refers to newborn brain injury resulting from severely deficient oxygen supply; it occurs in approximately 3 of every 1000 term US births. Olympic provided funding for clinical trial equipment to test a hypothesis by researchers at the University of Auckland that extended cooling of the brain soon after birth minimizes the severity of neurologic injury. [8] Based on initial positive results, Olympic then developed a system that simultaneously cools the brain through a fitted cap while maintaining a stable rectal temperature. In a randomized, controlled study of 218 infants over 18 months, within the subgroup of 172 infants with less severe injury, cooling resulted in significantly fewer deaths and less severe neuromotor disability [9].

I would like to welcome the readers of this book who are completing their educations in science and engineering to our profession. Bioengineering is a rewarding field that has saved and improved the quality of countless lives. If your research specialty is related to medical devices, please consider a career in industry first. Academia does wait. In later years, after honing your translational research skills, you can return, as I recently did, to continue your research endeavors. Then the payoff becomes translational research on the topic of your choosing.

References

1. NIH, *Translational Research: Overview.* www.nihroadmap.nih.gov/clinicalresearch/overview-translational.asp.
2. Dahms, A.S., "New Input Mechanisms into the Commercial Cycle from Universities and Research Institutes: Facilitated Movement of University IP, Translational Research and an Emerging Applied Research Model Leading to Enhanced Delivery of Healthcare." *OECD WPB/NESTI Workshop on Biotechnology Impacts and Outputs*, Paris, France, December 2006.
3. Wallace H. "Coulter Foundation," *The Foundation.* www.whcf.org/WHCF_TheFoundation.htm.
4. Hait, W.N., "Translating research into clinical practice: deliberations from the American Association for Cancer Research," *Clin Cancer Res*, 2005, 11:4275–77.
5. Greatbatch, W., and Holmes, C. F., "History of implantable devices: entrepreneurs, bioengineers and the medical profession, in a unique collaboration, build an important new industry," *IEEE EMB Magazine*, 1991, 10:38–41, 49.
6. The Deep-Brain Stimulation for Parkinson's Disease Study Group, Deep-brain stimulation of the subthalamic nucleus or the pars interna of the globus pallidus in Parkinson's disease, *NEJM*, 2001, 345:956–963.
7. Benabid, A,.L., Pollak, P., Louveau, A., Henry, S., and de Rougemont, J., "Combined (thalamotomy and stimulation) stereotactic surgery of the VIM thalamic nucleus for bilateral Parkinson disease," *Appl Neurophysiol*, 1987, 50:344–346.
8. Swain, E., "Transforming medical practice," *MD&DI*, 2007, 29:53–57.
9. CoolCap Study Group, "Selective head cooling with mild systemic hypothermia after neonatal encephalopathy: multicentre randomized trial," *Lancet*, 2005, 365:663–670.

63

Research Paving the Way for Therapeutics and Diagnostics

Dieter Falkenhagen, Dr. Med. Habil., Dipl. Phys.

Department for Environmental and Medical Sciences, Center for Biomedical Technology and Christian Doppler Laboratory, Danube University, Krems, Austria; International Faculty for Artificial Organs, University of Bologna, Italy; International Faculty for Artificial Organs, University of Strathclyde, Glasgow, UK

There is no progress without research. This slogan is frequently quoted when it comes to showing ways into the future of human society. One of the most prominent expectations that society has is the realization of a modern health system that incorporates the latest findings of research–an expensive undertaking which should, simultaneously be unexpensive for the patient. The problem is that research and science are costly processes, very costly! This is especially true for medical research because the newly developed technologies and techniques are becoming ever more sophisticated.

New findings in research often result in complicated consequences, which allow us to elaborate more detailed diagnostic differentiation of the particular symptoms. Such exact differentiation is key to evaluating a disease in regard to its prognosis and course, and is especially valuable for the development of the proper therapy.

There is vast potential to make new, essential discoveries. Let us just think of the concepts of "genomics" or "proteomics." With the help of so-called "ship-away readers," it has become possible to gain profoundly detailed information about the up-and down-regulation of genes in the various types of cells. This up-and down-regulation of genes shows the increasing or decreasing activity of individual genes. Consequently, different proteins are synthesized by the respective cell in an either strengthened or reduced fashion, which can lead to a pathophysiologically relevant rise or fall of active mediators or receptors, enzymes, or hormones. This is the point where the proteomics techniques become relevant. Proteomics, in fact will become an important factor in future diagnostics because proteomics techniques allow us to determine hundreds of different agents and substances, in suppressed, normal, or increased concentration, by means of especially developed protein chips. From these findings, we can form essential conclusions about the disease.

G. Madhavan et al. (eds.), *Career Development in Bioengineering and Biotechnology*, DOI: 10.1007/978-0-387-76495-5_63, © Springer Science+Business Media, LLC 2008

The next question is: How can we financially justify and appropriately select parameters from this flood of biological information? The answer is bioinformatics. By means of "systems biology," and thanks to large scale computer terminals with adequate software, bioinformatics fulfill the task of selecting the optimum parameters. Interestingly, it is currently the case that biology is "chasing" the field of bioinformatics forward because the latter have to develop computer programs and suitable terminals before biology can resolve the problem of optimizing the parameters.

One could even claim that these new tools have only just triggered the future of medicine, and in this case especially, the future of diagnostics. Therapeutics seem to be standing at the beginning of a new era as well: the era of cell therapy or tissue engineering. Latest findings in the field of cell therapy, including the vast and diverse field of stem cells, give ground for numerous speculations about the use of cell therapy or tissue engineering in many different conditions, such as Alzheimer's disease, stroke, or heart attack. The new techniques will give new momentum to the development of biotechnologically-produced pharmaceuticals with a highly specific result.

Nevertheless, two things must never be forgotten, despite all the scenarios of the future. First, every bioengineer and biotechnologist must know their anamnesis and retain their technical knowledge, since this is important to guarantee targeted and economically reasonable diagnostic and therapeutic technologies in the future. Second, it is equally relevant for research and for the future of all of us to find out how this field can be turned into a goldmine through the tools of innovation.

64

Interdisciplinary Collaboration and Competency Development

Joaquin Azpiroz Leehan, PhD

Center for Medical Imaging and Instrumentation, Department of Electrical Engineering, Universidad Autonoma Metropolitana-Iztapalapa, Col. Vicentina, Mexico D.F, Mexico.

Most modern engineering and technology projects are large and complex, and require the seamless interaction of a great many people at different levels. These projects often require the assembly of several multidisciplinary teams, each of which works on a specific area of the project.

One example of successful collaborative engineering is the Apollo Project. Here, the collaboration of a space agency and thousands of contractors allowed men to complete an extremely large project that, even today, almost five decades later, we find daunting. The fact that these teams of people carried out this project with "simple" tools and technologies makes the feat even more awesome. The teamwork that allowed the crew of Apollo 13 to return to earth unharmed after an in-flight explosion is also an example of successful collaboration that will not easily be forgotten. However, for all of these successes, we must not forget that the Apollo project was also beset by failures, such as the Apollo 1 capsule fire that resulted in the death of three astronauts, and the events that led to the explosion in the service module that initiated the Apollo 13 crisis.

Another notable failure is the delay in the construction of the Superjumbo Airbus A380, the largest passenger airliner in the world to date. This project, which is primarily being conducted in four different countries in Europe, and employs thousands of people, has run into delays now lasting more than two years, due mostly to failures in configuration management and change control. (Several sources have also cited unexpected wiring complexity and problems due to using different versions of the computer aided design software). In any case, this is an example of how large multidisciplinary projects can fail when specifications are not standardized across all working groups.

G. Madhavan et al. (eds.), *Career Development in Bioengineering and Biotechnology*,
DOI: 10.1007/978-0-387-76495-5_64, © Springer Science+Business Media, LLC 2008

Success depends on the right combination of leadership, teamwork, and collaboration. This applies to all fields of work. Several key elements of this type of work are:

1. Coordination among different teams
2. Effective communication of relevant information
3. Respect for the competencies of diverse groups in an organization, working toward the same goal.

Bioengineering and biotechnology are clearly multidisciplinary, so individuals in this field must know how to work with people who come from diverse backgrounds, such as nursing, medicine, basic sciences, and various fields of engineering design. Real life experiences show that bioengineers and biotechnologists do not need to know *everything* within a multidisciplinary project. However, the interdisciplinary outlook that they have after working with professionals from different disciplines makes them a valuable addition to a team that works in terms of technology design and development. This is especially relevant when they demonstrate they are able to help minimize the communication and coordination problems in large projects, such as the ones mentioned previously.

While developing projects at a university-industry technology transfer center, we have learned that the most efficient way to approach a project is to assemble teams of people with specific competencies who contribute their strengths to the final result. In the fields of bioengineering and biotechnology, research and development projects are not carried out exclusively by bioengineers or biotechnologists. These engineers, however, are capable of understanding a large part of the work that is being carried out by other members of the team. The most important issue that respect for other people's competencies is fundamental to the success of the project.

An example of this multidisciplinary approach to a project in our field is the design and evaluation of a ventricular assist device. In a project of this magnitude, a team of over thirty members must be assembled, with competencies in industrial design, mechanical engineering, materials science, computer science, and electronic instrumentation design. There is also a need for people with experience in human anatomy and physiology (physicians and nurses). In the testing and evaluation phases, there is a need for veterinarians (animal studies) and cardiovascular surgeons.

When I was interviewing this team to write this piece, the first thing that was evident was the respect that was granted to all of the team members' competence. There was never a case of "this is something that only bioengineers can do." On the contrary, as the project got off the ground, there was a lot of information and knowledge interchange among the team members who were amazed at the depth of knowledge of different specialists. The core of the team consisted of three bioengineers who interacted with specialists at different levels.

For this project to work, the device had to be tested under controlled conditions. Bioengineers were struggling to build a mock loop to simulate pressures and flow conditions in the human body, so a mechanical engineer was invited to collaborate on this section of the project. The result was that the mock loop was built in record time, thanks to the knowledge of fluid mechanics by the mechanical engineer. Even if the bioengineers were knowledgeable in the field, they were not up to the level of the specialist, in terms of implementation. So this brings us to the first conclusion on working with a multidisciplinary team: *Even if you can do the job, ask for help when you are struggling with a piece of your assignment.*

In another section of the design, the computer control of the actuators that power the device, many people in the team were very capable of working with microprocessors, microcomputers, and PIC devices, so a bioengineer with a little experience was chosen as the main designer. When it was time to seriously develop the software, a computer engineer from the Netherlands was hired. The interaction and understanding between these two people was so productive that the bioengineer ended up as the head of product development in the firm, and the computer engineer was promoted as head of the software engineering division. So the second conclusion is that *working together is rewarding*!

Yet another example of multidisciplinary work in this project was the development of the mechanical device itself, where the main idea was turned over to mechanical engineers and industrial designers to design the first prototypes. An expert in materials sciences soon joined these people to look at biocompatibility and reliability issues. This section was under the supervision of the "principal scientist," who had a doctorate in bioengineering. In this section of the project it became evident that expertise was needed from all fronts.

The last section of the system was the cart containing the consoles, which were built by a team that included industrial designers, mechanical engineers, and bioengineers. In this group, it became evident that a key to success was the development of the systems together with the input from the industrial designers. Nowadays it is no longer helpful to build a prototype and then find a box to fit it onto (witness the success of products such as iPods and iPhones that have integrated design into the construction process). We need to integrate industrial design into the early phases of development.

So, summing up, a successful project (design and development) has proven to be made up of teams that work together with people and with other teams with different competencies across the spectrum, from industrial design to component placement and human factor analysis. These successful team members respect the competencies of others and are willing to add their knowledge and input to the project under development. In addition to this, team members learn from the experience of others. This is most important for bioengineers and biotechnologists, who are normally expected to know a little bit (or a great deal, sometimes!) about many different subjects.

65

The 21st Century Mind: The Roles of a Futures Institute

Rick L. Smyre

Center for the Communities of Future, Gastonia, North Carolina, USA

> Evolutionary change is not something that only happened in the past, it is going on now from second to second, leading to the appearance of more and more complex systems, by which organisms come to live in every more diverse ways.
>
> —*John Zachary Young, "Programs of the Brain"*

We live in a time of historical transformation, requiring new ways of thinking and new concepts of perceiving reality. The age of a clockwork universe which maintains a static structure of predictability and control is giving way to one of changes and continuous dynamics.

Recent magnetic resonance imaging studies of brain activity for people who live in different industrialized countries have revealed something interesting: Whereas in other industrialized countries, brain activity was more equally distributed between left and right hemispheres, residents of the US sample were found to have developed their left brain's (analytical, problem solving, and linear thinking abilities) much more than their right brain's connective and intuitive capacities. So does this really mean that we are just action-oriented problem solvers, and not thinkers? Are not our great achievements in science and technology based on the ability to solve problems using linear, cause and effect rationality and the scientific method?

Of course they are. So let us begin this dialogue by recognizing and affirming the great achievements of science and the importance of traditional learning to our present society. The last century has witnessed amazing advances in products and inventions that have changed our very way of life. The integration of machinery and electronics has taken us to the moon, reduced the need for manual labor through automation, and reduced the effective size of the world as planes fly beyond Mach 2.

However, as progress accelerates, reality changes, increasing the strain on traditional ideas until the dam of convention breaks and we are left with apparent

G. Madhavan et al. (eds.), *Career Development in Bioengineering and Biotechnology*,
DOI: 10.1007/978-0-387-76495-5_65, © Springer Science+Business Media, LLC 2008

chaos. Look around us. Are our institutions not struggling to deal with today's complex issues? Our educational system seems to be falling apart; our political system often grinds in gridlock; our leaders and their experience can no longer point the way; and our local and national economies are sucked into a global competitive system that takes no prisoners. It dawns on us that many traditional methods no longer work. Yet, some decide to do more of the same, attempting to get away from the chaos of increasing ambiguity and uncertainty.

We search for the new right path to the future. Some advocate a return to traditional values, some push for more efficiency with what exists, some predict a new age, and all ask for meaning, wherever it appears.

Wait a minute! Do you sense that something is important to our future in that phrase, *wherever it appears?* It is such a simple phrase, but seemingly with such potential for transforming how we think about our society in the 21st century. I sense synchronicity at work.

Rethinking How We Think

Of all the changes, there is none greater than the change in the context of our society. The impact of speed does more than rush us ahead at a faster pace; it seems to redefine everything about our lives.

It is here that we need to stop and think. Think about what, you say? About speed, and its impact. Why? What do we need to think about? We just need to learn how to go faster, doing the same things we have always done more efficiently. Just like speeding up a manufacturing process or using technology to provide more information to governmental decision makers.

But what does this assume? We are not assuming, we are just becoming more efficient. So let us get on with it. We need to do something. If we do not act, we cannot get anything done. But there "is" an assumption at work in just the way you respond to my suggestion about thinking about speed, and it is the product our educational system. We teach achievement, content, outcomes, and results. The way we are taught to learn imbues us with the assumption that outcomes and actions are the most important factors. As a result, when we evaluate learning, we test content and knowledge. Why? So that we can get on with it, so that we can do something, so that we can achieve.

Our educational system does not emphasize conceptual thinking: knowing "why" things work and knowing how to ask appropriate questions. Historically, we have not put a premium on ideas. In the past, we have not seen "talking about concepts" as a good use of time. So we hold underlying assumptions constant, and seek improvements in what we do. We just decide to speed things up and make them more efficient. Such "thinking" has led us to set standards to get to the bottom line quickly. The premium has been on doing, not thinking.

Such an approach has served us well over the years because the assumptions undergirding our society and its institutions have not changed in many years. In education, lecturing based on standard curricula has served us well. In government, representative democracy is seen as the model of citizen decision making. In economics, capitalism has produced the height of affluence for industrialized countries.

So what is different as we approach a new millennium? Is the increased speed of society creating a new context in some way? And as a result, are we going to have to think differently?

Let us see what some key people who have been very successful in our traditional society say about these questions:

> "My critique of the global capitalist system falls under two main headings. One concerns the defects of the market mechanism. Here I am talking primarily about the instabilities built into financial markets. The other concerns the deficiencies of what I have to call, for lack of a better name, the non market sector. By this I mean primarily the failure of politics and the erosion of moral values on both the national and international level."
>
> —*George Soros, The Crisis of Capitalism*

> "The question is not whether the transformation to instant public feedback through electronics is good or bad, or politically desirable or undesirable. Like a force of nature, it is simply the way our political system is heading. The people are being asked to give their own judgment before major governmental decisions are made. Since personal electronic media, the teleprocessors and computerized keypads that register public opinion are inherently democratic . . . some fear too democratic . . . and their effect will be to stretch our political system toward more sharing of power, at least by those citizens motivated to participate."
>
> —*Lawrence Grossman, The Electronic Republic*

> "Knowledge is at the heart of a dynamic civilization . . . but so is surprise. A dynamic civilization maximizes the production and use of knowledge by accepting widespread ignorance. At the simplest level, only people who know they do not know everything will be curious enough to find things out. To celebrate the pursuit of knowledge, we must confess our ignorance. Dynamism gives individuals both the freedom to learn and the incentives to share what they discover. A dynamic civilization allows its members to gain from the things they themselves do not know but other people do. Its systems and institutions evolve to let people develop, extend, and act on their particular knowledge without asking permission of a higher, but less informed, authority."
>
> —*Virginia Postrel, The Future and Its Enemies*

In each case, an underlying assumption has changed. As society moves away from hierarchies to webs, broad participation becomes possible. As telecommunications speed up connections, top down decision-making becomes inefficient. As knowledge explodes, innovative thinking replaces standard answers.

And so we come to a new understanding as society transforms. A web society requires new concepts and new underlying assumptions. With a society in transformation, we are faced with a future where we need to *think* and *do* simultaneously. We must create the capacity for communities to build parallel processes to act on present issues, while building new capacities for a constantly changing, interconnected, and increasingly complex society.

The Need for Futures Institutes

If context has emerged as a key concept for education, what do we do to help people learn how to build capacities for transformation? If the underlying assumptions are changing, how do we coach people to think within a futures context? And possibly the most important question: how do we introduce into educational curricula the need to think about the impact of future trends, as well as transforming underlying assumptions? How can schools, community colleges, and universities begin to create learning environments so that issues are considered within an evolving futures context?

There are probably many answers to these questions. However, let us return to the phrase wherever it appears, and consider its implication. Present curricula are usually based on standard answers. Our society is increasingly fast-paced, interdependent, and complex. As a result, a problem exists in how to adjust to prepare a different kind of learning experience that will prepare learners to think differently, and build skills of innovations for a constantly changing society. Therefore, a concept of focusing on the future within a context of a transformation would allow new ideas to appear as generative dialogue occurs.

What is needed is a separate and parallel structure that can be attached to existing educational structures to think about future trends, but that remains autonomous. Such a futures institute could be established to provide community research and development, and allow new trends and knowledge to appear in the thinking and operations of educational institutions and local communities.

The Target: Community Colleges

Although any educational institution can establish a futures institute, there is no better place to position one than at a local community college. Not only is there potential to prepare existing students to use a futures focus and to think differently, but there exists the opportunity to introduce 21st century ideas into the thinking and operations of organizations in the local community.

Local community colleges can establish new structures and outreach efforts more easily than public schools and universities as a result of their mission. There are several approaches any community college can take when creating a curricula for a futures institute.

- Establish a day of futures orientation, where all students are introduced to future trends of all types.
- Develop a set of futures modules for all areas of knowledge. Use these modules in many different ways, both within the structure of the curricula as well as for community seminars.
- Create a series of courses which can be used during any quarter or semester.

There are a set of ideas and concepts that should serve as foundation principles for any futures institute. All students should be schooled in the simple concepts of chaos and complexity theory. With the inability to predict specific outcomes and control standard processes (as in the past), all people must learn how to think about and participate in nonlinear processes of uncertainty, using generative dialogue to help solutions "appear." This is called *emergence*, and is a key concept of *integral science.* Other ideas related to chaos and complexity become important for people to understand how to deal with a constantly changing society.

In addition (other than a list of future trends), the following concepts should be a part of any futures institute:

- openness to new ideas
- focus on interdependency
- transforming, not reforming
- collaborative individualism
- building connections as a part of web of people, organizations, and ideas.
- choices, not standardization
- generative dialogue, not debate
- process leader, not leader-follower
- webs, not hierarchies
- a futures context
- living systems as models for society
- capacities for transformation
- focused on synthesis
- connective listener
- self-organizing
- emerging truth
- parallel processes

Conclusion

A futures institute can be many things, taking many different forms, and offering continuously new ideas. It is very important not to search for the one perfect model to use, for there will be no such thing in the 21st century. Just as important as it is not to build a standard curricula with a fixed structure, it is very important to establish key principles which undergird the development of such a dynamic institution. There is no greater challenge for educational institutions and engineers than to find appropriate ways to build futures context capacity in the thinking of all local students, citizens, and organizations. The idea of a futures institute become a necessity for learning in all communities in a 21st century society. This is a critical area, where engineers, technologists, and scientists need to join hands and get involved.

66

Accelerating Innovation in the 21st Century

Ralph W. Wyndrum Jr., ScD, MBA

The Institute of Electrical and Electronics Engineers-USA Innovation Institute, Washington, District of Columbia, USA

In its 2006 report, *Rising Above the Gathering Storm: Energizing and Employing America for a Brighter Economic Future,* the National Academies of Engineering and Science offer what has become the definitive summary of the challenges facing the United States and its technology driven prosperity in an increasingly global economy. At its core, the report emphasizes the need to strengthen human capital by being even more productive and innovative. Legislative activity in the United States Congress followed to support this goal, and wide public, academic, and industrial support surfaced. Implicit in these activities was the notion that innovation is a process that can be optimized and taught.

The findings of the Academies do not just apply to the United States, but globally to the developed world. Generalizing broadly, the decade preceding the 21st century and the first few years of the 21st century had been years of optimizing productivity and maximizing profits in the corporate world. Much of the world's industrial production moved to low-cost countries in Asia and Central America. Focus on research and development, and consequently on innovation (and particularly its front end, invention) was deemphasized. Whole research and development laboratories were shrunk or discontinued. Organizations such as Bell Laboratories became shells of their former selves, and the innovation process became lost in much of the industrial productive sector.

In response, a large body of technology assessments have been developed in the last several years, driven by various government organizations, as well as scientific and engineering professional organizations and universities. These assessments propose renewed funding and increased engineering, scientific, and mathematical university enrollments to counteract these troubling trends. It would appear that broad increases in innovative development require a jump start, and this activity may best be provided by universities and professional engineering and scientific organizations over the next decade. Universities can develop, through research into innovation practices and processes, curricula that emphasize those practices

G. Madhavan et al. (eds.), *Career Development in Bioengineering and Biotechnology,*
DOI: 10.1007/978-0-387-76495-5_66, © Springer Science+Business Media, LLC 2008

that enhance the likelihood of innovation. At a high level, integrated into their course structure for both undergraduate and masters level programs, the subject matter should stress:

1. Creating a culture of customer oriented innovation
2. Best innovation practices in large and small organizations
3. The innovation process, from invention to customer delivery
4. Innovation: Capitalizing on new technologies – what to look for and leverage

In addition to inculcating innovation in graduating students, we all need to reach the practicing engineering and scientific work forces, with a series of related tools that will rely heavily on in-person workshops of several days duration, and on online tools integrated into a unified program. Some areas for us to focus on as professionals, and as volunteers in engineering and scientific professional societies, may include:

1. **Innovation Forums and Case Study Series:** Workshops on specific aspects of the innovation process, using case study methodology, focused on the four items cited above.
2. **Innovation Network:** In-person or online innovation networks, for interactive mentoring and networking discussion programs supported by an established virtual community tool. These would serve to further connect fellow professionals with the findings and results of all of the forums, and to continually evolve the forum curricula.
3. **Innovation Library:** All participants and associates would have the opportunity to access a web-based clearinghouse of innovation resources. These would include a catalog and description of the increasing body of courses in accredited degree programs and workshops globally that focus on innovation. Web accessible programs would also be included.
4. **Annual Innovation Conference:** A convention that would include tutorials, invited paper sessions, and open paper sessions and panels focusing on innovation policy. Important government policy leaders would play a role.

In each global entity, country, or organization, such as the European Community, there needs to be an overseeing organization to lead and coordinate the activities described above. The scientific and engineering community needs to drive this to happen within the existing political process.

67

Benign Application of Knowledge through Evolutionary Theory

David Sloan Wilson, PhD

Departments of Biology and Anthropology, State University of New York, Binghamton, New York, USA

The Panama Canal was the largest engineering project of its time. It almost failed because of yellow fever and malaria. In their ignorance, the French and Americans attributed these diseases to moral weakness. As author David McCullough writes in *The Path Between The Seas*, "Nearly everyone was profoundly shaken whenever the death of some notably upright person seemed to make a mockery of such views. 'Certainly his moral quality was above reproach,' wrote one bewildered, grieving French engineer of another who had died of yellow fever the first year." Eventually it was discovered that both yellow fever and malaria are transmitted by mosquitoes. In fact, mosquitoes were breeding under the very beds of the hospital patients, in bowls of water that were placed under each leg to deter ants. Once the true causes were understood, practical measures could be taken to bring the diseases under control.

The message of this story to the readers is that *some problems require practical solutions.* All of the moralizing in the world could not prevent yellow fever and malaria. An ounce of factual knowledge could.

We need factual knowledge today more than ever. In addition to our continuing battle against disease organisms, we must battle new problems of our own creation, from compounds in plastic that mimic hormones to global climate change. In each case, there is something comparable to mosquitoes under the bed that has nothing to do with moral weakness, and must be discovered to achieve a practical solution. Yet it would be naïve to portray factual knowledge as unambiguously good. It is the cause of our problems as much as the solution. In Kurt Vonnegut's novel *Cat's Cradle*, an ounce of factual knowledge in the form of a newly discovered stable configuration of water combines with human folly to bring about the end of the world.

At least three problems make factual knowledge a mixed blessing. The first is *unforeseen consequences.* Plastic appeared unambiguously good when it was invented, and nobody had the slightest idea that it might release hormone mimics

G. Madhavan et al. (eds.), *Career Development in Bioengineering and Biotechnology*,
DOI: 10.1007/978-0-387-76495-5_67, © Springer Science+Business Media, LLC 2008

into the environment. The second is *unethical use.* Facts are powerful and can be used as weapons by some against others, in addition to their more benign uses. The third is the *erosion of moral values.* It is easy to smile at the idea of malaria as a moral weakness, but some problems *are* caused by moral weakness and solved by implementing a strong moral community.

These problems might be enormously difficult to solve in practice, but the way to begin is relatively straightforward, at least as far as I can see. To solve the problem of unforeseen consequences, we need to be suitably humble about what we know, cautious about implementing new technologies, and diligent about discovering unforeseen consequences. The ultimate solution to partial knowledge is more complete knowledge.

To solve the problem of unethical use, we need ethical social systems that prevent the exploitation of some by others. There is nothing special about factual knowledge *per se* when it comes to ethical conduct. If we can create ethical social systems, then the use of factual knowledge will become more benign, along with everything else.

To prevent the erosion of moral values, we must think carefully about the relationship between factual knowledge of the world and the beliefs required for strong moral systems. Is there *necessarily* a tradeoff between these? We know about tradeoffs for particular moral systems, such as religious beliefs that depart from factual reality, but is there a more general tradeoff that cannot be avoided?

Evolutionary theory is deeply relevant to each of these three problems and their solutions. Let's begin by asking whether the theory leads to factual knowledge that can be used for solving the practical problems of life. If I were to pose this question to a biologist, you can imagine the incredulous look that I would receive in return. The mantra "*nothing in biology makes sense except in the light of evolution*" was coined 35 years ago. Yet, if I were to pose the same question to people in a bar or a supermarket, I would receive another kind of incredulous look. A few might mumble something about antibiotic resistance, but the vast majority wouldn't have the slightest idea how to reply. The most extraordinary fact about public awareness of evolution is not that 50% do not believe it, but that nearly 100% have not connected it to anything of importance in their lives.

The reason that we believe so firmly in the physical sciences is not because they are better documented than evolution, but because they are so essential to our everyday lives. We cannot build bridges, drive cars, or fly airplanes without them. In my opinion, evolutionary theory will prove just as essential to our welfare, and we will wonder, in retrospect, how we lived in ignorance for so long. If we value factual knowledge at all, we must place a very high value on evolutionary theory, and work hard to expand it beyond the boundaries of the biological sciences, where it has been restricted for most of the 20th century.

With respect to the ethical use of factual knowledge, evolutionary theory has often been misused, but is it *more* prone to misuse than other scientific theories? A fascinating book by Rebecca Lemov titled *World as Laboratory: Experiments with Mice, Mazes and Men* recounts the history of the social sciences in America, including the "blank slate" tradition of behaviorism. Grandiose expectations about controlling human behavior were combined with blind trust in authority and a willingness to inflict suffering on individuals to benefit society as a whole that appear shocking to us today. Some of the experiments conducted by American scientists during the Korean War, involving ice-pick lobotomies that leave no scar or the total breakdown of a person prior to "reconstruction," are as ghoulish as anything that can be imagined in fiction or the dark recesses of a totalitarian state. If behaviorism and evolutionary theory have both been misused, then the real problem is to avoid the misuse of *any* theory, not just evolutionary theory.

The main reason that the term "social engineering" has earned such a bad name in the past is because it lacked a vital ingredient—mutual trust. People dread being controlled by others without their consent, for the best of reasons. When people agree upon their social priorities by consensus, then the practical application of scientific knowledge becomes benign. When knowledge is used without mutual consent, it becomes sinister. It does not matter whether the knowledge is based on evolutionary theory or its polar opposite, radical behaviorism.

The question then becomes: How can we build more durable bonds of mutual trust? Evolutionary theory has much to say about this subject. The social sciences are full of scenarios about the lives and minds of our ancestors prior to the advent of civilization. Rousseau imagined a noble savage corrupted by society. Hobbes imagined a brutish savage that must be tamed by society. Freud imagined a guilty savage whose patricidal act somehow became embedded in racial memory. Economists imagine a selfish savage, sometimes even referred to as *Homo economicus*, who became civilized only by appealing to his self-interest. It is worth asking why these origin myths are necessary when they have no more basis in fact than the Garden of Eden. My guess is that they play a practical role in the belief systems that create and sustain them, much as the distorted versions of history in the four Gospels. In any case, we are on the verge of replacing these scientific creation myths with more authentic knowledge about our species as a product of genetic and cultural evolution. This knowledge can almost certainly help us build bonds of mutual trust, which in turn can promote the benign application of all knowledge by mutual consent.

Finally, there is the question of whether a belief system can combine the best of religion and science, enabling people to flourish in sustainable communities while remaining fully committed to factual realism. It is important to realize that this would be a new cultural adaptation, never before seen on the face of the earth. Factual realism has always been the servant of practical realism, showing up when useful and excusing itself otherwise. This has been true starting with the perceptual systems of bacteria. Our minds are genetically designed to encode instructions for how to behave as factual statements. It is as natural for us as having sex and demonizing our enemies. Only now, in highly differentiated modern societies, has it become important to create a large body of factual knowledge that can be trusted, to solve practical problems at an unprecedented social, spatial, and temporal scale. Fortunately, human moral systems are flexible enough to embody anything that is deemed good and right, even if it demands discipline and self-restraint, as it usually does. Hence, our first step is to decide that factual knowledge is a virtue—sacred, if you like—and that value systems must treat statements of fact more respectfully in the future than the past.

> "To defy the authority of empirical evidence is to disqualify oneself as someone worthy of critical engagement in a dialogue"
>
> —*His Holiness the XIV Dalai Lama Tenzin Gyatso*

Honor Thy Profession

Max E. Valentinuzzi, PhD

Universidad Nacional de Tucumán and Instituto Superior de Investigaciones Biológicas, Tucumán, Argentina

"*Honor thy father and thy mother,* for they gave you at least your biological life;
Honor thy teachers, too for they showed you the way;
Honor thy disciples, assuming you had the rare privilege of having them, for they put content and projection to what you have offered them;
Honor the soil thou werest born... for there you got education, probably at the expense of those who could not get what you got;
Honor thy profession, for somehow it condenses at least partially all four previous commandments;
Honor life as the supreme gift, which easily you can destroy but to be sure, you can never recall."

My personal life is interwoven with social, political, economic, and scientific events; there is no way to separate them out. After all, my generation was born with the vacuum tube, raised with the transistor, and in adulthood watched development of the operational amplifier, integrated circuits, the microprocessor, and the personal computer. Point-to-point telecommunications were superseded by satellites, the World Wide Web, and cellular phones. In October, 1957, the new Sputnik satellite, and a few months later, the dog astronaut Laika, left us speechless. Technological advancement after advancement came in a waterfall of developments: cardiac pacemakers, implantable defibrillators, open heart surgery, cardiac transplantation, myoelectric prostheses, implants of different types, artificial organs, virtual surgery, the human genome, molecular biology, cellular engineering, biomaterials, and nanoscience. These developments have been so many and so revolutionary that our life styles, ways of thinking, and philosophical attitudes have been profoundly changed. In the realm of science and technology, and in biomedical knowledge, mankind has advanced more in the 20th century than in all the previous centuries. Yes, undoubtedly scientifically and technically the challenges faced by bioengineers and biotechnologists branch off in many directions.

G. Madhavan et al. (eds.), *Career Development in Bioengineering and Biotechnology*,
DOI: 10.1007/978-0-387-76495-5_68, © Springer Science+Business Media, LLC 2008

During the Middle Ages, all knowledge was thought of as a unified set, with science being considered a relatively smaller portion called Natural Philosophy (as opposed to the higher Theologic and Human Philosophy). Our current doctoral degrees stem from these old ideas. The *Theory of Creativity*, introduced by Arthur Koestler in the 1960s, clearly holds that mental collisions, facilitated by inter- and multi-disciplinary activities, favor the understanding of problems, discovery, creativity, and the generation of fresh knowledge.

However, despite tremendous interdisciplinary scientific and technological progress, the true and cruel problem of mankind is still open: inequality, as manifested by deep and growing socio-economic differences that lead to insufficient education, malnutrition, endemic diseases, and violence. Terrorist tragedies in many places take an immeasurable toll in human lives and destruction, sowing nothing but hatred for the future.

The 20th century started with about one billion inhabitants. The 21st century has begun with six billion. We know that before 2050, the number will climb up to perhaps 12 billion. In a timeframe of less than 30 years, there will be widespread and severe limitations in the availability of fresh potable water and land. Science and technology in general, and bioengineering in particular—what can they offer to alleviate the pressures emerging from such demands? Certainly, more and better weapons and more powerful armies are not the proper and sensible way. Perhaps we should simply remember that before anything else, we are simply men and women. Being is much better than having, and independently of how much richness, power, knowledge, or worldly glories we might collect and store, the important fact is how much we love and how much we have loved. And the scientific endeavor calls for a lot of love.

The road to that love perhaps might be found through the epigraph heading this chapter. Perhaps a few comments about these profound thoughts might best conclude this brief article.

Honor thy father and thy mother advise the Holy Scriptures (*Exodus* 20:12), for your parents gave you at least your biological life. Some confusion is more likely to happen nowadays, as with the young man who on Mother's Day bought three bouquets of beautiful flowers: one for the woman who donated an ovum, a second one for the woman in whose uterus the fertilized ovum was implanted, and a third for the woman who lovingly brought him up, through the modern progresses in biotechnology! Even so, honor them.

Honor thy teachers, too because they gave you direction. Whether they were good teachers, or sometimes perhaps not so good, it is always best to honor them. And honor them with actions rather than words, remembering that a disciple's success is the best reward a teacher can have. Besides, those who had students are relatively few, and those who had disciples are even fewer, meaning that having disciples is a true privilege. Disciples bring content and projection to what they received from their teachers, often incomplete, precarious, and even plagued with errors, but still, honor them for that reason.

Honor also thy disciples, had you the privileged luck of enjoying them, recognizing that alone, very likely, you would have not done anything, or very little, for in this complex world, we strongly set our feet on others' shoulders. For a society, not having schools is bad; however, it is much worse to have bad schools. Thus, good teachers and good schools are mandatory requirements for a growing society, from the elementary levels to the highest scientific hierarchies.

Honor also the soil thou werest born for there you have gotten education, probably at the expense of those who could not get what you got. Sometimes, many immigrate to other regions where they freely deliver their intellects' fruits, occasionally complain of, or even renege on their origins. John F. Kennedy's motto of the 1960s appears as fresh as ever: *Do not ask what your country can do for you but what you can do for your country*, let me add, *constructively, peacefully, respectfully, humbly.*

Honor thy profession as well, as somehow it condenses at least partially all four previous influences: parents, teachers, students, and countries, all integratively leading to the maximum commandment.

Honor life as the supreme gift offered to you as an engineer and scientist; easily you can destroy it, but surely you cannot recall and recreate it.

Technical Leadership: An International Imperative

Colonel Barry L. Shoop, PhD

Joint Improvised Explosive Device Defeat Organization, Office of the Deputy Secretary of Defense, Army Pentagon, US Department of Defense, Washington, District of Columbia; and Department of Electrical Engineering and Computer Science, United States Military Academy, West Point, New York, USA

"To lead people, walk beside them . . .
As for the best leaders, the people do not notice their existence.
The next best, the people honor and praise.
The next, the people fear;
and the next, the people hate . . .
When the best leader's work is done the people say,
"We did it *ourselves!*"

—*Lao-tsu*

Leadership can be defined as the ability of an individual to influence, motivate, and enable others to contribute to the effectiveness and success of the organizations of which they are members[1]. In the early stages of a career, most people naturally focus on the technical aspects of their career. They begin their careers applying the technical disciplinary knowledge they acquired in college. However, technical leadership is something that should be considered and fostered early and often throughout our entire career. This applies to both leadership within our chosen technical discipline, as well as leadership of our profession.

There are a number of different theories about leadership – three are described below:

1. **Trait Theory.** Some personality traits may lead people naturally into leadership roles.

[1] House, R. J. (2004) *Culture, Leadership, and Organizations: The GLOBE Study of 62 Societies*, SAGE Publications, Thousand Oaks, 2004

G. Madhavan et al. (eds.), *Career Development in Bioengineering and Biotechnology*, DOI: 10.1007/978-0-387-76495-5_69, © Springer Science+Business Media, LLC 2008

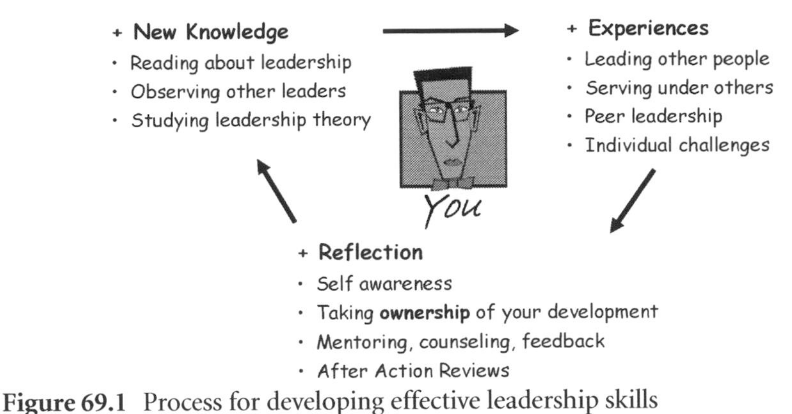

Figure 69.1 Process for developing effective leadership skills

2. **Great Events Theory.** A crisis or important event may cause a person to rise to the occasion, which brings out extraordinary leadership qualities in an ordinary person.
3. **Transformational Leadership Theory.** People can choose to become leaders, and can learn leadership skills. This is the most widely accepted theory today.

Effective leaders are made, not born. If you have the desire and willpower, you can become an effective leader. Good leaders develop through a continuous process of self-study, education, training, experience, and reflection. Although your position may give you the authority to accomplish certain tasks and objectives in your organization, power or authority does not make you a leader . . . it simply makes you the boss. Figure 69.1 describes a cycle of development that can help you learn and improve effective leadership skills.

Through this process, you acquire new knowledge about leadership theory and application, gain leadership experience, and then reflect on what you have learned to improve your leadership skills. To be effective, this developmental cycle should continue throughout your entire career – effective leaders are life-long learners.

You should consciously seek to acquire new knowledge about leadership. This involves reading articles and books on leadership, studying leadership theory, and observing other leaders. You can gain tremendous insight from observing both effective and ineffective leaders. Leadership experience is equally important in developing effective leadership skills. Leadership experience at work, in volunteer organizations, and in professional societies provides the opportunity to practice the leadership theory and new knowledge acquired during the first phase of the developmental process. Reflection is arguably the most important element in this overall developmental process. Conducting an honest inventory of one's leadership strengths and weaknesses, and then reviewing your leadership experiences through the lens of the newly acquired knowledge, will help you further refine your leadership skills. This developmental cycle continues throughout your entire career.

The basis of good leadership is honorable character and selfless-service to your organization. Your leadership is everything you do that affects the organization's objectives and the well being of those who follow you. Respected leaders concentrate on what they are [**be**] (such as beliefs and character), what they [**know**] (such as job, tasks, and human nature), and what they [**do**] (such as implementing, motivating, and providing direction).

To help you *be, know,* and *do;* follow these eleven principles of leadership:[2]

1. **Know yourself and seek self-improvement** – To know yourself, you have to understand your own attributes. Seek self-improvement, continually strengthening your attributes. This can be accomplished through self-study, formal classes, reflection, practice, and interacting with others.
2. **Be technically proficient** – As a leader, you must know your job and those of your subordinates.
3. **Seek responsibility, and take responsibility for your actions** – Search for ways to guide your organization to new heights. When things go wrong, take responsibility and do not blame others. Analyze the situation, take corrective action, and move on to the next challenge.
4. **Make sound and timely decisions** – Use good problem solving, decision making, and planning tools.
5. **Set the example** – Be a good role model – show by your own actions what you want others to do.
6. **Know your people, and look out for their well-being** – Know human nature and the importance of sincerely caring for your people.
7. **Keep everyone informed** – Know how to communicate not only with your subordinate, but also with seniors and other key people.
8. **Develop a sense of responsibility in those you lead** – Help to develop good character traits that will help those who work for you to carry out their professional responsibilities.
9. **Ensure that tasks are understood, supervised, and accomplished** – Communication is the key to this responsibility.
10. **Train as a team** – Although many so-called leaders term their organization, departments, or sections a team; such units are not necessarily teams . . . they may just be a group of people doing their jobs.
11. **Use the full capabilities of your organization** – By developing a team spirit, you will be able to employ your organization, department, or section, to its fullest capabilities.

The following suggests a leadership framework based on Be-Know-Do:

- **BE** a professional – be loyal to the organization, perform selfless service, take personal responsibility.
- **BE** a professional who possess good character traits – honesty, competence, candor, commitment, integrity, and courage.
- **KNOW** the four factors of leadership – follower, leader, communication, situation.
- **KNOW** yourself – strengths and weakness of your character, knowledge, and skills.

[2] US Army Handbook (1973). *Military Leadership.*

- **KNOW** human nature – human needs, emotions, and how people respond to stress.
- **KNOW** your job – be proficient, and be able to train others in their tasks.
- **KNOW** your organization – where to go for help, your organization's climate and culture, and who the unofficial leaders are.
- **DO** provide direction – goal setting, problem solving, decision making, and planning.
- **DO** implement – communicating, coordinating, supervising, and evaluating.
- **DO** motivate – develop moral and esprit in the organization. Train, coach, and counsel.

Being an effective leader will serve you well throughout your entire professional career. Regardless of where you are in your organizational structure, effective technical leadership can make you a more effective contributor, and can make your organization more productive. Practicing your leadership skills in volunteer positions in your professional societies will contribute to and strengthen the entire profession.

> "If your actions inspire others to dream more, learn more, do more and become more, you are a leader."
>
> —*John Quincy Adams*

70

The Art of Achieving the Menschhood

Guy Kawasaki, MBA

Garage Technology Ventures and Nononina, Inc., Palo Alto, California, USA

"The true measure of a man is how he treats someone who can do him absolutely no good."

—*Samuel Johnson*

I have a theory (as opposed to a dream) that Heaven is a three-class Boeing 777. You can sit in a narrow seat that does not recline and eat chicken-like substances next to a screaming baby in coach class. Alternatively, you can sit in a slightly wider seat that reclines a bit more and eat a beef-like substance in business class. But the goal is to spend eternity in first class—specifically, Singapore Airlines first class. Here your seat reclines to a completely flat position, and there is a power outlet, personal video player, wireless access to the Internet, and noise-canceling headphones. There are also chefs, not microwave ovens.

In Heaven, you cannot buy your way into first class; nor can you use frequent flyer miles. The only way to earn an upgrade is to be a *mensch.* Mensch is the Yiddish term for someone who is ethical, decent, admirable, and emulable. It is the highest form of praise one can receive from the people whose opinions matter. The key to being "a real mensch" is nothing less than character, rectitude, dignity, and a sense of what is right, responsible, decorous. Here is my humble attempt to help you achieve menschdom.

1. ***Help people who cannot help you.*** A mensch helps people who cannot ever return the favor. He or she does not care if the recipient is rich, famous, or powerful. This does not mean that you should not help rich, famous, or powerful people (indeed, they may need the most help), but you should generally try to help those who are simply in need.
2. ***Help without the expectation of return.*** A mensch helps people without the expectation of return—at least in this life. What is the payoff? Not that there has to be a payoff, there is a pure satisfaction in helping others. Nothing more, nothing less.

G. Madhavan et al. (eds.), *Career Development in Bioengineering and Biotechnology,*
DOI: 10.1007/978-0-387-76495-5_70, © Springer Science+Business Media, LLC 2008

3. ***Help many people.*** Menschdom is a numbers game: you should help many people, so you do not hide your generosity under a bushel. (Of course, not even a mensch can help everyone. Trying to do so would mean failing to help anyone).
4. ***Do the right thing the right way.*** A mensch always does the right thing – not the easy thing, the expedient thing, the money-saving thing, or the I-can-get-away-with-it thing. Right is right, and wrong is wrong. There are absolutes in life, and mensches heed and exemplify this truth.
5. ***Pay back society.*** A mensch realizes that he or she is blessed. For example, entrepreneurs are blessed with vision and passion, plus the ability to recruit, raise money, and change the world. These blessings come with the obligation to pay back society. The baseline is that we *owe* something to society—we are not a doing a *favor* by paying back society.

I hope this stimulates the readers of this book to become mensches and solidify their stance to make *meaningful* contributions to the world. No need to thank me if it does—helping you is reward enough. In other words, don't menschion it.

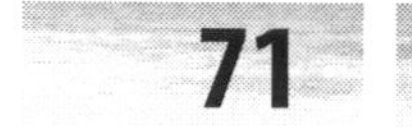

Ten Questions for Individual Leadership Development

Reverend John C. Maxwell, PhD

The INJOY Group and Maximum Impact, Atlanta, Georgia, USA

As we age, it seems our childlike curiosity diminishes. Yet, after all I have learned, as a leader, I realize there is even more that I have not yet discovered. Through my leadership journey, I have tried to keep my mind open to growth by continuing to probe for new ideas. In this readership forum, consisting of bioengineers, biotechnologists, and cross-disciplinary scientists and professionals, I would like to share ten broader questions that I regularly ask myself as a leader.

1. *Am I investing in myself?*
   This is a **personal growth** question.
   Lifelong learners have a common set of characteristics:
   - They develop a personal growth plan.
   - They possess a teachable attitude.
   - They invest in growth-oriented resources and relationships.
   - They continually leave their comfort zone.
   - They capture what they learn by applying their knowledge.
   - They reflect on what they learn and turn experience into insight.
   - They pass what they learn on to others.
2. *Am I Genuinely Interested In Others?*
   This is a **motive** question.
   Leaders see *before* others see, and they see *more* than others see. Since leaders "figure it out" first, they can be tempted to take advantage of others. Self-centered leaders manipulate when they move people for personal benefit. Mature leaders motivate by moving people for mutual benefit. They place what's best for others above themselves.
3. *Am I Doing What I Love and Loving What I Do?*
   This is a **passion** question.
   You will never fulfill your destiny doing work you despise. You are nothing unless it comes from your heart. If you go to work only to cycle through rote

G. Madhavan et al. (eds.), *Career Development in Bioengineering and Biotechnology*, DOI: 10.1007/978-0-387-76495-5_71, © Springer Science+Business Media, LLC 2008

processes and functions, then you are effectively retired. It scares me when most people I see are retired by the age of 28. To be a difference-maker, you have to bring passion, commitment, and caring to your career. Passion gives you the energy advantage over others.

4. *Am I Investing My Time with the Right People?*
   This is a **relationship** question.
   Most people can trace their successes and failures to the relationships in their lives. Be selective about who you join with on the leadership journey. Choose companions with a commitment to personal growth, a healthy attitude, and high potential.

5. *Am I Staying in My Strength Zone?*
   This is an **effectiveness** question.
   Effective leaders stop working on their weaknesses and diligently develop their strengths. You do not have to be a jack of all trades. Delegation frees you to focus on what only you can offer to your organization.

6. *Am I Taking Others to a Higher Level?*
   This is a **mission** question.
   My success is determined by the seeds I sow, not the harvest I reap. My life mission is to add value to leaders who will multiply value to others. Leaders add value to others rather than accumulating value for themselves. Dr. Martin Luther King, Jr. said it best: "Life's most urgent question is: what are you doing for others?"

7. *Am I Taking Care of Today?*
   This is a **success** question.
   The secret of your success is determined by your daily agenda. Are the habits in your life steering your toward success or simply frittering away your time? Be serious about making each day count.

8. *Am I Taking Time To Think?*
   This is a **leadership** question.
   A minute of thought is greater than an hour of talk. Taking time to think allows you to live life purposefully. Do not let life's circumstances dictate your path or allow the expectations of others to determine your course. Author your own life by clearing your schedule for thinking.

9. *Am I Developing Leaders?*
   This is a **legacy** question.
   "The ultimate test for a leader is not whether he or she makes smart decisions and takes decisive action, but whether he or she teaches others to be leaders and builds an organization that can sustain its success even when he or she is not around. True leaders put ego aside and strive to create successors who go beyond them." ~ Lorin Woolfe

10. *Am I Living Honorably?*

This is an **eternity** question.

In the light of history, our years are short and our days are few. Yet, our lives have greater significance than we can imagine. As the Roman general Maximus, exhorts his men in the motion picture *Gladiator*, "What we do in life echoes in eternity." Live your life honorably and with a clean conscience. Focus your effort on worthwhile causes that will outlast your time on this planet.

Afterword

Bioengineering and biotechnology have undergone rapid growth to become major scientific, technological, and humanitarian fields that enrich human health and welfare. Surges in the number of people involved in bioengineering and biotechnology are now evident worldwide. All of those working in these exciting areas have the responsibility of promoting and enhancing these fields, and using their knowledge to improve social wellbeing and economic development. An instrumental element in such endeavors will be to foster the career development of bioengineers and biotechnologists to further enhance the visibility of fields that are crucial to the future of society.

Career Development in Bioengineering and Biotechnology is an important work to meet such a critical need, and I am honored and delighted to pen the afterword. The book focuses not only on the career development aspects, but also on the broader social and professional development facets required for engineering and technology professionals working at the intersection of life sciences, medicine, and engineering, and in many closely related fields such as agricultural, ecological, and environmental sciences. Because of the important social roles of bioengineering and biotechnology, engineers and technologists now need to be knowledgeable about non-engineering and non-technology subjects, which include sociology, business, law, government, and international affairs. It is wonderful that the chapters of this book have addressed a wide set of knowledge and life skill that have become increasingly relevant to our profession.

I agree with Drs. Joachim Nagel, Robert Langer, and Bruce Alberts in their *Editorial, Foreword*, and *Introduction* of this book, respectively, that this is truly an outstanding book, and the first of its kind. The publication of this book is most timely for meeting the critical needs of the rapidly developing, frontier fields of bioengineering and biotechnology. The editors are to be congratulated for having the vision to select such a broad range of pertinent topics, and such a seminal group of experts to write these informative and valuable chapters. I have no doubt that this superb book will reach and help a broad international audience within the domains of bioengineering and biotechnology, including undergraduates, graduate students, and postdoctoral fellows in engineering, medical, and life sciences, as well as professionals in industry (including those seeking career transitions), entrepreneurs, educators, and policy makers. This book is certainly a pioneering contribution to bioengineering and biotechnology.

Shu Chien, MD, PhD
President, Biomedical Engineering Society and the International Society of Biorheology; Past-President, American Institute of Medical and Biological Engineering, and the American Physiological Society; University Professor of Bioengineering and Medicine, Director, Whitaker Institute of Biomedical Engineering, University of California, San Diego, California, USA

Credits and Permissions

Chapter 33 – *Holistic Engineering: The Dawn of the New Era for the Profession,* previously appeared under the title "Holistic Engineering" in *The Chronicle of Higher Education,* Vol. 53, No. 28, March 16, 2007. The chapter is presented here with some modifications and kind permission from both the authors and publisher, and is gratefully acknowledged.

Chapter 34 – *On Searching for New Genes: A 21st* Century DNA for Higher Education previously appeared under the same title in *New Horizons for Learning* at www.newhorizons.org/ (January 2006). The chapter is presented here with some modifications and kind permission from both the author and publisher, and is gratefully acknowledged.

Chapter 35 – *Protean Professionalism and Career Development* previously appeared under the title "Continual Career Change" in the American Society of Mechanical Engineering's *Mechanical Engineering* Magazine, Vol. 129, No.7, July 2007. The chapter is presented here with some modifications and kind permission from both the author and publisher, and is gratefully acknowledged.

Chapter 36 – *Leadership and Social Artistry* previously appeared under the title "Social Artistry" in *New Horizons for Learning* at www.newhorizons.org/ (June 2004). The chapter is presented here with some modifications and kind permission from both the author and publisher, and is gratefully acknowledged.

Chapter 41 – *Science, Ethics, and Human Destiny* was delivered as a speech at the *Couchiching Institute on Public Affairs Conference* on August 6, 1999 and is presented as a chapter with kind permission from the author, and is gratefully acknowledged.

Chapter 49 – *From War to Law Via Science* previously appeared under the same title in *Toronto Star* on February 19, 2006, and is presented as a chapter with kind permission from the author, and is gratefully acknowledged.

Chapter 50 – *Science and Technology for Sustainable Well-Being* previously appeared as an editorial titled "Sustainable Well-Being" in *Science,* Vol. 315, No. 5814, p.913, on February 16, 2007, published by the American Association for Advancement of Science. The chapter is presented here with some modifications and kind permission from both the author and publisher, and is gratefully acknowledged.

Chapter 53 – *Feeding the Hungry* previously appeared as an editorial under the same title in *Science,* Vol. 318, No. 5849, p.359, on October 19, 2007, published by the American Association for Advancement of Science. The chapter is presented here with some modifications and kind permission from both the author and publisher, and is gratefully acknowledged.

Chapter 57 – *Energy and Sustainability in the 21st* Century previously appeared as an editorial under the title "Energy and Sustainability" in *Science*, Vol. 315, No. 5813, p.737, on February 9, 2007, published by the American Association for Advancement of Science. The chapter is presented here with some modifications and kind permission from both the author and publisher, and is gratefully acknowledged.

Chapter 61 – *Enhancing Humanity* previously appeared under the same title in Philosophy Now (A Magazine of Ideas), May/June 2007 issue. The chapter is presented here with some modifications and kind permission from both the author and publisher, and is gratefully acknowledged.

Chapter 63 – *Research Paving the Way for Diagnostics and Therapeutics* previously appeared under the title "Research Paving the Way for Therapy and Diagnostics" in the *Course Catalogue of Medicine & Health* 2005, pp. 196–197, edited by Gerhard Polak, and published by Going International Information Services, Vienna (www.goinginternational.org). The chapter is presented here with some modifications and kind permission from both the author and publisher, and is gratefully acknowledged.

Chapter 65 – *The 21st Century Mind: The Roles of a Future Institute* previously appeared under the title "Transforming the 20th Century Mind: The Roles of a Futures Institute" in *New Horizons for Learning* at www.newhorizons.org/ (April 2000). The chapter is presented here with some modifications and kind permission from both the author and publisher, and is gratefully acknowledged.

Chapter 67 – *Benign Application of Knowledge through Evolutionary Theory* has been adapted from D.S. Wilson, *Evolution for Everyone: How Darwin's Theory Can Change the Way We Think About Our Lives* (Bantam Press, 2007) with modifications and input from the author, and is gratefully acknowledged.

Chapter 71 – *Questions for Individual Leadership Development* previously appeared under the title "Questions I Ask Myself" in Maximum Impact's *Leadership Wired*, Vol. 10, No. 3, February 16, 2007. The chapter is presented here with some modifications and kind permission from the publisher, and is gratefully acknowledged.

Index